DEVELOPMENTS IN SEDIMENTOLOGY 46

Carbonate Diagenesis and Porosity

This volume has been published in cooperation with OGCI and has been based on training courses organised by OGCI and taught by Dr. Moore.

FURTHER TITLES IN THIS SERIES
VOLUMES 1–11, 13–15 and 21–24 are out of print

12 R.C.G. BATHURST
CARBONATE SEDIMENTS AND THEIR DIAGENESIS
13 H.H. RIEKE III and G.V. CHILINGARIAN
COMPACTION OF ARGILLACEOUS SEDIMENTS
17 M.D. PICARD and L.R. HIGH Jr.
SEDIMENTARY STRUCTURES OF EPHEMERAL STREAMS
18 G.V. CHILINGARIAN and K.H. WOLF, Editors
COMPACTION OF COARSE-GRAINED SEDIMENTS
19 W. SCHWARZACHER
SEDIMENTATION MODELS AND QUANTITATIVE STRATIGRAPHY
20 M.R. WALTER, Editor
STROMATOLITES
25 G. LARSEN and G.V. CHILINGAR, Editors
DIAGENESIS IN SEDIMENTS AND SEDIMENTARY ROCKS
26 T. SUDO and S. SHIMODA, Editors
CLAYS AND CLAY MINERALS OF JAPAN
27 M.M. MORTLAND and V.C. FARMER, Editors
INTERNATIONAL CLAY CONFERENCE 1978
28 A. NISSENBAUM, Editor
HYPERSALINE BRINES AND EVAPORITIC ENVIRONMENTS
29 P. TURNER
CONTINENTAL RED BEDS
30 J.R.L. ALLEN
SEDIMENTARY STRUCTURES
31 T. SUDO, S. SHIMODA, H. YOTSUMOTO and S. AITA
ELECTRON MICROGRAPHS OF CLAY MINERALS
32 C.A. NITTROUER, Editor
SEDIMENTARY DYNAMICS OF CONTINENTAL SHELVES
33 G.N. BATURIN
PHOSPHORITES ON THE SEA FLOOR
34 J.J. FRIPIAT, Editor
ADVANCED TECHNIQUES FOR CLAY MINERAL ANALYSIS
35 H. VAN OLPHEN and F. VENIALE, Editors
INTERNATIONAL CLAY CONFERENCE 1981
36 A. IIJIMA, J.R. HEIN and R. SIEVER, Editors
SILICEOUS DEPOSITS IN THE PACIFIC REGION
37 A. SINGER and E. GALAN, Editors
PALYGORSKITE-SEPIOLITE: OCCURRENCES, GENESIS AND USES
38 M.E. BROOKFIELD and T.S. AHLBRANDT, Editors
EOLIAN SEDIMENTS AND PROCESSES
39 B. GREENWOOD and R.A. DAVIS Jr., Editors
HYDRODYNAMICS AND SEDIMENTATION IN WAVE-DOMINATED COASTAL ENVIRONMENTS
40 B. VELDE
CLAY MINERALS — A PHYSICO-CHEMICAL EXPLANATION OF THEIR OCCURRENCE
41 G.V. CHILINGARIAN and K.H. WOLF, Editors
DIAGENESIS, I
42 L.J. DOYLE and H.H. ROBERTS, Editors
CARBONATE-CLASTIC TRANSITIONS
43 G.V. CHILINGARIAN and K.H. WOLF, Editors
DIAGENISIS, II
44 C.E. WEAVER
CLAYS, MUDS AND SHALES
45 G.S. ODIN, Editor
GREEN MARINE CLAYS

DEVELOPMENTS IN SEDIMENTOLOGY 46

Carbonate Diagenesis and Porosity

Clyde H. Moore
Basin Research Institute, Louisiana State University, Baton Rouge, LA 70903-4101, U.S.A.

ELSEVIER
Amsterdam — Oxford — New York — Tokyo   1989

ELSEVIER SCIENCE PUBLISHERS B.V.
Sara Burgerhartstraat 25
P.O. Box 211, 1000 AE Amsterdam, The Netherlands

Distributors for the United States and Canada:

Oil & Gas Consultants International, Inc.
4554 South Harvard
Tulsa, Oklahoma 74135-2980, U.S.A.

ISBN 0-444-87415-1 (Vol. 46) (Hardbound)
ISBN 0-444-87416-X (Vol. 46) (Paperback)
ISBN 0-444-41238-7 (Series)

© Elsevier Science Publishers B.V., 1989

All rights reserved. No part of this publication may be reproduced, stored in a retrieval system or transmitted in any form or by any means, electronic, mechanical, photocopying, recording or otherwise, without the prior written permission of the publisher, Elsevier Science Publishers B.V./ Physical Sciences & Engineering Division, P.O. Box 330, 1000 AH Amsterdam, The Netherlands.

Special regulations for readers in the USA — This publication has been registered with the Copyright Clearance Center Inc. (CCC), Salem, Massachusetts. Information can be obtained from the CCC about conditions under which photocopies of parts of this publication may be made in the USA. All other copyright questions, including photocopying outside of the USA, should be referred to the publisher.

No responsibility is assumed by the Publisher for any injury and/or damage to persons or property as a matter of products liability, negligence or otherwise, or from any use or operation of any methods, products, instructions or ideas contained in the material herein.

Printed in The Netherlands

PREFACE

This book is an outgrowth of an annual seminar delivered to the Industrial Associates of the Applied Carbonate Research Program for the past 12 years here in Baton Rouge, and a number of public short courses given at various localities in the U.S. and Europe under the auspices of Oil and Gas Consultants of Tulsa, Oklahoma. The aim of these courses and the purpose of this book is to provide the working geologist, and the university graduate student, with a reasonable overview of carbonate diagenesis and its influence on the evolution of carbonate porosity. It is an enormously complex subject, incorporating large dollops of petrography, geochemistry, hydrology, mineralogy, and some would say, witchcraft. I do hope that in my effort to make carbonate diagenesis and porosity evolution understandable to the "normal geologist" that I have not damaged the subject by drifting too far toward over-simplification. In the discussion of tools useful for the recognition of diagenetic environments, I have stressed the value of basic petrology and geologic setting because of the perceived audience of this book. In the case of geochemical techniques I attempt to give a balanced view concerning the strengths and constraints of each technique at its present state of development, with the purpose of alerting the reader to the pits and traps that abound in modern high tech geoscience, and to help the reader to evaluate modern published studies in carbonate diagenesis. Case histories must be an integral part of any work treating the interrelationships of diagenesis and porosity. The case histories presented here were chosen on the basis of how well the details of porosity evolution could be tied to specific diagenetic evironments, the level of documentation of the case study, the demonstrated economic importance of the sequence, and finally, the familiarity of the author with the area and the rocks. No particular effort was made to ensure that specific regions, or geological ages were represented, because it is strongly felt that the concepts being developed in the book, and documented in the case histories are, in general neither site nor age specific. Nevertheless it is realized, principally because of the author's experience, that the book is heavily weighted toward the Americas. Perhaps this unintended provincialism can be rectified in the future.

A number of people have assisted in the preparation of this work. The manuscript was read in its entirety by Ellen Tye, Emily Stoudt, and Chad McCabe. Their comments and observations were very helpful and many found their way into the final version. Jay Banner read the sections on trace elements and isotopes and his suggestions were particularly beneficial. Marlene Moore constructed and maintained the bibliographic data base, and together with Brenda Kirkland, produced the reference section. Mary Lee Eggert, Clifford Duplechain, and James Kennedy produced the figures, while Dana Maxwell printed the photos. The manuscript was edited by Lynn Abadie with the able assistence of Marlene Moore and Brenda Kirkland . Dr. Douglas Kirkland served as consulting grammarian. Brenda Kirkland organized the "Index Gang" consisting of Brian Carter, Allison Drew, Pete George, Ezat Heydari, Marilyn Huff, Kasana Pitakpaivan, Tinka Saxena, Bill Schramm, Chekchanok Soonthornsaratul, Bill Wade, and Paul Wilson. This group produced the index. The Industrial Associates of the Applied Carbonate Research Program, the Basin Research Institute, and Louisiana State University are thanked for support during the project. The manuscript was produced camera ready in house on a Macintosh SE.

This book is dedicated to Marlene Mutz Moore, a lady of courage, class, and a ready smile. She has touched us, her friends, family, and compatriots, and we are the better for it.

February 16, 1989 Clyde H. Moore
Baton Rouge, La.

CONTENTS

PREFACE V

CHAPTER 1. THE NATURE OF CARBONATE DEPOSITIONAL SYSTEMS-COMPARISON OF CARBONATES AND SILICICLASTICS 1
Introduction 1
Consequences of Biological Influences over Carbonate Sediments 1
    Introduction 1
    Origin of carbonate sediments 1
    The reef, a unique depositional environment 2
    Unique biological control over the texture and fabric of carbonate
    sediments 3
    Carbonate grain composition 5
Sedimentary Processes and Depositional Environments Common to
Both Carbonates and Siliclastics 6
Carbonate Rock Classification 7
Sedimentation Style-The Ubiquitous Carbonate Shoaling Upward Sequence
and Cyclicity 9
Carbonate Shelf Evolution-Response to Sea Level 11
The Changing Nature of Carbonate Shelf Margins in Response to Global
Tectonics 14
Consequences of High Chemical Reactivity of Carbonates Relative
to Siliclastics 15
    Carbonate precipitation in the marine environment 15
    Susceptibility of shallow marine carbonates to early diagenetic
    overprint 16
    Susceptibility of carbonates to burial diagenesis 17
Summary 18

**CHAPTER 2. THE CLASSIFICATION AND NATURE OF CARBONATE
POROSITY** 21
Introduction 21
The Classification of Carbonate Porosity 21
    Introduction 21
    Fabric selectivity 22

Primary porosity	24
Secondary porosity	24
The utilization of the Choquette-Pray porosity classification	26
The Nature of Primary Porosity in Modern Carbonate Sediments	27
Interparticle porosity	27
Intraparticle porosity	28
Depositional porosity of mud-bearing sediments	29
Framework and fenestral porosity	30
Secondary Porosity	33
Introduction	33
Secondary porosity formation by dissolution	33
Secondary porosity associated with dolomitization	35
Secondary porosity associated with breccias	39
Secondary porosity associated with fractures	39
Summary	40
CHAPTER 3. DIAGENETIC ENVIRONMENTS OF POROSITY MODIFICATION AND TOOLS FOR THEIR RECOGNITION IN THE GEOLOGIC RECORD	43
Introduction	43
The Diagenetic Environments of Porosity Modification	43
Introduction	43
Marine environment	44
Meteoric environment	45
Subsurface environment	45
Tools for the Recognition of Diagenetic Environments of Porosity Modification in the Geologic Record	46
Introduction	46
Petrography-cement morphology	47
Petrography-cement distributional patterns	51
Petrography-grain-cement relationships relative to compaction	53
Trace element geochemistry of cements and dolomites	56
Stable isotopes	61
Strontium isotopes	68
Fluid inclusions	71
Summary	73
CHAPTER 4. NORMAL MARINE DIAGENETIC ENVIRONMENTS	75
Introduction	75
Shallow Water Normal Marine Diagenetic Environments	76

Introduction to the shallow marine cementation process 76
Recognition of ancient shallow marine cements 80
Diagenetic setting in the intertidal zone 83
Modern shallow water submarine hardgrounds 87
Recognition and significance of ancient hardgrounds 88
Diagenetic setting in the modern reef environment 89
Recognition of reef-related marine diagenesis in the ancient record ... 93
Porosity evolution of reef-related Lower-Middle Cretaceous shelf margins:
the Golden Lane of Mexico and the Stuart City of south Texas 97
Porosity evolution of Middle Devonian reef complexes: Leduc, Rainbow,
"Presqu'ile", and Swan Hills reefs, Western Canadian Sedimentary Basin 102
Deep Marine Diagenetic Environments 108
Introduction to diagenesis in the deep marine environment 108
Diagenesis within the zone of aragonite dissolution 109
Dolomitization below the calcite compensation depth 112
The thermal convection model of marine water dolomitization 114
Summary 116

CHAPTER 5. EVAPORATIVE MARINE DIAGENETIC ENVIRONMENTS ... 120
Introduction 120
Introduction to diagenesis in evaporative marine environments 120
The Marginal Marine Sabkha Diagenetic Environment 123
Modern marginal marine sabkhas 123
Diagenetic patterns associated with ancient marginal marine sabkhas . 127
Ordovician Red River marginal marine sabkha reservoirs, Williston
basin, U.S.A. 130
Mississippian Mission Canyon marginal marine sabkha reservoirs,
Williston basin, U.S.A. 134
Ordovician Ellenburger marginal marine sabkha-related dolomite
reservoirs, west Texas, U.S.A. 136
Criteria for the recognition of ancient marginal marine sabkha
dolomites 141
Marginal Marine Evaporative Lagoons and Basins (Reflux Dolomitization) ... 143
The marginal marine evaporative lagoon as a diagenetic environment .. 143
The MacLeod salt basin 145
The Upper Permian Guadalupian of west Texas, U.S.A.: an ancient
marginal marine evaporative lagoon complex 148
Ferry Lake Anhydrite, central Gulf of Mexico basin, U.S.A. 150
Upper Jurassic Smackover platform dolomitization, east Texas, U.S.A.:
a reflux dolomitization event 151

The Elk Point Basin of Canada 155
Michigan Basin, U.S.A. 157
Criteria for recognition of ancient reflux dolomites 158
Summary 159

CHAPTER 6. INTRODUCTION TO DIAGENESIS IN THE METEORIC ENVIRONMENT 161
Introduction 161
Chemical and Mineralogical Considerations 161
Geochemistry of meteoric pore fluids and precipitates 161
Isotopic composition of meteoric waters and carbonates precipitated
from meteoric waters 165
Mineralogic drive of diagenesis within the meteoric environment 167
Implications of the kinetics of the $CaCO_3$-H_2O-CO_2 system to grain
stabilization and to porosity evolution in meteoric diagenetic
environments 168
Hydrologic setting of the meteoric diagenetic environment 172
Summary 175

CHAPTER 7. METEORIC DIAGENETIC ENVIRONMENTS 177
Introduction 177
The Vadose Diagenetic Environment as Developed in Metastable Carbonate
Sequences 177
Introduction 177
Upper vadose soil or caliche zone 177
Lower vadose zone 179
Petrography of vadose cements 179
Trace element composition of vadose cements 180
Isotope composition of vadose cements 180
Porosity development in the vadose diagenetic environment 181
The Meteoric Phreatic Diagenetic Environment as Developed in Metastable
Carbonate Sequences 181
Introduction 181
Local floating meteoric water lens 182
Petrography of meteoric phreatic cements 183
Trace element composition of meteoric phreatic cements from
a local meteoric lens 184
Stable isotopic composition of meteoric phreatic cements from
a local meteoric lens 184
Porosity development in a local meteoric lens 185

Local island model of diagensis 185
The local island model through time 186
The Walker Creek field: a Jurassic example of the local island model? 187
Regional meteoric aquifer system 193
Regional meteoric aquifer diagenetic model 194
Porosity development and predictability in regional meteoric aquifer
environments 196
Geochemical trends characteristic of a regional meteoric aquifer system .. 196
The Jurassic Smackover Formation, U.S.Gulf of Mexico: a case history
of economic porosity evolution in a regional meteoric aquifer system 199
Mississippian grainstones of southwestern New Mexico, U.S.A.:
a case history of porosity destruction in a regional meteoric
aquifer system 204
The Meteoric Diagenetic Environment in Mature, Mineralogically Stable
Systems 209
    Introduction 209
    Karst processes and products 210
    Solution, cementation, and porosity evolution in a diagenetically mature
    system 211
    Karst-related porosity in the Permian San Andres Formation at the
    Yates field, west Texas, Central basin platform, U.S.A. 212
Summary 216

CHAPTER 8. DOLOMITIZATION ASSOCIATED WITH METEORIC AND MIXED METEORIC AND MARINE WATERS 219
Introduction 219
    Meteoric-marine mixing, or Dorag model of dolomitization 219
    Concerns relative to the validity of the Dorag, or mixing model of
    dolomitization, and its application to ancient rock sequences 220
    Mississippian North Bridgeport field, Illinois basin, U.S.A.: mixed water
    dolomite reservoirs 225
    Dolomitization by continental waters, Coorong Lagoon, south Australia .. 230
Summary 234

CHAPTER 9. BURIAL DIAGENETIC ENVIRONMENT 237
Introduction 237
The Burial Setting 237
    Introduction 237
    Pressure 239
    Temperature 240

    Deep burial pore fluids 241
    Hydrology of subsurface fluids 243
Compaction 243
    Introduction 243
    Mechanical compaction and dewatering 244
    Chemical compaction 247
    Factors affecting the efficiency of chemical compaction 251
    The North Sea Ekofisk field: a case history of porosity preservation
    in chalks 254
Burial Cementation 260
    The problem of source of $CaCO_3$ for burial cements 260
    Petrography of burial cements 262
    Geochemistry of burial cements 262
    Impact of late subsurface cementation on reservoir porosity 266
Subsurface Dissolution 267
Subsurface Dolomitization 268
    Introduction 268
    Petrography and geochemistry 269
    Impact of burial dolomitization on reservoir porosity 270
    Upper Devonian dolomitized sequences of Alberta, Canada: a case
    history of burial dolomitization 271
The Role of Early, Surficial Depositional and Diagenetic Processes Versus
    Burial Processes in Shaping Ultimate Porosity Evolution 277
Predicting Changes in Porosity with Depth 279
Summary 283

REFERENCES 285

INDEX 317

Chapter 1

THE NATURE OF CARBONATE DEPOSITIONAL SYSTEMS- COMPARISON OF CARBONATES AND SILICICLASTICS

INTRODUCTION

While this book is concerned primarily with diagenesis and porosity evolution in carbonate reservoirs, the reader and author must ultimately share a common understanding of the fundamental characteristics of the carbonate realm. Therefore, this introductory chapter is designed to compare and contrast carbonates and siliciclastics, and to highlight general concepts unique to the carbonate regime, such as: 1) the biological origin enjoyed by most carbonate sediments; 2) the complexity of carbonate rock classifications; 3) the ubiquitous cyclicity of autochthonous carbonate rock sequences, and their response to sea level fluctuations and global tectonics; and 4) the diagenetic consequences of the high chemical reactivity of carbonates.

Readers desiring an in-depth review of carbonate depositional environments should read the extensive compilation edited by Scholle, Bebout, and Moore (1983) and the review of carbonate facies by James (1984).

CONSEQUENCES OF BIOLOGICAL INFLUENCE OVER CARBONATE SEDIMENTS

Introduction

The striking differences between carbonate and siliciclastic sediments can generally be traced to the overwhelming biological origin of carbonate sediments, and the influences that this origin exerts on sediment textures, fabrics, and depositional processes such as the ability of certain organisms to build a rigid carbonate framework. The following section outlines these broad influences on carbonate sediments and sedimentation.

Origin of carbonate sediments

Well over 90% of the carbonate sediments found in modern environments are biological in origin and are forming under marine conditions (Milliman, 1974; Wilson, 1975; Sellwood, 1978). Distribution of carbonate sediments is directly controlled by environmental parameters favorable for the growth of the calcium carbonate organisms. These parameters include temperature, salinity, substrate, and the presence of siliciclastics (Lees, 1975). Carbonate sediments generally are deposited near the site of their origin. In contrast, siliciclastics are generally formed outside the basin of deposition, and are transported to the basin, where physical processes control their distribution. For siliciclastics, climate is no constraint, for they are found worldwide, and are abundant at all depths, in fresh water as well as marine environments.

The reef, a unique depositional environment

The ability of certain carbonate-secreting organisms to dramatically modify their environment by encrusting, framebuilding, and binding leads to the depositional environment unique to the carbonate realm--the reef (Fig. 1.1). In this discussion, we shall use the term *reef* in its genetic sense--a solid organic framework that resists waves (James, 1983).

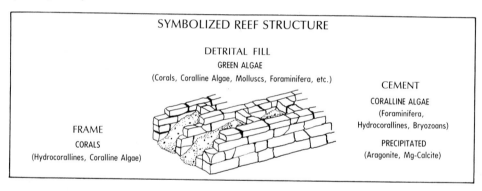

Fig. 1.1. Diagram showing symbolized reef structure emphasizing interaction between frame, detrital fill, and cement. Modified from Ginsburg and Lowenstam, 1958, Journal of Geology, v.66, p.310. Reprinted with permission of the University of Chicago Press. Copyright (C) 1958, Journal of Geology.

In a modern reef, there is an organism-sediment mosaic that sets the pattern for framework ecologic reef sequences. There are four elements: the *framework organisms*, including encrusting, attached, massive, and branching metazoa; *internal sediment*, filling primary growth as well as bioeroded cavities; the *bioeroders*, which break down reef elements by boring, rasping, or grazing, thereby contributing sediment to peri-reef as well as internal reef deposits; and *cementation*, which actively lithifies, and perhaps contributes to internal sediment (Fig. 1.1). While the reef rock scenario is complex, it is consistent. Chapter 4 presents a comprehensive treatment of marine cementation associated with reef depositional environments.

Today, the reef frame is constructed by corals and red algae. Ancillary organisms such as green algae contribute sediment to the reef system. Reef organisms have undergone a progressive evolution through geologic time, so that the reef-formers of the Lower and Middle Paleozoic (i.e., stromatoporoids) are certainly different from those of the Mesozoic (rudistids and corals) and from those that we observe today (James, 1983). Indeed, there were periods, such as the Upper Cambrian, Mississippian, and Pennsylvanian, when the reef-forming organisms were diminished, or not present, and major reef development did not occur (James, 1983).

Reef complexes, then, while certainly influenced by physical processes, are dominated by a variety of complex, unique, biological, and diagenetic processes that have no siliciclastic counterpart.

Unique biological control over the texture and fabric of carbonate sediments

The biological origin of most carbonate sedimentary particles places severe constraints on the utility of textural and fabric analysis of carbonate sediments and rocks. Sediment size and sorting in siliciclastics are generally indicators of the amount and type of physical energy (such as wind, waves, directed currents and their intensity) influencing sediment texture at the site of deposition (Folk, 1968). Size and sorting in carbonate sediments, however, may be more influenced by the population dynamics of the organism from which the particles were derived, as well as the peculiarities of the organism's ultrastructure. Folk and Robles (1964) documented the influence of the ultrastructure of coral and *Halimeda* on the grain-size distribution of beach sands derived from these organisms at Alacran Reef in Mexico (Fig. 1.2).

In certain restricted environments, such as on a tidal flat, it is common to find carbonate grains composed entirely of a single species of gastropod (Shinn and others, 1969). The mean size and sorting of the resulting sediment is controlled by natural size distribution of the gastropod population, and therefore, grain size tells little about the physical conditions at the site of deposition. For example, large conchs are commonly encountered in and adjacent to mud-dominated carbonate lagoons in the tropics. Upon death, these conchs become incorporated as large clasts in a muddy sediment. This striking textural inversion does not necessitate, as it might in siliciclastics, some unusual transport mechanism such as ice rafting, but simply means that the conch lived and died in the environment of deposition.

Other textural and fabric parameters, such as roundness, also suffer biological control. Roundness in siliciclastic grains is generally believed to indicate distance of transport, and/or the intensity of physical processes at the site of deposition (Blatt,

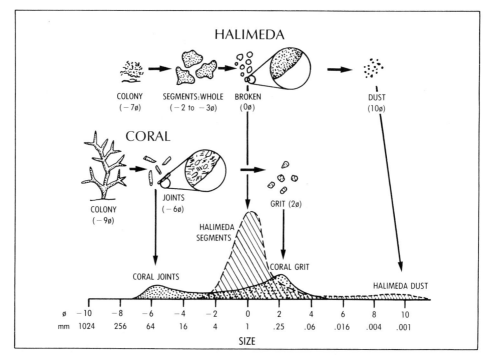

Fig. 1.2. Bioclast ultrastructural control of grain size in beach sediments from Alacran Reef, Mexico. Halimeda breaks down into two main grain-size modes; coarse sand composed of broken segments, and micrite-sized dust composed of the ultimate ultrastructural unit of the organism, single needles of aragonite. Corals also break down into two modes; large gravel-sized joints, and sand-sized particles which represent the main ultrastructural unit of the coral (Fig. 1.3). From Folk and Robles, 1967, Journal of Geology, v.72, p.9. Used with permission of University of Chicago Press. Copyright (C) 1967, Journal of Geology.

Middleton, and Murray, 1972). Roundness in carbonate grains, however, may well be controlled by the initial shape of the organism from which the grain is derived (for example, most Foraminifera are round). In addition, an organism's ultrastructure, such as the spherical fiber fascioles characteristic of the coelenterates, may also control the shape of the grains derived from the coral colony (Fig. 1.3).

Finally, some grains such as oncoids, rhodoliths, and ooids are round because they originate in an agitated environment where sequential layers are acquired during the grain's travels over the bottom, with the final product assuming a distinctly rounded shape (Bathurst, 1971). Therefore, great care must be used when interpreting the textures and fabrics of carbonate sediments and rocks as a function of physical conditions at the site of deposition.

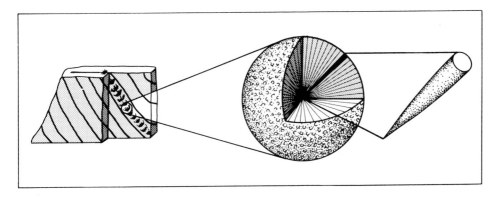

Fig. 1.3. Ultrastructure of a coral septum composed of spherical bodies stacked one on top of the other in inclined rows. Each sphere is about 2 phi sized (grit) and consists of aragonite fibers arranged about a point center. See Fig. 1.2. (Taken from Majeweski, 1969.). Copyright, E. J. Brill, the Netherlands. Used with permission.

Carbonate grain composition

Since skeletal remains of organisms furnish most of the sediments deposited in carbonate environments, the grain composition of carbonate sediments and rocks often directly reflects their environment of deposition because of the general lack of transport in carbonate regimens and the direct tie to the biological components of the environment. A number of researchers have documented the close correlation between biological communities, depositional environment, and subsequent grain composition in modern carbonate depositional systems (Ginsburg,1956; Swinchatt,1965; Thibodeaux, 1972) (Fig. 1.4).

The ability to determine the identity of the organism from which a grain originates by its distinctive and unique ultrastructure (Bathurst,1971; Milliman, 1974; Majewske, 1969; and Horowitz and Potter, 1971) is the key to the usefulness of grain composition for environmental reconstruction in ancient carbonate rock sequences. Wilson (1975) and Carrozi and Textoris (1967) based their detailed microfacies studies on the thin section identification of grain composition, including detailed identification of the biological affinities of bioclasts.

In contrast, grain composition in siliciclastics is related to the provenance of the sediment, climate, and stage of tectonic development of the source, rather than to conditions at the site of deposition (Krynine, 1941; Folk, 1954 and 1968; Pettijohn,1957; and Blatt, Middleton and Murray, 1972).

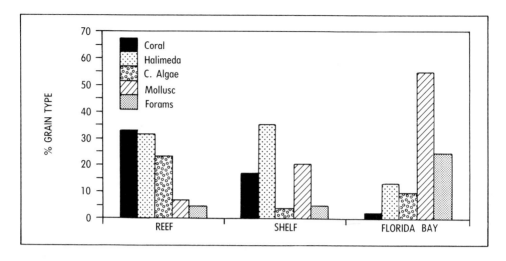

Fig. 1.4. Graph showing percentage of major grain types in the dominant depositional environments of south Florida (Data from Thibodeaux, 1972).

SEDIMENTARY PROCESSES AND DEPOSITIONAL ENVIRONMENTS COMMON TO BOTH CARBONATES AND SILICICLASTICS

Once formed, carbonate sediments not bound by organisms react to physical processes of transport and deposition in the same manner as their siliciclastic counterparts. Carbonate sands under the influence of directed currents will exhibit appropriate bedforms and cross-stratification as a function of current characteristics (Ball, 1967). Carbonate muds will be winnowed if enough wave or current energy is present in the environment. A carbonate turbidite, under appropriate conditions, will exhibit the classic Bouma sequence of sedimentary structures (Reinhardt, 1977; Rupke, 1978), thus indicating similar processes at the site of deposition regardless of origin of the sediment.

It follows that carbonates and siliciclastics share some common marine depositional environments and that general paths to the recognition of these environments in ancient sequences, such as type and sequences of sedimentary structures, textures, and fabrics, will be similar (Inden and Moore, 1983). These analogous environments include: shoreline complexes consisting of beach, dune, tidal flat, tidal channel, and tidal delta environments; submarine tidal bar complexes; submarine canyon fills and slope deposits; and basinal turbidites (see Scholle, Bebout, and Moore, 1983 for a comprehensive discussion of these and other carbonate depositional environments).

CARBONATE ROCK CLASSIFICATION

Carbonate rock classification parallels, in some respects, the classifications commonly used to characterize siliciclastics (McBride, 1964; Folk, 1968). Siliciclastics are normally classified on the basis of composition or texture, or both. Compositional classifications of sandstones generally are based on three end-members: Quartz+Chert, Feldspars, and Unstable Rock Fragments (Blatt, Middleton and Murray, 1972). As noted above, sandstone compositional classes generally reflect the tectonic setting, provenance, and climate of the extrabasinal source of the sandstones. In his textural classification, Folk (1968) uses three end-members: sand, mud, or gravel, or sand, silt, and clay, in the absence of gravel. The sediment and rock types recognized by such a

VOLUMETRIC ALLOCHEM COMPOSITION		>10% Allochems ALLOCHEMICAL ROCKS (I AND II)		<10% Allochems MICROCRYSTALLINE ROCKS (III)		UNDISTURBED BIOHERM ROCKS (IV)
		Sparry calcite cement > Microcrystalline ooze Matrix SPARRY ALLOCHEMICAL ROCKS (I)	Microcrystalline Ooze Matrix > Sparry Calcite Cement MICROCRYSTALLINE ALLOCHEMICAL ROCKS (II)	1-10% Allochems (Most Abundant Allochem)	<1% Allochems	
>25% Intraclasts (i)		Intrasparrudite (Ii:Lr) Intrasparite (Ii:La)	Intramicrudite (IIi:Lr) Intramicrite (IIi:La)	Intraclasts: Intraclast-bearing Micrite (IIIi:Lr or La)	Micrite (IIIm:L); if disturbed, Dismicrite (IIImX:L); if primary dolomite, Dolomicrite (IIIm:D)	Biolithite (IV:L)
<25% Intraclasts	>25% Oolites (o)	Oosparrudite (Io:Lr) Oosparite (Io:La)	Oomicrudite (IIo:Lr) Oomicrite (IIo:La)	Oolites: oolite-bearing Micrite (IIIo:Lr or La)		
<25% Intraclasts	<25% Oolites; Volume Ratio of Fossils to Pellets >3:1 (b)	Biosparrudite (Ib:Lr) Biosparite (Ib:La)	Biomicrudite (IIb:Lr) Biomicrite (IIb:La)	Fossils: Fossiliferous Micrite (IIIb: Lr, La, or L1)		
<25% Intraclasts	<25% Oolites; Volume Ratio of Fossils to Pellets 3:1-1:3 (bp)	Biopelsparite (Ibp:La)	Biopelmicrite (IIbp:La)	Pellets: Pelletiferous Micrite (IIIp:La)		
<25% Intraclasts	<25% Oolites; Volume Ratio of Fossils to Pellets <1:3 (p)	Pelsparite (Ip:La)	Pelmicrite (IIp:La)			

Fig. 1.5. Carbonate rock classification of Folk (1959), as modified by Folk (1962). This classification is compositional as well as textural. Reprinted by permission of the American Association of Petroleum Geologists.

texturally-based classification are believed to reflect the general level of energy present at the site of deposition (Folk, 1968). This concept of interdependency of sediment texture and energy at the site of deposition has been incorporated into the two most widely used carbonate classifications: those by Folk (1959), and Dunham (1962).

Folk's classification is more detailed, encompassing a textural scale that incorporates grain size, roundness, sorting, and packing, as well as grain composition (Fig. 1.5). The complexity of the Folk classification makes it more applicable for use with a petrographic microscope. Dunham's classification, on the other hand, is primarily

CLASSIFICATION OF CARBONATE ROCKS ACCORDING TO DEPOSITIONAL TEXTURE

DEPOSITIONAL TEXTURE RECOGNIZABLE					DEPOSITIONAL TEXTURE NOT RECOGNIZABLE
Original Components Not Bound Together During Deposition				Original components were bound together during deposition... as shown by intergrown skeletal matter, . lamination contrary to gravity, or sediment-floored cavities that are roofed over by organic or questionably organic matter and are too large to be interstices.	Crystalline Carbonate (Subdivide according to classifications designed to bear on physical texture or diagenesis.)
Contains mud (particles of clay and fine silt size)		Lacks mud and is grain-supported			
Mud-supported		Grain-supported			
Less than 10 percent grains	More than 10 percent grains				
Mudstone	Wackestone	Packstone	Grainstone	Boundstone	

Fig. 1.6. Dunham's carbonate classification (1962). This classification is primarily textural, and depends on the presence of recognizable primary textural elements. Reprinted by permission of American Association of Petroleum Geologists.

textural in nature, is simple, and is easily used in the field. This classification will be used exclusively in this book.

Dunham's major rock classes are based on the presence or absence of organic binding, presence or absence of carbonate mud, and the grain versus matrix support (Fig. 1.6). Four of the major rock classes, mudstone, wackestone, packstone, and grainstone, represent an energy continuum (Dunham, 1962). The choice of these names by Dunham emphasizes the close affinity of his classification to commonly used siliciclastic terminology. The term *Boundstone* emphasizes the major role of organic binding and framework formation in carbonates, and is a rock type unique to the carbonate realm. As with Folk's classification, these rock names are modified by the most abundant constituents (such as ooid, pellet grainstone), that further define the biological and physical conditions at the site of deposition.

SEDIMENTATION STYLE-THE UBIQUITOUS CARBONATE SHOALING UPWARD SEQUENCE AND CYCLICITY

Shallow marine carbonate shelf limestones are generally dominated by multiple, shoaling upward sequences (Wilson, 1975; James, 1979). A typical sequence consists of three basic units (Fig. 1.7): a basal normal marine or subtidal lagoonal unit whose base may be a high energy lag deposit, or a marine corrosion surface representing the return to marine conditions; an intertidal unit that may exhibit both low energy (such as tidal flats) (Fig. 1.7A) or high energy conditions (such as beach shoreface) (Fig. 1.7C); and a supratidal unit, generally above normal high tide, consisting of storm-tide and wind-blown deposits that exhibit evidence of subaerial exposure (Figs. 1.7A and C). This shoaling upward sequence is similar to, and often equated with, siliciclastic progradational shoreline sequences. Its occurrence in the record of carbonate platform deposition, however, is distinctly different from its siliciclastic counterpart within a clastic shelfal complex.

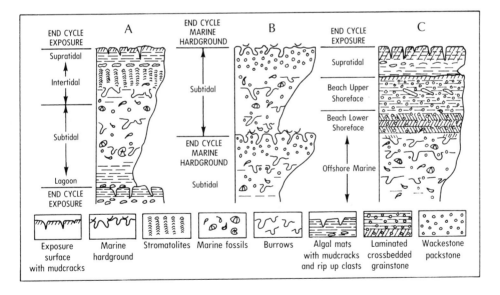

Fig. 1.7. Conceptual models of common shoaling upward sequences. Models A and C have a terminal subaerial exposure phase, while model B type cycles are punctuated by marine hardgrounds developed on grainstones or packstones.

The siliciclastic progradational shoreline sequence generally occurs during the regressive phase of large-scale, transgressive-regressive cycles driven by global sea level fluctuations, regional tectonics, or a combination of the two. These shoreline sequences most often occur as a relatively thin, but geographically widespread progradational shoreline complex associated with the terminal phase of a major siliciclastic cycle (Fig. 1.8) (Selley, 1970; Molenaar, 1977). This style of occurrence is dictated by the extrabasinal origin of siliciclastics where variable sediment availability as driven by tectonics, weathering, and indirectly by sea level, is a major factor in determining the ultimate architecture of the siliciclastic cycle.

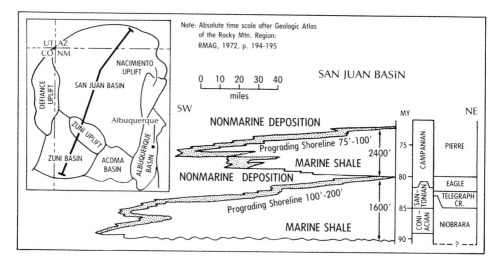

Fig. 1.8. *Conceptual stratigraphic cross section of Upper Cretaceous siliciclastic sequences, San Juan and Zuni basins, New Mexico and Arizona. This section illustrates regional shoreline progradation associated with the regressive phase of major marine cycles (Molenaar, 1977). Original diagram published in New Mexico Geological Society Guide Book to the San Juan Basin, used with permission.*

In direct contrast, the carbonate realm is marked by the *in situ* production (biological as well as chemical) of such enormous volumes of carbonate sediment that sedimentation can generally track or outstrip even the most rapid sea level rise (Wilson, 1975; Schlager, 1981). Rapid vertical sedimentation to or above base level, therefore, is the norm for carbonate shelves, and the result is the ubiquitous, stacked, shoaling upward sequence that gives the carbonate shelf its typical internal cyclical architecture. The nature of these small-scale internal cycles, such as the relative thickness of subtidal versus intratidal-supratidal caps, may depend on the complex interplay between the lag time before reestablishment of carbonate sedimentation between cycles (Schlager, 1981), and the period and amplitude of short-term sea level change (Reed and others, 1986). The thickness of the total cycle, however, seems to be controlled by the rate of shelf subsidence (Reed and others, 1986).

CARBONATE SHELF EVOLUTION-RESPONSE TO SEA LEVEL

The major difference in patterns of shelfal evolution in siliciclastic systems versus carbonate-dominated systems again lies in sediment source. In siliciclastic settings, the shelf is fed at the shoreline, and gross shelfal architecture through time is marked by distinct coastal onlap and offlap that correspond to fluctuations of sea level (Vail and Hardenbol, 1979), by changes in the rate of sea level fluctuation, and by the availability of sediment coming into the system, as driven by source area climate and tectonism (Pittman, 1978).

While coastal onlap and offlap are certainly a reality for carbonate sequences during shelf development, the continuous, massive *in situ* generation of biologically-derived carbonate sediment across the shelf dramatically affects the patterns of carbonate shelf evolution during fluctuations of sea level. Fig. 1.9 summarizes the general response of shallow marine carbonate sedimentation during various possible permutations of sea level. In this discussion, it is assumed that climate and, hence, rate of carbonate production remain the same, and that sea level change is relative, being an actual eustatic sea level fluctuation, a tectonically related subsidence/rise of the sediment-water interface, or all three.

During a rapid rise in sea level (Fig.1.9A), where rate of relative sea level rise outstrips carbonate sedimentation, a carbonate ramp is generally established (Wilson, 1975). A transgressive, high-energy shoreline forms along the innermost margins of the shelf at any one instant in time, and moves across the ramp as sea level rises without leaving behind a significant deposit, other than a thin, high-energy lag. Under these conditions, there is no well-defined shelf margin, and facies tracts tend to be bathymetrically controlled. During periods of high evaporation, the basin could become density stratified, and euxinic, leading to the preservation of organic material and the development of significant carbonate hydrocarbon source facies. As an example, Upper Jurassic source rocks across the Gulf of Mexico are dark, euxinic limestones deposited during a rapid sea level rise in conjunction with the development of a ramp in the Oxfordian lower Smackover (Moore, 1984; Sassen and Moore, 1988).

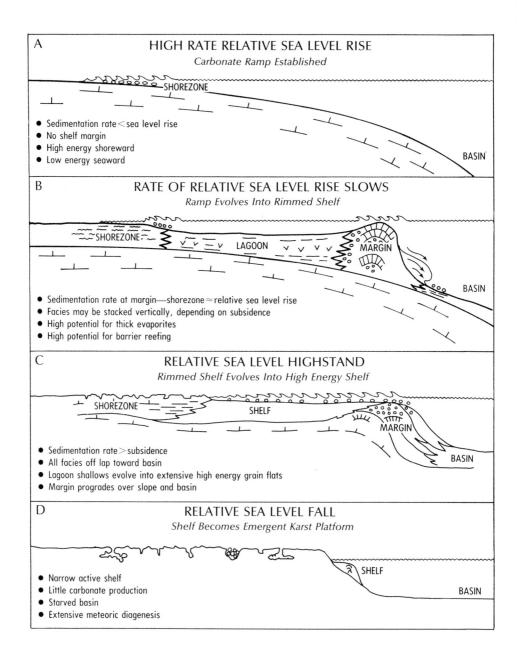

Fig. 1.9. Summary diagram showing the general response of shallow marine carbonate sedimentation during various possible changes of sea level. It is assumed that climate and hence rate of carbonate production remain the same, and that sea level change is relative.

As relative sea level rise slows (Fig. 1.9B), carbonate sedimentation begins to catch up to and match the rate of sea level rise, the shelf builds into the euphotic zone, the inner high energy zone of the ramp is translated across the shelf toward the basin, and the ramp evolves into a rimmed shelf as a result of the higher rate of sediment accumulation in the distal basin margin, high-energy zone (Wilson, 1975; Moore, 1987). If sea level continues to rise, shorezone, lagoonal, and shelf margin facies tracts will be maintained and stacked vertically through time with no apparent onlap or offlap. This configuration is ideal for the development of barrier-type reef complexes, with patch and pinnacle reefs in the lagoon. Shelf margin relief is potentially the greatest, the basin is generally starved, and there is a high potential for margin-to-basin transport by gravity processes (Moore and others, 1976; Enos and Moore, 1983; James, 1983). During times of dry climate, the lagoon can become the site for thick evaporite deposits, such as have been described by Loucks and Longman (1982) for the Lower Cretaceous Ferry Lake Anhydrite, and Moore (1984, 1987) for the Upper Jurassic Buckner Anhydrite.

As sea level reaches a highstand, the relative accumulation rates between margin, lagoon, and shorezone all equal or exceed subsidence, and the lagoon aggrades into the zone of high energy, resulting in the deposition of high energy facies from the shelf margin to the shorezone. If sedimentation is balanced by subsidence, a high energy shelf or platform like the Great Bahama Bank may be maintained. If sedimentation in all zones is greater than subsidence, then all facies tracts will offlap toward the basin, and the shelf margin will prograde over basinal deposits, much like the Permian reef complex of New Mexico and Texas (Wilson, 1975).

During a relative sea level fall (Fig. 1.9C), the shelf becomes emergent, undergoes karstification, and a new, very narrow shelf is established on the slope of the previous shelf margin. In this setting, little carbonate sediment is produced, and the adjacent basin is carbonate starved (Moore and others, 1976; James, 1979). The emergent platform is commonly the site of extensive meteoric vadose and phreatic diagenesis. The north Jamaica coast, centered on Discovery Bay is an example of this situation, except the sea level fall is relative, since the north coast is actively being tectonically uplifted. The present shelf is very narrow, and presently, little reef-derived sediment is being transported toward the basin beyond 350 m water depth. During interglacial sea level highstands in the Pleistocene, relatively large volumes of reef-derived sediment was transported into the adjacent basin as turbidites (Moore and others, 1976).

THE CHANGING NATURE OF CARBONATE SHELF MARGINS IN RESPONSE TO GLOBAL TECTONICS

Two contrasting scales of carbonate shelves have developed through geologic time: large epicontinental platforms (1000's to 100's of kms wide) with isolated carbonate platforms in open-ocean basins, and small (100's to 10's of kms wide) carbonate platforms and buildups associated with intracratonic basins (Fig. 1.10) (James and Mountjoy, 1983). While the record is not clear, it would appear that plate tectonics has played a major role in the temporal distribution of these platform types. During times of continental breakup epicontinental platforms were formed on passive margins facing large, open seas. Epicontinental platforms are characteristically found, therefore, in the

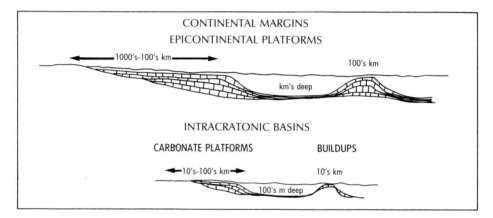

Fig. 1.10. Contrasting scales of carbonate platforms associated with continental versus cratonic margins (James and Mountjoy, 1983). Used with permission of SEPM.

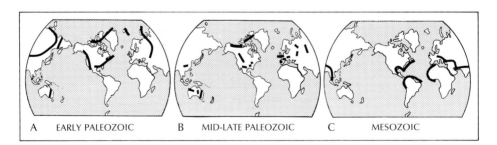

Fig. 1.11. Temporal distribution of carbonate platform types. Long lines represent epicontinental platform margins, short discontinuous lines represent cratonic-based platform margins (James and Mountjoy, 1983). Used with permission of SEPM.

Early Paleozoic (Cambrian-Ordovician) and the Mesozoic (such as the Lower Cretaceous of the Gulf of Mexico). This trend continues on into the Cenozoic (Figs. 1.11A and C) (James and Mountjoy, 1983). In contrast, the mid-Late Paleozoic saw the development of intracratonic basins, small carbonate platforms and buildups (such as the Central Basin Platform of the Midland Basin) during periods of continental accretion (Fig. 1.11B) (James and Mountjoy, 1983).

CONSEQUENCES OF HIGH CHEMICAL REACTIVITY OF CARBONATES RELATIVE TO SILICICLASTICS

Carbonate precipitation in the marine environment

Most surface tropical marine waters are supersaturated with respect to calcium carbonate (Berner, 1971). The presence of enormous volumes of potential carbonate precipitational nuclei derived from biological activity gives rise to the high potential for carbonate precipitation across carbonate platforms. This precipitation takes three forms: spontaneous precipitation of tiny carbonate crystals in the water column (so-called whitings), the precipitation of calcium carbonate on moving nuclei in a high energy situation (ooid formation), and the precipitation of calcium carbonate as cement in voids and intergranular pore spaces (see Chapter 4, this book).

Whitings have long been a controversial subject (Bathurst, 1974), and the controversy continues today (Morse and others, 1985; Shinn and others, 1985). If they are indeed an important source of fine-grained carbonate, it is obvious that the total carbonate production across carbonate platforms has been consistently underestimated. Much of the sediment potentially formed in this manner would find its way onto the shoreline as intertidal and supratidal mud flats, or be swept off the platform during storms to be deposited on the slope and in adjacent basins.

Ooids generally form in shallow, tidally active waters along the margins of carbonate platforms, such as seen in the Bahamas today (Bathurst, 1974). This is an ideal setting for carbonate precipitation because of the large volume of supersaturated marine waters fronting the platform and the potential for the upwelling of colder waters into a rapidly warming high-energy environment full of photosynthesizing plants. All these factors tend to enhance the evolution of carbon dioxide from the water with the resultant precipitation of calcium carbonate (Berner, 1971).

Carbonate marine cementation is concentrated both at the platform margin, associated with reefs and ooid shoals, and at high-energy shorelines associated with beach environments (James and Choquette, 1983). Cementation occurs where there is the potential for a large flux of calcium carbonate-saturated waters through porous sediment. Marine cementation acts as a sediment stabilizer, enhancing the potential for accumulation, rather than transportation after deposition. At the shelf margin, massive carbonate cementation allows development of steep-to-vertical profiles, and helps to overcome the constraint of the angle of repose for bioclastic and ooid sands (Halley and others, 1983). At the shoreline, marine cementation as beach rock or in association with tidal channels drastically affects normal sedimentation patterns, such as tidal channel migration and beach destruction by storm activity. In addition, massive, early marine cementation preserves delicate sedimentary structures such as burrows and keystone vugs (Inden and Moore, 1983). If ripped up by storms, beach rock will form distinctive chaotic conglomerates and breccias within beach shoreface sequences (Inden and Moore, 1983) (see Chapter 4, this book, for a comprehensive discussion of marine cementation).

Siliciclastic beach and shallow shelf sediments in tropical regions are also bathed by carbonate-saturated marine waters, but are seldom cemented because of lack of available carbonate nuclei, and the general progradational nature of siliciclastic beach deposits. Darwin (1841) described a notable exception along the coast of Brazil."The most curious object which I saw in this neighbourhood, was the reef that forms the harbour. I doubt whether in the whole world any other natural structure has so artificial an appearance. It runs for a length of several miles in an absolutely straight line, parallel to, and not far distant from, the shore. It varies in width from thirty to sixty yards, and its surface is level and smooth; it is composed of obscurely-stratified hard sandstone."

Susceptibility of shallow marine carbonates to early diagenetic overprint

Carbonate platform sediments are particularly susceptible to drastic, early diagenetic modification. Marine carbonate sediments consist of metastable carbonate phases, such as aragonite and magnesian calcite (Milliman, 1971; Bathurst, 1974), which are easily dissolved and recrystallized by fresh meteoric waters or mixtures of meteoric and marine waters, such as are encountered in surface and shallow subsurface conditions (Longman, 1980; James and Choquette, 1984). Because carbonate platforms can be maintained near sea level for extended periods, the possibility of fresh water flushing of the platforms, with attendant massive diagenesis, is significant.

One of the major impacts of meteoric water diagenesis is the basic rearrangement of calcium carbonate by dissolution of grains and the reprecipitation of calcium carbonate as cement in pore spaces. There is a distinct tendency for primary porosity in carbonate sequences to be destroyed early by cementation with the attendant generation of secondary porosity during the dissolution phase (Longman, 1980; Moore and Druckman, 1981; James and Choquette, 1984) (See Chapter 6 for a comprehensive discussion of meteoric diagenesis).

In contrast, coarse siliciclastics are practically untouched diagenetically during early burial because of the low solubility and stability of most of the grains relative to surface waters and surface-related diagenetic environments. Porosity in siliciclastic sequences in the subsurface, therefore, is generally primary, can be tied directly to

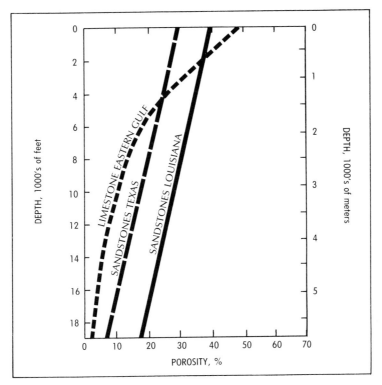

Fig. 1.12. Graph showing porosity-burial depth relationships for carbonates and sandstones across the Gulf of Mexico. (Limestone data from Halley and Schmoker, 1983; sandstone data from Loucks and others, 1979.) Used with permission of AAPG and SEPM.

variations in depositional environment, and hence is highly predictable.

Susceptibility of carbonates to burial diagenesis

Carbonates not extensively altered during early diagenesis are particularly susceptible to chemical compaction processes such as grain-to-grain pressure solution and stylolitization during burial because of the relatively high solubility of calcium carbonate (Bathurst 1975, 1983). The contrast between porosity preservation in limestones and siliciclastics is particularly vivid when comparing porosity versus burial depth relationships (Loucks and others, 1979; Halley and Smocker, 1983). Primary porosity in siliciclastic sequences is obviously retained to much greater burial depths than could be expected for carbonates (Fig. 1.12).(see Chapter 7, this book).

Aggressive pore fluids are common in the deep subsurface because of high temperatures, dissolved organic acids, carbon dioxide, hydrogen disulfide, and other components released into pore fluids during mineral phase changes and thermal degradation of organic matter and hydrocarbons (Sassen and Moore, 1988). While carbonates are particularly susceptible to dissolution by these brines, the more unstable siliciclastic phases such as feldspars are also attacked, potentially forming significant secondary porosity (Schmidt and McDonald, 1979; Surdam and others, 1984). Again, however, the potential for formation of late burial secondary porosity should be higher for carbonates than siliciclastics. (A comprehensive discussion of burial diagenesis can be found in Chapter 9.)

SUMMARY

The biological influence over the origin and nature of carbonate sediments is reflected in distinct climatic controls over their distribution, their general authocthonous nature leading to cyclicity, and shape/skeletal architectural control over sediment texture. The reef is seen as an important depositional system unique to the carbonate realm.

The high chemical reactivity of carbonates relative to siliciclastics leads to extensive early carbonate diagenesis including marine cementation in the environment of deposition, and dissolution and cementation under meteoric water influences. These early diagenetic processes strongly influence porosity type and quality in the resulting limestones. Upon burial, carbonates react to pressure solution more readily than their siliciclastic counterparts, leading to a more rapid loss of porosity with depth.

Table 1.1 summarizes the differences between carbonate and siliciclastic rocks outlined in Chapter 1(see also Choquette and Pray, 1970; Wilson, 1975).

SUMMARY

TABLE 1.1
Summary of differences between carbonates and siliciclastics

CARBONATE SEDIMENTS, ROCKS	SILICICLASTIC SEDIMENTS, ROCKS
Most occur in tropics	Climate, water depth no constraint
Most are marine	May be marine or non-marine
Organisms erect structure	No analogous process
Sediment texture controlled by growth form and ultrastructure of organisms	Sediment texture reflects hydraulic energy in environment of deposition
Grain composition directly reflects environment of deposition	Grain composition related to provenance of sediment, climate and tectonics of source
Shelf limestones often consist of numerous stacked shoaling upward sequences	Shelf clastics do not generally show cyclicity
Shelf undergoes predictable evolution in response to sea level because rate of carbonate production constant across shelf	Shelf evolution in response to sea level more complex because of potential changes in sediment availability through tectonisim and climate at the source
Often cemented in marine environment	Seldom cemented in marine environment
Muds and grains may be formed by chemical precipitation	Muds and grains formed by breakdown of pre-existing rocks
Susceptible to early diagenetic overprint, porosity difficult to predict	Less susceptible to early diagenesis, porosity related to depositional environment, predictable
More susceptible to burial diagenesis, porosity basement relatively shallow	Less susceptible to burial diagenesis, porosity basement relatively deep

Chapter 2

THE CLASSIFICATION AND NATURE OF CARBONATE POROSITY

INTRODUCTION

Pore systems in carbonates are much more complex than siliciclastics (Choquette and Pray, 1970) (Table 2.1). This complexity is a result of the overwhelming biological origin of carbonate sediments that results in porosity within grains, growth framework porosity within reefs, and the common development of secondary porosity due to pervasive diagenetic processes such as solution and dolomitization affecting the more chemically reactive carbonates throughout their burial history. This tie between diagenetic processes and porosity is the focus of the remainder of this book.

In this chapter, however, the basic classification of carbonate porosity and the nature of the main carbonate porosity types are discussed in order to develop a framework within which to unfold this complex story.

THE CLASSIFICATION OF CARBONATE POROSITY

Introduction

Choquette and Pray (1970) developed a workable carbonate porosity classification that has received wide acceptance. The validity and usefulness of the classification is attested to by the fact that no other viable scheme has been put forward since its introduction. The classification is illustrated in Fig. 2.1. The following sections discuss the various concepts upon which the classification is based, such as fabric selectivity. The utilization of the classification and the construction of porosity categories are covered. The nature of the main classes of porosity recognized in this classification, such as intergranular, moldic, framework, and fenestral are discussed and illustrated.

TABLE 2.1. *Comparison of porosity in sandstone and carbonate rocks. Reprinted by permission of the American Association of Petroleum Geologists.*

Aspect	Sandstone	Carbonate
Amount of primary porosity in sediments	Commonly 25-40%	Commonly 40-70%
Amount of ultimate porosity in rocks	Commonly half or more of initial porosity; 15-30% common	Commonly none or only small fraction of initial porosity; 5-15% common in reservoir facies
Type(s) of primary porosity	Almost exclusively interparticle	Interparticle commonly predominates, but intraparticle and other types are important
Type(s) of ultimate porosity	Almost exclusively primary interparticle	Widely varied because of postdepositional modifications
Sizes of pores	Diameter and throat sizes closely related to sedimentary particle size and sorting	Diameter and throat sizes commonly show little relation to sedimentary particle size or sorting
Shape of pores	Strong dependence on particle shape—a "negative" of particles	Greatly varied, ranges from strongly dependent "positive" or "negative" of particles to form completely independent of shapes of depositional or diagenetic components
Uniformity of size, shape, and distribution	Commonly fairly uniform within homogeneous body	Variable, ranging from fairly uniform to extremely heterogeneous, even within body made up of single rock type
Influence of diagenesis	Minor; usually minor reduction of primary porosity by compaction and cementation	Major; can create, obliterate, or completely modify porosity; cementation and solution important
Influence of fracturing	Generally not of major importance in reservoir properties	Of major importance in reservoir properties if present
Visual evaluation of porosity and permeability	Semiquantitative visual estimates commonly relatively easy	Variable; semiquantitative visual estimates range from easy to virtually impossible; instrument measurements of porosity, permeability and capillary pressure commonly needed
Adequacy of core analysis for reservoir evaluation	Core plugs of 1-in. diameter commonly adequate for "matrix" porosity	Core plugs commonly inadequate; even whole cores (~3-in. diameter) may be inadequate for large pores
Permeability porosity interrelations	Relatively consistent; commonly dependent on particle size and sorting	Greatly varied; commonly independent of particle size and sorting

Choquette and Pray, 1970

Fabric selectivity

The solid depositional and diagenetic constituents of a sediment or rock are defined as its fabric. These solid constituents consist of: various types of primary grains, such as ooids and bioclasts; later-formed diagenetic constituents, such as calcite, dolomite, and sulfate cements; and recrystallization or replacement components, such as dolomite and sulfate crystals. If a dependent relationship can be determined between

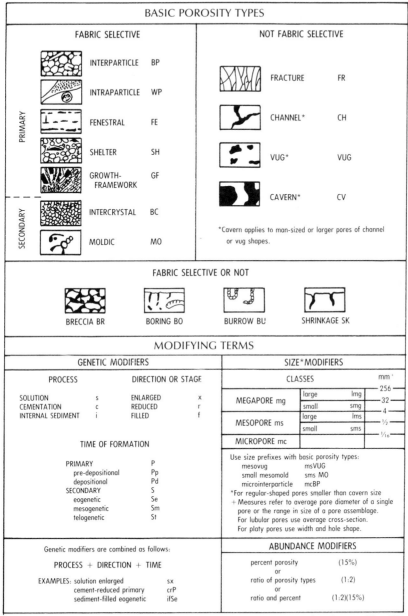

Fig. 2.1. Classification of carbonate porosity. Classification consists of basic porosity types, such as moldic. Each type is represented by an abbreviation (MO). Modifying terms include genetic modifiers, size modifiers and abundance modifiers. Each modifier also has an abbreviation. Reprinted by permission of the American Association of Petroleum Geologists.

porosity and fabric elements, that porosity is referred to as *fabric-selective*. If no relationship between fabric and porosity can be established, the porosity is classed as *not fabric-selective* (Fig. 2.1).

It is important to assess fabric selectivity in order to better describe, interpret, and classify carbonate porosity. Two factors determine fabric selectivity: the configuration of the pore boundary, and the position of the pore relative to fabric (Choquette and Pray, 1970). In most primary porosity, pore boundary shape and location of the pore are determined completely by fabric elements. Primary intergranular pore space in unconsolidated sediments, then, is obviously fabric-selective because its configuration is determined solely by depositional particles, as is primary intragranular porosity, which is controlled by the shape and location of cavities determined by the nature of growth of the organism giving rise to the particle.

In secondary pore systems, however, porosity may be either fabric-selective or not, depending primarily on diagenetic history. As an example, moldic porosity is commonly fabric-selective because of the preferential removal of certain fabric elements from the rock, such as aragonitic ooids and bioclasts early during mineral stabilization (Fig. 2.10), or anhydrite, gypsum, or even calcite from a dolomite matrix later in the diagenetic history of a sequence. On the other hand, phreatic cavern development commonly cuts across most fabric elements, is not fabric-selective, and is controlled primarily by joint systems.

Primary porosity

Primary porosity is any porosity present in a sediment or rock at the termination of depositional processes. Primary porosity is formed in two basic stages, the *predepositional stage* and the *depositional stage*. The predepositional stage begins when individual sedimentary particles form and includes intragranular porosity such as is seen in forams, pellets, ooids, and other nonskeletal grains. This type of porosity can be very important in certain sediments.

The depositional stage is the time involved in final deposition, at the site of final burial of a sediment or a growing organic framework. Porosity formed during this stage is termed *depositional porosity* and is important relative to the total volume of carbonate porosity observed in carbonate rocks and sediments (Choquette and Pray, 1970). The nature and quality of primary porosity will be covered more fully later.

Secondary porosity

Secondary porosity is developed at any time after final deposition. The time involved in the generation of secondary porosity relative to primary porosity may be enormous (Choquette and Pray, 1970). This time interval may be divided into stages based on differences in the porosity-modifying processes occurring in shallow surficial diagenetic environments versus those encountered during deep burial. Choquette and Pray (1970) recognized three stages: *eogenetic, telogenetic, and mesogenetic* (Fig. 2.2).

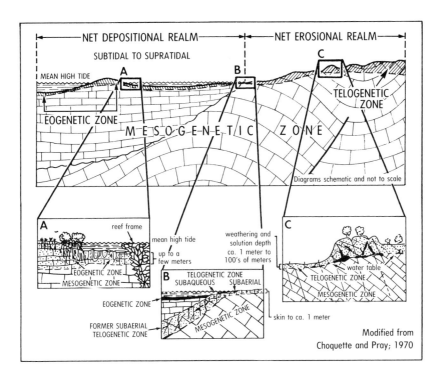

Fig. 2.2. Major environments of porosity evolution as developed by Choquette and Pray, 1970. Reprinted by permission of the American Association of Petroleum Geologists.

The eogenetic stage is the time interval between when sediments have been deposited until when they are buried below the influence of surficial diagenetic processes. The upper limit is generally a depositional interface, which can be either subaerial or subaqueous. In this book, the lower limit of the eogenetic zone is considered to be that point at which surface recharged meteoric waters, or normal (or evaporated) marine waters, cease to actively circulate by gravity or convection.

Generally, the sediments and rocks of the eogenetic zone are mineralogically unstable or are in the process of stabilization. Porosity modification by dissolution, cementation, and dolomitization is quickly accomplished and volumetrically very important. Diagenetic environments that are active within the eogenetic zone include: *meteoric phreatic, meteoric vadose, shallow and deep marine, and evaporative marine.* Each of these diagenetic environments and their importance to porosity development and quality will be discussed in Chapters 4-9.

The mesogenetic stage is the time interval during which the sediments are buried at depth below the major influence of surficial diagenetic processes. In general, the mesogenetic zone is one of rather slow porosity modification and is dominated by compaction and compaction-related processes. While rates are slow, the time interval over which diagenetic processes are operating is enormous, and hence porosity modification (generally destruction) may well go to completion (Scholle and Halley, 1985). The burial diagenetic environment coincides with the mesogenetic stage, and will be discussed in detail in Chapter 9.

The telogenetic stage is the time interval during which carbonate sequences that have been in the mesogenetic zone are exhumed in association with unconformities to once again be under the influence of surficial diagenetic processes. The term *telogenetic* is reserved specifically for the erosion of old rocks, rather than the erosion of newly-deposited sediments during minor interruptions in depositional cycles. As such, sequences affected in the telogenetic zone are mineralogically stable limestones and dolomites which are less susceptible to surficial diagenetic processes. While most surficial diagenetic environments may be represented in the telogenetic zone, meteoric, vadose, and phreatic diagenetic environments are most common. Telogenetic porosity modification will be discussed as a special case under the meteoric diagenetic environment in Chapter 7.

The utilization of the Choquette-Pray porosity classification

The Choquette-Pray porosity classification consists of four elements (Fig. 2.1): *basic porosity types, genetic modifiers, size modifiers, and abundance modifiers.*

Choquette and Pray recognize 15 basic porosity types as shown in Fig. 2.1. Each type is a physically or genetically distinctive kind of pore or pore system that can be separated by characteristics such as pore size, shape, genesis, or relationship to other elements of the fabric. Most of these pore types are illustrated later in this chapter.

Genetic modifiers are used to provide information concerning the processes responsible for the porosity, or modification of the porosity (i.e., solution, cementation), the time of formation of the porosity (i.e., primary or secondary-eogenetic), and whether or not the porosity has been reduced or enlarged during its travels through the burial cycle.

Size modifiers are utilized to differentiate various size classes of pore systems, such as large pores (megapores) from small pores (micropores), while abundance modifiers are used to characterize percentages or ratios of pore types present in a carbonate rock sequence.

Porosity designations are constructed by linking these four elements in the

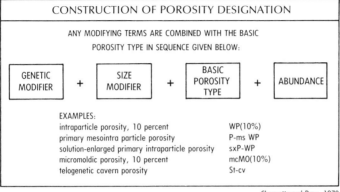

Fig. 2.3. Technique for constructing a porosity designation. Reprinted by permission of the American Association of Petroleum Geologists.

sequence shown in Fig. 2.3. The use of the shorthand designations shown on Fig. 2.1, such as BP for interparticle and MO for moldic, allows rapid, precise, porosity designation. Accurate porosity description utilizing a well known scheme such as is described here, is probably the single most important task that an exploration or exploitation geologist can perform in a carbonate province.

THE NATURE OF PRIMARY POROSITY IN MODERN CARBONATE SEDIMENTS

Interparticle porosity

Mud-free carbonate sediments, like their siliciclastic counterparts, are dominated by intergranular porosity at the time of deposition. These sediments exhibit porosities from 40- 50% (Fig. 2.4) (Enos and Sawatsky, 1981), which is near the upper limit of 48%

expected in spherical particles with minimum packing (Graton and Frazer, 1935). The excess porosity over the 27-30% expected in spherical particles showing close, maximum packing and commonly observed in siliciclastic sediments (Graton and Frazer, 1935) is because of the wide variability of particle shape seen in carbonates. This shape variation seems to be a function of their biological origin (Dunham, 1962), and the common presence of intraparticle porosity that may occupy a significant percentage of the bulk volume of the sediment (Enos and Sawatsky, 1981).

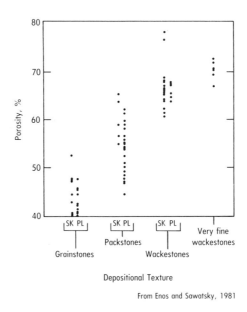

Fig. 2.4. Primary depositional porosity exhibited by various Holocene carbonate sediment textural types. Used with the permission of SEPM.

Intraparticle porosity

Intraparticle depositional porosity is one of the fundamental differences between carbonate and siliciclastic porosity. Intraparticle porosity may originate in a variety of ways. The living chambers of various organisms such as Foraminifera, gastropods, rudists, and brachiopods often provide significant intraparticle porosity (Fig. 2.7). The ultrastructure of the tests and skeletons of organisms--such as the open fabric of the *Halimeda* segment, consisting of a felted framework of aragonite needles, or the open structure of many coral polyps--can also provide intraparticle porosity (Fig. 2.7). The ultrastructure of some abiotic grains, such as ooids, and composite grains, such as pelloids, which consist of packed, needle-shaped crystals, may also lead to significant intraparticle porosity (Fig. 2.7) (Robinson, 1967; Loreau and Purser, 1973; Enos and

Sawatsky, 1981). Finally, the activity of microboring algae and fungi may significantly increase the intraparticle porosity of carbonate grains, before, during, and shortly after deposition (Perkins and Halsey, 1971).

Depositional porosity of mud-bearing sediments

Carbonate sediments containing mud range in porosity from 44 to over 75%. Grain-supported muddy sediments such as packstones show the lowest porosity range (44 to 68%), mud-supported sediments (wackestones) show porosities from 60-78% (Fig. 2.4) (Enos and Sawatsky, 1981), while deep marine oozes can have porosities of up to 80% (Schlanger and Douglas, 1974). The high porosities seen in the mud-supported shelf sediments are surely the effect of shape and fabric (the mud fraction in modern sediments is dominated by elongate needles which pack like jackstraws), and perhaps the effect of oriented sheaths of water molecules responding to the strongly polar aragonite crystals that result from the common orientation of the carbonate radical (Enos and Sawatsky, 1981). The exceptionally high porosities reported for deep marine oozes, however, are undoubtedly the result of the high intragranular porosity found in the dominant organic components of these sediments, such as the inflated chambers of

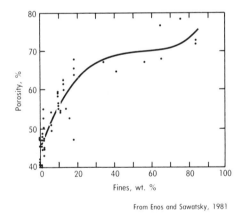

From Enos and Sawatsky, 1981

Fig. 2.5. Weight % of carbonate fines (silt plus clay) in Holocene sediments plotted against % porosity. Used with permission of SEPM.

pelagic foraminifera. Schlanger and Douglas (1974) reported 45% intragranular porosity, and 35% intergranular porosity for pelagic oozes encountered during Deep Sea Drilling Program operations in the Pacific. The high porosities reported for packstones by Enos and Sawatsky (1981), however, remain an enigma. Logically, one would assume that as the pores of a grain-supported sediment are filled with mud, the porosity of the resulting mixture (packstone) should decrease. In reality, if the porosity of modern mud-bearing sediment is plotted against weight percent fines, one sees a steady increase in porosity with increasing weight percent fines, rather than the predicted porosity decrease in the grain-supported field (Fig. 2.5). Enos and Sawatsky (1981) believed that the sediments used in their study might have contained small, isolated domains of mud-supported sediment that could have masked the expected drop in porosity in the packstone field.

Permeability characteristics of modern grain-supported versus mud-supported sediments show an inverse relationship to porosity, with the lower porosity grain-supported sediments exhibiting the highest permeability, and the higher porosity mud-supported sediments, the lowest permeability, as might be expected (Fig. 2.6) (Enos and Sawatsky, 1981).

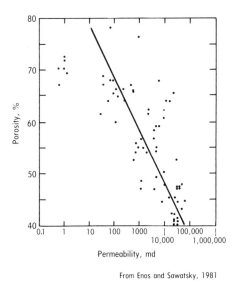

From Enos and Sawatsky, 1981

Fig. 2.6. Porosity-permeability plot of Holocene carbonate sediments. Used with permission of SEPM.

Framework and fenestral porosity

Framework porosity, associated with the activity of reef-building organisms, can be an important depositional porosity type in the reef environment. Framebuilders, such as scleractinian corals, can construct an open reef framework, potentially enclosing enormous volumes of pore space during the development of the reef (Fig. 2.8). It is this depositional porosity potential that has long attracted the economic geologist to the study of reefs. Framework porosity potential is, however, dependent on the type of framebuilding organism. While scleractinian corals construct an open framework reef, coralline algae, and in the past, stromatoporoids, and sponges have tended to erect a more closed framework structure with significantly less framework porosity because of the general encrusting mode of growth of these organisms (Fig. 2.9A) (James, 1983) . The borings of a number of organisms, such as clionid sponges and pelecypods, can generate a significant volume of porosity in the reef framework during the development of the reef complex (Fig. 2.8B) (Moore and others, 1976; Land and Moore, 1980; James, 1983). Framework porosity tends to become quickly filled during early reef development by both coarse, and fine-grained, internal sediments, leading to a complex depositional pore system (Schroeder and Zankl,1974; Land and Moore, 1980; James, 1983).

Fenestral porosity (Fig. 2.9B), commonly associated with supratidal, algal-related, mud-dominated sediments can be locally important (Shinn, 1968). Enos and Sawatsky (1981) report depositional porosities from supratidal, algal-related wackestones of up to 65 % (Fig. 2.4). One would expect low permeabilities from such a mud-dominated system after lithification. Subsequent dolomitization, however, would tend to increase the permeability significantly, opening communication between the larger fenestral pores through the intercrystalline porosity developed in the matrix dolomite.

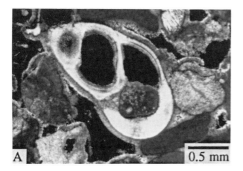
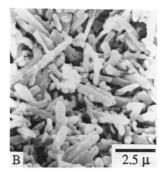

Fig. 2.7. Primary intragranular porosity. A. Living chamber of a Holocene gastropod, Florida. B. Ultrastructure of Halimeda sp. showing porosity between aragonite elements of the segment.

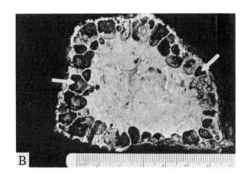

Fig. 2.8. A. Framework porosity developed within a modern coral reef, Grand Cayman, West Indies. Vertical dimension of the pore is 3 m. B. Boring sponge (Clionid) galleries (arrows) within modern coral framework, Jamaica. Galleries are being filled with pelleted internal sediment and marine cement.

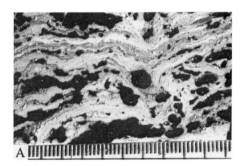
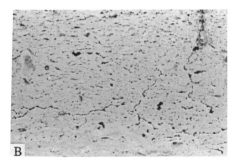

Fig. 2.9. A. Framework porosity developed in Holocene algal cup reef, Bermuda. Coralline algae show as white layers, voids are black. Scale in cm, photo by R.N.Ginsburg. Reprinted by permission of the American Association of Petroleum Geologists. B. Fenestral porosity developed in semi-lithified muddy supratidal sequence, Sugarloaf Key, Florida. Photo by Gene Shinn. Reprinted by permission of the American Association of Petroleum Geologists.

SECONDARY POROSITY

Introduction

Primary depositional porosity is progressively lost during burial through a number of interrelated processes (Chapter 9). During this same evolutionary journey, porosity generating processes, which are dominantly diagenetic in nature, may also operate to increase the total pore volume of the evolving limestone. These processes are mainly of two types: 1) porosity generation by dissolution, and 2) porosity as a result of dolomitization. Fracturing, another important process, generally acts to increase permeability, rather than total pore volume.

Secondary porosity formation by dissolution

Dissolution of limestones and sediments may occur at any point in the burial history of the sequence (Chapter 9, Fig. 9.30). The dissolution event, with its attendant

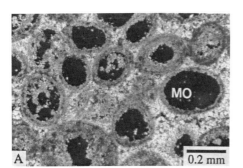 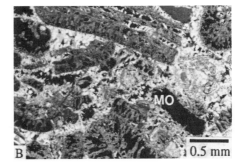

Fig. 2.10. Moldic porosity. A. Oomoldic porosity (MO), Jurassic Smackover Formation, Miller Co. Arkansas, U.S.A. Crossed polars. B. Biomoldic porosity (MO), Miocene limestone, subsurface core, Enewetak Atoll. Aragonite Halimeda segments partially dissolved. Crossed polars.

increase in pore volume, generally occurs in response to a significant change in the chemistry of the pore fluid, such as a change in salinity, temperature, or partial pressure of CO_2. These changes are most likely to occur early in the history of burial (eogenetic stage), such as the development of a meteoric water system in a shallow shelf sequence; late in the history of burial (mesogenetic stage), where hydrocarbon maturation or shale dewatering may provide aggressive fluids; or finally, anytime during burial history, when limestones have been exhumed in association with an unconformity (telogenetic stage) and placed into contact with meteoric waters (James and Choquette, 1984).

If original marine pore fluids are replaced by meteoric waters early in the burial history of a carbonate sequence, before mineral stabilization, the process of dissolution leading to porosity enhancement can be distinctly fabric-selective, and controlled by the mineralogy of individual grains. As was mentioned previously, modern shallow-marine carbonate sediments are composed of a metastable mineral suite consisting of aragonite and magnesian calcite. Both are unstable under the influence of meteoric waters but generally take contrasting paths to the stable phases, calcite and dolomite (James and Choquette, 1984) (Chapter 6, this book, Fig. 6.6). Both magnesian calcite and aragonite stability are achieved by the dissolution of the unstable phase and precipitation of the stable phase. Magnesian calcite dissolves incongruently, giving up its magnesium without the transport of significant $CaCO_3$, and generally does not exhibit an empty moldic phase (Land, 1967; James and Choquette, 1984). Aragonite, on the other hand, dissolves, and depending on the rate of fluid flow and the state of saturation of the transient pore fluid, often shows significant transport of $CaCO_3$ away from the site of dissolution and thus, the formation of moldic porosity (Fig. 2.10) (Matthews, 1974; Longman, 1980). It is this early moldic porosity formed during the rapid stabilization of aragonitic limestones in meteoric water systems that may provide much of the early calcite cement that often occludes the intergranular depositional porosity of shallow-marine carbonate sequences (Longman, 1980; Moore and Druckman, 1981; James and Choquette, 1984) (Chapters 6 and 7, this book).

As will be seen later in Chapters 6 and 7, there is some disagreement as to whether there is any net effect on the total porosity of carbonate sequences by eogenetic processes, such as solution and cementation.

Dissolution occurring later in the burial history of a carbonate sequence, after mineral stabilization, will generally be characterized by non-fabric-selective dissolution, where the resulting pores cut across all fabric elements such as grains, cement, and matrix (Fig. 2.11), rather than being controlled by preferential dissolution of certain fabric elements, such as grains, that were originally aragonite. These pore types are commonly called vugs, channels, and caverns, depending on size (Choquette and Pray, 1970). This type of porosity development in exhumed limestones associated with unconformities (telogenetic) is the direct result of exposure to meteoric vadose and phreatic conditions where high partial pressures of CO_2 are common, and waters are undersaturated relative to most carbonate phases, including dolomite (James and Choquette, 1984). Vuggy or cavernous late secondary porosity associated with unconformities can be extensive, as well as economically important (Chapter 6) (James and Choquette, 1984; Choquette and James, 1987).

Late subsurface secondary porosity (mesogenetic) is a bit more difficult to explain because most subsurface fluids are believed to be supersaturated with respect to most

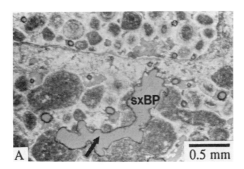
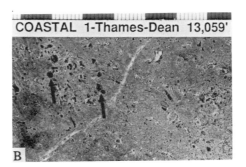

Fig. 2.11. A. Solution-enlarged, intergranular porosity (sxBP) developed in the subsurface in Jurassic upper Smackover ooid grainstone reservoir rocks, Arkansas, U.S.A. Plain light. Reprinted by permission of the American Association of Petroleum Geologists. B. Vuggy porosity developed in Smackover ooid grainstones, subsurface, western Alabama. Scale in cm.

carbonate phases because of long-term, extensive rock-water interaction (Choquette and James, 1987). However, high pressures, temperatures, hydrocarbon maturation, and thermal degradation have been linked to the development of aggressive subsurface fluids that have the capability of developing significant subsurface secondary pore space in both siliciclastics as well as carbonates (see Chapter 7). In some cases, the porosity development seems to have been substantial, such as is commonly seen in the Upper Jurassic Smackover Formation of southern Arkansas (Figure 2.11) (Moore and Druckman, 1981). This type of late porosity development is often described as a simple expansion of early secondary moldic, or primary intergranular porosity, to vuggy or cavernous porosity (Fig. 2.11) (Choquette and Pray, 1970; Moore and Druckman, 1981; Choquette and James, 1987).

In both subsurface and unconformity-related secondary porosity, the ultimate distribution of porosity is controlled either by preexisting porosity established earlier by diagenesis or original depositional environments, or by fractures, by faults, and perhaps even by distribution of stylolites.

Secondary porosity associated with dolomitization

Intercrystalline porosity associated with dolomites form an important reservoir type in a number of settings ranging from supratidal/sabkha to normal marine sequences (Murray, 1960; Roehl and Choquette, 1985). There has long been discussion concerning

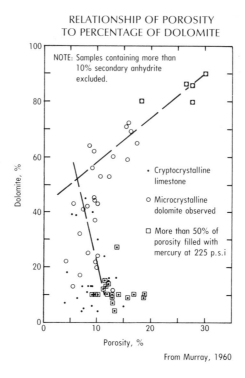

Fig. 2.12. Relationship of porosity to % dolomite in the Midale beds of the Charles Formation, Midale field, Saskatchewan, Canada. Used with permission of SEPM.

the role of dolomitization in porosity development and destruction (Fairbridge, 1957). Murray (1960) documented the close relationship between percentage of dolomite and porosity in the Midale beds of the Charles Formation in Saskatchewan (Fig. 2.12). In this example, porosity initially dropped as dolomite percentage increased until 50% dolomite was reached, at which point, porosity increased with increasing dolomite percentage (Fig. 2.12). Murray (1960) explained the dolomitization control of porosity by concurrent dissolution of calcite to provide the carbonate for dolomitization. In Murray's example, the Midale beds were originally lime muds. In those samples with less than 50% dolomite, the undolomitized mud compacted during burial, and the floating dolomite rhombs occupied porosity, causing porosity to decrease with increasing dolomitization. At 50% dolomite, the rhombs act as a framework, thus preventing compaction, and as dolomite percent increases, porosity also increases. The mode by which interrhomb calcite is lost is critical to the understanding of dolomite-related porosity. If the interrhomb calcite is lost by dissolution unrelated to dolomitization, the increase in porosity is related to a specific diagenetic event that will affect only those dolomitic

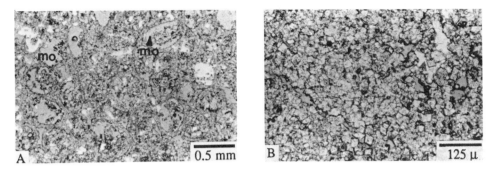

Fig. 2.13. A. Moldic (MO), intercrystalline (BC) porosity in Jurassic upper Smackover reservoir rocks, Bryans Mill field, Texas. Plain light. B. Lower Cretaceous central Texas tidal flat dolomite. Porosity (BC) approximately 35%. Plain light.

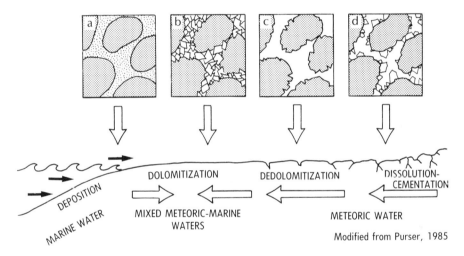

Fig. 2.14. Geologic setting for the dolomitization-dedolomitization of Jurassic ooid packstones from the Paris Basin. Final porosity consists of cement-reduced, dolomite crystal-moldic porosity. Reprinted from Carbonate Petroleum Reservoirs, with permission. Copyright, Springer-Verlag, New York.

sequences with appropriate geologic and burial histories, such as exposure along an unconformity with related meteoric phreatic dissolution. On the other hand, if the loss of calcite is associated with the dolomitization process itself (calcite providing the CO_3 for dolomitization), then most such dolomites should exhibit significant porosity.

Weyl (1960) built a compelling case for local production of CO_3 by concurrent dissolution of calcite during dolomitization as the dominant source for dolomite-related porosity based on conservation of mass requirements. In addition, he noted that if dolomitization was truly a mole-for-mole replacement with local sourcing of carbonate, then the greater specific gravity of dolomite would lead to an increase of 13% porosity in the conversion from calcite to dolomite. If the dolomitization occurs early, prior to compaction, such as on a tidal flat, original porosity of the mud could exceed 60% (Enos and Sawatsky, 1981). These high initial porosities combined with the above solid volume decrease during dolomitization could easily lead to the 30-40% porosity exhibited by many sucrosic dolomite reservoirs (Murray, 1960; Choquette and Pray, 1970; Roehl and Choquette, 1985).

While it is uncertain whether it is actually local sourcing of CO_3 that takes place during most dolomitization events, the potential for fresh water flushing, and hence enhancement of porosity by dissolution of relict calcite or aragonite (Murray, 1960), is great in most shallow marine-transitional dolomitization environments, such as the tidal flat, sabkha, or zones of fresh-marine water mixing (see Chapters 5 and 8).

An interesting variation on dolomite-related porosity is provided by Purser (1985), who describes relatively early dolomite dissolution by fresh meteoric water associated

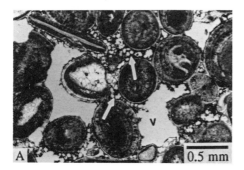 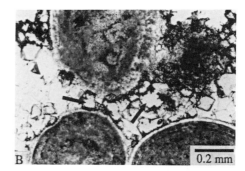

Fig. 2.15. Dolomite moldic (MO) porosity in Jurassic Great Oolite at Bath, United Kingdom. A. General view showing vuggy (v) porosity development in the main pores, and dolomite molds (arrows) concentrated between the grains. Plain light. B. Closer view showing dolomite molds (arrows) outlined in bitumen. Plain light.

with the landward margins of a prograding coastal plain developed in middle Jurassic sequences of the Paris Basin. The dolomites were formed either in tidal flats, or in a mixed fresh-marine water-mixing zone, and subsequently were flushed by a major meteoric water system that dissolved the dolomite rhombs as progradation proceeded (Fig. 2.14). Reservoirs at Coulommes field in the central Paris Basin, 35 km east of Paris, exhibit an average of 15% porosity developed in rhombic molds and interparticle porosity. A similar occurrence was described by Sellwood and others (1987) in the Jurassic Great Oolite reservoir facies in the subsurface of southern England (Fig. 2.15).

Secondary porosity associated with breccias

Brecciation of carbonate rock sequences can occur in a number of situations, including: evaporite solution collapse, limestone solution collapse (Fig. 2.16A), faulting, and soil formation (Blount and Moore, 1969). Limestone breccias, particularly those associated with evaporite or limestone solution collapse, often result in enhanced porosity that may form either a reservoir for hydrocarbons or a host for mineralization. Loucks and Anderson (1985; see Chapter 7) describe porous zones developed in evaporite solution collapse sequences in the Ordovician Ellenburger at Puckett field in west Texas. These breccia zones developed during periodic subaerial exposure of the landward portions of the lower Ellenburger shoreline. The influence of meteoric waters, then, contributed significantly to the productivity of Puckett field.

 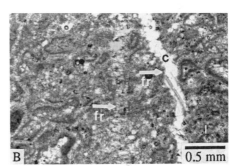

Fig. 2.16. A. Breccia porosity (BR) associated with evaporite solution collapse breccia, L.Cretaceous, central Texas, U.S.A. B. Calcite cement (c) reduced fracture porosity (FR) in Jurassic Smackover limestone in the Sun Bd. of Supervisors S-768, 10768' (3283 m), Clarke Co. Mississippi, U.S.A. Plain light.

Porous breccia sequences developed during karstic limestone collapse are often associated with major unconformities and can produce reservoirs of prodigious size, such as the Mississippian Northwest Lisbon field, Utah (Miller, 1985), and the Permian Yates field (Craig, 1988; see Chapter 7).

Secondary porosity associated with fractures

Intense fracturing is present and affects the reservoir characteristics of some of the world's largest oil fields (Roehl and Choquette, 1985). While it is not always clear how much actual porosity is gained during the fracturing of carbonate reservoir rocks, because of the difficulty in measuring this type of porosity, there can be little doubt concerning the benefits that fractures can bring to ultimate reservoir production. Consider, for example, the case of Gaschsaran oil field in Iran, where a single well can produce up to 80,000 barrels of oil per day from fractured Oligocene Asmari Limestone with a matrix porosity of just 9% (McQuillan, 1985). Another example is the Monterrey Shale, at West Cat Canyon field, California, where matrix porosity is not measurable, but effective fracture porosity averages 12% and the field has 563 million barrels of oil in place (Roehl and Weinbrandt, 1985).

Fracturing is particularly effective and common in carbonate reservoirs because of the brittle nature of carbonates relative to the more ductile fine-grained siliciclastics with which they are often interbedded (Longman, 1985). Fracturing can take place at practically any time during the burial history of a carbonate sequence starting with shallow burial because of common early lithification. Fracturing can be associated with faulting, folding, differential compaction, salt dome movement, and hydraulic fracturing within overpressured zones (Roehl and Weinbrandt, 1985; Longman, 1985; McQuillan, 1985).

Fractures in carbonates are commonly filled with a variety of mineral species including, calcite, dolomite, anhydrite, galena, sphalerite, celestite, strontianite, and fluorite (Fig. 2.16B). These fractures are, however, generally dominated by carbonate phases. Fracture fills are precipitated as the fracture is being used as a fluid conduit. CO_2 degassing during pressure release associated with faulting and fracturing in the subsurface can result in extensive, almost instantaneous calcite and dolomite precipitation in the fracture system (Woronick and Land, 1985; Roehl and Weinbrandt, 1985). These late carbonate fracture fills commonly have associated hydrocarbons as stains, fluid inclusions, or solid bitumen, partial fracture fills (Moore and Druckman, 1981; Roehl and Weinbrandt, 1985).

SUMMARY

Carbonate sediments and rocks generally have a much more complex pore system than do siliciclastics because of the wide variety of grain shapes common in carbonates, the presence of intragranular, framework, and fenestral porosity in carbonates, and the potential for the development of moldic and highly irregular dissolution-related porosity in carbonates.

These complex pore systems necessitate the utilization of a comprehensive carbonate porosity classification based on four elements: basic pore types (fabric-selective or not, primary versus secondary); genetic modifiers (process, direction, time of formation); size modifiers (megapore, mesopore, micropore); and abundance modifiers (percentage or ratio, or combination).

Primary porosity in carbonate sedimentary sequences is often higher than similar siliciclastic sequences because of the common occurrence of intragranular, fenestral, and framework porosity in carbonate sediments.

Secondary porosity in carbonates is dominated by dissolution processes that may affect the sequence at any time during burial. If early (eogenetic), fabric-selective, secondary porosity is common with its configuration controlled by individual grain mineralogy. If late (telogenetic or mesogentic), porosity is generally non-fabric-selective, but its distribution is controlled by porosity existing at the time of the dissolution event.

Non-fabric-selective breccia and fracture porosity may be exceedingly important porosity types in the subsurface. While brecciation may significantly enhance the total porosity of a sequence, fractures generally enhance permeability rather than total porosity.

Chapter 3

DIAGENETIC ENVIRONMENTS OF POROSITY MODIFICATION AND TOOLS FOR THEIR RECOGNITION IN THE GEOLOGIC RECORD

INTRODUCTION

As discussed in Chapter 2, Choquette and Pray (1970) sketched a broad temporal and spatial framework within which they considered the nomenclature and classification of carbonate porosity. They differentiated between primary porosity, developed before and during deposition, and secondary porosity, developed after deposition. It is post-depositional porosity evolution, and the ultimate predictability of this porosity that will concern us for much of the remainder of this book.

Choquette and Pray (1970) recognized three zones (eogenetic, mesogenetic and telogenetic) in which post-depositional porosity modification and evolution occur (Fig. 2.1). While their scheme is useful in the discussion of the nomenclature and classification of carbonate porosity, it is really an inadequate framework within which to consider the diagenetic processes and products responsible for the porosity development and evolution in carbonate rock sequences after deposition, because it ignores the basic fuel of diagenesis—water. The chemical characteristics of carbonate pore fluids, the rate of flux through the pore system, and the temperature and pressure regimen under which the resulting rock-water interactions are effected control the diagenetic processes that affect and modify carbonate porosity through dissolution of existing carbonate phases and/or the precipitation of new phases. Therefore, the environmental framework within which porosity development and evolution will be discussed is based on three broad families of waters and their distribution either on the surface or in the subsurface, as adapted from Folk (1974).

THE DIAGENETIC ENVIRONMENTS OF POROSITY MODIFICATION

Introduction

There are three major diagenetic environments in which carbonate porosity is formed or modified: *meteoric, marine,* and *subsurface* (Fig. 3.1). Two surface or near-surface environments, the meteoric and marine (eogenetic and telogenetic zones of

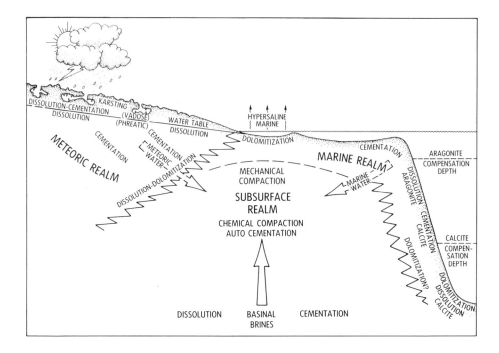

Fig. 3.1. Schematic diagram illustrating the common diagenetic environments within which post-depositional porosity modification and evolution occurs. Active diagenetic processes are identified in each of the three main diagenetic realms.

Choquette and Pray, 1970), are marked by the presence of pore fluids that are distinctly different from each other, and a third, the subsurface environment (mesogenetic zone of Choquette and Pray, 1970), is characterized by mixtures of marine-meteoric waters or complex basinal-derived brines.

Marine environment

The marine environment in which most carbonate sediments originate is characterized by normal or modified marine pore fluids generally supersaturated with respect to most carbonate mineral species (Bathurst, 1975) (Chapter 4, this book). The marine environment, therefore, is potentially the site of extensive porosity destruction by marine cements (James and Choquette, 1983). This environment has little potential for the formation of secondary porosity by dissolution, except for the deep marine environment where supersaturated surface waters become progressively undersaturated with depth (Fig. 3.1). The distribution of marine cementation is generally controlled by the rate of fluid movement through the sediment pore systems, and hence is dramatically affected by conditions at the site of deposition, such as energy levels, sediment porosity and permeability, and rate of sedimentation. Cementation, therefore, is not ubiquitous in the marine environment, but occurs only in certain favorable subenvironments, such as within the shelf margin reef and the intertidal zone (see Chapter 4, this book).

Secondary porosity development through dolomitization is common in association with evaporative marine waters (Chapter 5, this book) and may also occur where normal marine waters are flushed through limestone sequences by thermal convection (Chapter 4, this book).

Meteoric environment

The meteoric environment is characterized by common exposure to subaerial conditions and the presence of relatively dilute waters that exhibit a wide range of saturation states, from the strongly undersaturated to the supersaturated. The supersaturated refer to the stable carbonate mineral species, calcite and dolomite. However, meteoric waters are generally strongly undersaturated with respect to the metastable carbonate species, aragonite and magnesian calcite (Bathurst, 1975; James and Choquette, 1984; Chapter 6, this book). The ready availability of soil zone carbon dioxide to meteoric fluids within the vadose zone (that zone above the water table where both air and water are present) greatly influences the saturation state of these waters relative to carbonate mineral phases. The meteoric environment, therefore, has a high potential for the generation of secondary porosity by dissolution, as well as the potential for porosity destruction by passive cementation. The porosity modification potential of the meteoric environment is often enhanced by relatively rapid rates of fluid flow through the phreatic zone of the system (Hanshaw and others 1971; Back and others, 1979).

Subsurface environment

The subsurface environment is characterized by pore fluids that may either be a mixture of meteoric and marine waters (Folk, 1974), or a chemically complex brine resulting from long-term, rock-water interaction under elevated temperatures and pressures (Stossell and Moore, 1983). Because of this extensive rock-water interaction, these fluids are generally thought to be supersaturated with respect to most stable carbonate species such as calcite and dolomite (Choquette and James, 1987). However, under the high pressure and temperature regimens of the subsurface, pressure solution is an important porosity destruction process that is often aided by cement precipitation in adjacent pore spaces due to the general supersaturation of the pore fluids. Finally, local areas of undersaturation related to thermal degradation of hydrocarbons may result in secondary porosity generation by dissolution.

Most diagenetic processes operate slowly in the subsurface because of the relatively slow movement of fluids under conditions of deep burial (Choquette and James, 1987) (Chapter 9, this book). These processes, however, have enormous spans of geologic time within which to accomplish their work.

TOOLS FOR THE RECOGNITION OF DIAGENETIC ENVIRONMENTS OF POROSITY MODIFICATION IN THE GEOLOGIC RECORD.

Introduction

Each of the diagenetic environments briefly described above is a unique system in which porosity is created or destroyed in response to diagenetic processes driven by the chemical and hydrological characteristics of pore fluids and the mineralogical stability of the sediments and rocks upon which these fluids act. The uniqueness of these environments is the basis for the predictive conceptual models that have been developed over the years to aid the economic geologist in anticipating the nature, quality, and distribution of carbonate porosity on a regional and a reservoir level.

In order to fully utilize the predictive value of these models, the geologist must be able to place the dominant pore modification event, whether dissolution or cementation, into this environmental framework. If, for example, one is able to determine that a pore-occluding cement was precipitated early—during mineral stabilization in a regional meteoric phreatic environment associated with a sea level lowstand, or stillstand—it might be reasonable to expect, based on the knowledge of this environment, to find more favorable porosity development in an up-dip, or up-flow direction (Longman, 1980; Moore and Druckman, 1981).

Today, the geologist in search of answers to those questions concerning the conditions under which porosity modification events take place has a broad spectrum of technology available for the hunt. Fundamental *petrographic relationships* between various cements, and among grains, cements, and matrix, as well as the relationship of carbonate phases to other introduced phases, such as sulfates, allow the geologist to construct relative paragenetic sequences, and to time porosity modification events relative to other diagenetic phases. Cement morphology, cement distributional patterns, and cement size can be related to certain aspects of the precipitational environment, such as chemistry of the pore fluids, rate of cement precipitation, and the relative saturation of the pore system with water (vadose zone versus phreatic zone).

The *isotopic and trace element chemistry* of pore-filling cements, as well as the chemistry of porosity-enhancing dolomites can be related to the trace element and isotopic chemistry of the precipitating and dolomitizing fluids. Therefore, carbonate geochemistry can be used in conjunction with basic petrography to help identify the diagenetic environment, such as meteoric-phreatic, in which porosity modification was accomplished.

Oxygen isotopic composition in conjunction with *two-phase fluid inclusions* can be used to estimate the temperature of formation of diagenetic phases, such as cements and dolomites, and therefore further refine paragenetic sequences as well as place more realistic time constraints on porosity modification events.

The remainder of this chapter is devoted to a general discussion of some of the most useful tools available to the geologist seeking to determine the environment in which important porosity modification events have occurred. This discussion is not meant to be definitive, because, obviously, entire books could be and have been written on each subject mentioned above. Instead, it is hoped that the reader will have an introduction to each technique and observation, and, most importantly, an outline of the major problems that can seriously constrain their use. The observational petrographic techniques are stressed in this discussion because they are generally available to most working geologists in both industry and academia. The more sophisticated instrument-oriented techniques, such as isotopes and trace element geochemistry, are included in a most general way, so that the working geologist can become familiar with the manner in which they are currently used by the specialist. Familiarity with the technique and its constraints will allow subsequent evaluation of the appropriateness of the technique's application to specific problems of porosity evolution.

Petrography-cement morphology

Numerous authors have commented on the diverse morphology of calcite and aragonite pore-fill cements and have attempted to relate crystal morphology to the chemical environment of precipitation (Lippmann, 1973; Folk, 1974; Lahann, 1978; Lahann and Seibert, 1982; Given and Wilkinson, 1985). The conventional wisdom of the '60s and '70s was that most fresh water calcite cements tend toward equant shapes, while marine calcite and aragonite cements tend toward elongate fibrous shapes. Folk (1974) developed a model directly relating the morphology of calcite cements to the Mg/Ca ratio of the precipitational fluid. Folk's model was based on the concept of the sidewise poisoning of the growing calcite crystal by the substitution of the Mg ion for Ca. The smaller ionic radius of the Mg ion, relative to the Ca ion, causes lattice distortion at the edge of the growing crystal, stopping growth at the edge, and ultimately leading to the elongation of the crystal in the C-axis direction (Fig. 3.2). According to Folk's scheme, therefore, meteoric waters, generally characterized by low Mg/Ca ratios, would tend toward equant crystal shapes, while marine waters marked by high Mg/Ca ratios would produce the elongate calcite crystals noted in marine beach rocks and reef sequences (Fig. 3.3). Fig. 3.4 illustrates common calcite cement morphologies found in nature.

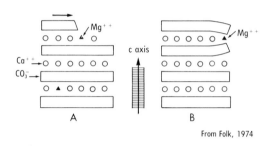

From Folk, 1974

Fig. 3.2. Morphology of calcite crystals as controlled by selective Mg-poisoning (Folk, 1974). If as is shown in A, a Mg ion is added to the end of a growing crystal it can easily be overstepped by the next succeeding CO_3 layer without harm to the crystal growth. If, however, as in B, the small Mg ion is added to the side of the crystal, the adjacent CO_3 sheets are distorted to accommodate it in the lattice, hampering further sideward growth, and resulting in the growth of small, fibrous crystals. Used with permission of SEPM.

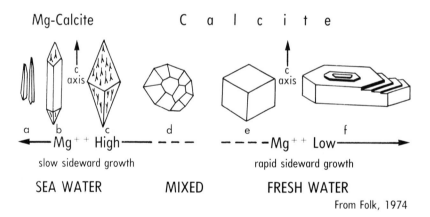

From Folk, 1974

Fig. 3.3. Calcite crystal growth habit as a function of Mg/Ca ratio. Used with permission of SEPM.

Lahann (1978), while agreeing with Folk that Mg could poison the growth of calcite due to lattice distortion, argued that the Mg poisoning effect could not be used to explain morphological differences in calcite crystals, because the Mg poisoning should affect all growing surfaces equally, including the C-axis direction. Instead, Lahann called upon the differences in surface potential that develop on the different calcite crystal faces due to calcite crystallography to explain the range of morphologies observed in naturally formed calcites. For example, a crystal edge parallel to the C-axis will expose both Ca and CO_3 ions to the precipitating fluid, while the growing face normal to the C-axis will expose either Ca or CO_3 ions to the precipitating fluids, but not both at the same time (Fig.3.2). In marine water, where there is a great excess of surface-active cations compared to surface-active anions, the C-axis face will normally be saturated with cations, creating a strong positive potential on the C-axis face that will be greater than that found on the edge parallel to the C-axis, because both CO_3 as well as Ca will be exposed to the fluid. This positive potential will then ensure a greater concentration of CO_3 ions at the C-axis face, and will result in elongation of the calcite crystal parallel to

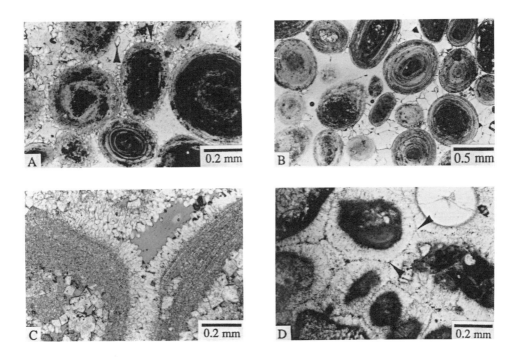

Fig. 3.4. Common calcite cement morphologies found in nature. A. Small equant to rhombic calcite crystals (arrows) forming a crust around grains. Meteoric phreatic zone, Joulters Cay, Bahamas. Plain light. B. Equant calcite mosaic showing irregular distribution pattern. Sample taken at water table, at Joulters Cay, Bahamas. Plain light. C. Bladed calcite circumgranular crust cement in ooid grainstone that has seen meteoric water conditions. Upper Jurassic Smackover Formation, Murphy Giffco #1, Bowie Co. Texas, 7784' (2373 m). Plain light. D. Fibrous magnesian calcite marine cement a Holocene grainstone collected from the Jamaica island slope at 260 m below the sea surface by Research Submersible Nekton. Note the polygonal suture patterns developed between adjacent cement crusts (arrows). Plain light.

the C-axis under normal marine conditions.

The Lahann model satisfactorily explains the common occurrence of equant calcite crystals in most meteoric water situations. In this case, both CO_3 and Ca ion concentrations are low; saturation with respect to most carbonate phases are at equilibrium or undersaturated; and differences in surface potential between faces are minimal, resulting in equant calcite morphologies. The major exception is in the meteoric vadose environment where rapid CO_2 outgassing would lead to elevated saturation conditions; the precipitation of elongate calcite crystals, such as are found in speleothems; and the whisker crystal cements of the soil zone (Given and Wilkinson, 1985).

Most subsurface calcite cements are also equant, and while Ca ion concentrations might be high, low CO_3 availability results in slow equant crystal growth. A parallel situation exists in the deep marine environment, close to the calcite lysocline, where Ca ion concentrations are close to those found in surface waters. However, CO_3 availability is limited, and therefore these calcites exhibit an equant morphology (Schlager and James, 1978; Given and Wilkinson, 1985).

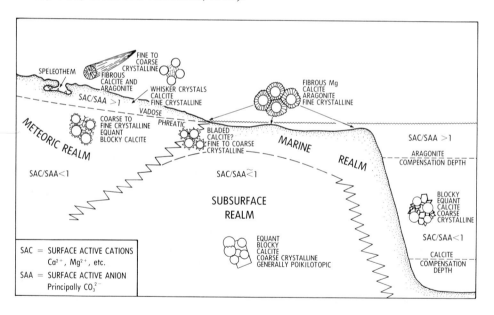

Fig. 3.5 Schematic diagram showing anticipated growth habits of pore-fill calcite cement in the principal diagenetic environments. as controlled by the ratio of surface active cations (SAC) to surface active anions (SAA).

In summary, calcite crystal morphology seems to be related, by way of the CO_3 ion, to the rate of precipitation, as controlled by the state of carbonate saturation within the environment. While each of the major diagenetic environments exhibits a wide range of saturation relative to calcite and can thus support a range of calcite morphologies, the mean water chemistry of each environment is distinctive enough so that meteoric-phreatic waters generally precipitate equant calcite, while surface marine waters generally precipitate fibrous-to-bladed calcite, and subsurface waters almost always precipitate equant-to-complex polyhedral calcite cements. Fig. 3.5 summarizes the general calcite cement morphology that can be expected within our major diagenetic environments and subenvironments.

This is obviously a simplistic view of a complex problem (Given and Wilkinson, 1985), especially considering the effect of organics on the growing surfaces of crystals (Kitano and Hood, 1965), as well as the effects of other components such as silica. Nevertheless, if used with caution within a solid geological framework, and in conjunction with other tools such as geochemistry, calcite cement morphology can be a useful index to the conditions of precipitation, and a valuable clue to the diagenetic history and porosity evolution of a rock.

Petrography-cement distributional patterns

Under most normal conditions of calcite cement precipitation, the sediment or rock pore is totally saturated with water, and cement is precipitated equally around the periphery of the pore (Figs. 3.6C and 3.7C). In the vadose zone, above the water table, however, precipitation takes place in the presence of two phases, air and water. Under these conditions, water is held at the grain contacts, water movement is by capillarity, and cement distributional patterns are controlled by the distribution of the liquid phase (Dunham, 1971; Longman, 1980; James and Choquette, 1984). Well within the zone of capillarity, calcite cement will tend to be concentrated at the grain contacts, and will exhibit a curved surface that reflects the meniscus surface of the water at the time of precipitation caused by surface tension (Figs. 3.6A and 3.7A and B). This curved pattern of cementation is called *meniscus cement* (Dunham, 1971). In the more saturated areas immediately below the zone of capillarity and above the water table, more fluid is available, and excess water accumulates as droplets beneath grains. Calcites precipitated in this zone often show this gravity orientation and are termed *microstalactitic cements* (Longman,1980) (Figs. 3.6B and 3.7D). The vadose cementation patterns discussed above may occur in both the meteoric environment and the marine environment (beach rock) (Moore and others, 1972) (see Chapters 4 and 6, this book).

While these cementation patterns are quite useful in differentiating the vadose and phreatic subenvironments in Holocene and Pleistocene sequences (Halley and Harris, 1979), they are more difficult to utilize in ancient rock sequences where subsequent generations of cement often overgrow the original vadose cements, muting and

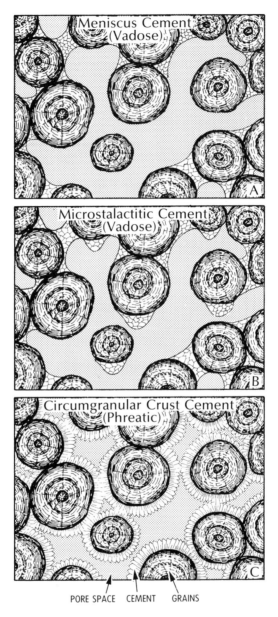

Fig. 3.6. Cement distributional patterns as a function of diagenetic environment. A. In the vadose zone, cements are concentrated at grain contacts and the resulting pores have a distinct rounded appearance due to the meniscus effect. The resulting cement pattern is termed meniscus cement. B. In the vadose zone immediately above the water table there is often an excess of water that accumulates at the base of grains as droplets. These cements are termed microstalactitic cements. C. In the phreatic zone, where pores are saturated with water, cements are precipitated as circumgranular crusts.

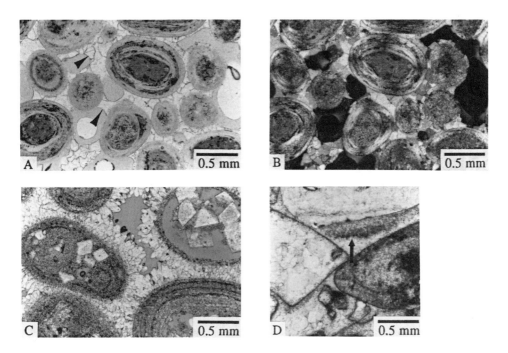

Fig. 3.7. Cement distribution patterns as a function of diagenetic environment. (A) Meniscus calcite cement characteristic of the vadose zone, Joulters Cay, Bahamas. Plain light. (B) Same as (A), crossed polars. (C) Circumgranular calcite crust cement characteristic of phreatic precipitation. Murphy Griffco #1, 7784' (2373 m), Miller Co., Arkansas, U.S.A. Plain light. (D) Microstalactitic calcite cement (Arrows) in Lower Cretaceous Edwards Formation, north central Texas. Cement precipitated as aragonite in the marine vadose zone (beach rock) associated with a prograding beach sequence (see Chapter 2 for a discussion of this beach complex). Plain light.

ultimately destroying the distinctive vadose features. In addition, later partial dissolution of fine crystalline phreatic calcite cements can mimic the rounded pores associated with vadose cements (Moore and Brock, 1982) (Fig. 2.11A; Chapter 9).

Petrography-grain-cement relationships relative to compaction

The timing of compaction events relative to cementation is both important and relatively easy to observe. If, for example, a cemented grainstone shows no evidence of compaction, such as high grain density, or grain interpenetration, the cementation event is most certainly early and before burial (Fig. 3.7C)(see Coogan and Manus, 1975, for an in-depth discussion on determination of grain compaction fabrics and compaction estimation). Cements that are broken and involved in compaction are at least contemporary with the compaction event (Figs. 3.8A and 3.9A-C), and are probably relatively early

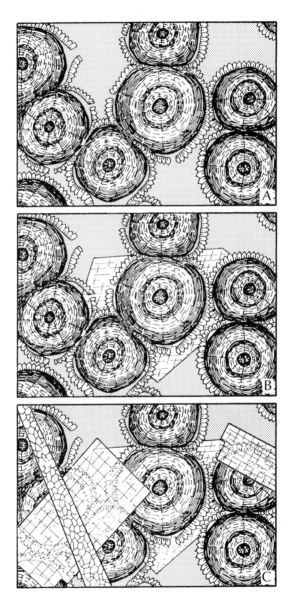

Fig. 3.8. Timing of diagenetic events by petrographic relationships. A. Involvement of isopachous cement in compaction indicates that cement was early. B. Cement that encases compacted grains and spalled isopachous cement is relatively late. C. Cross cutting relationships of features such as fractures and mineral replacements are useful. Calcite filling fracture is latest event, replacement anhydrite came after the late poikilotopic cement, but before the fracture, while the isopachous cement was the earliest diagenetic event.

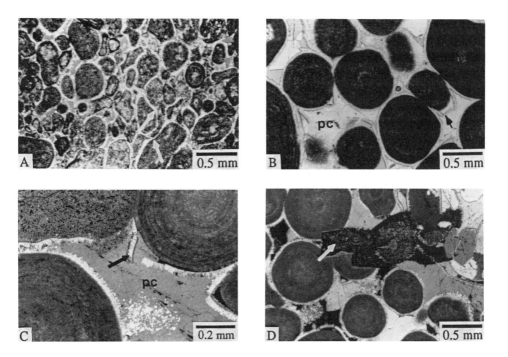

Fig. 3.9. Grain-cement relationships useful in determining relative timing of cementation events. (A) Compacted Jurassic Smackover pellet grainstone. Circumgranular crust cements obviously affected by compaction (arrows), so cement is early, before compaction. Walker Creek field, 9842' (3000 m), Columbia Co., Arkansas, U.S.A. Plain light. (B) Distorted, and spalled, circumgranular crust calcite cements in a Jurassic Smackover ooid grainstone. These cements are early, precompaction. Areas of spalled cement crust (arrows) have been engulfed by poikilotopic calcite cements (pc) that are obviously later, subsurface cements. Nancy field, 13550' (4130 m), Mississippi, U.S.A. Plain light. (C) Close view of spalled calcite crust cement (arrow) of (B) engulfed by coarse poikilotopic calcite cement (pc). Poikilotopic cement post-compaction. (Crossed polars). (D) Anhydrite (arrow) replacing both ooids and post-compaction poikilotopic calcite cement, suggesting that anhydrite is a late diagenetic event. Note inclusions of calcite that outline ghosts of grains and cements. Jurassic Smackover, 13613' (4149 m), Nancy field, Mississippi, U.S.A. (Crossed polars).

(Moore and Druckman, 1981; Moore, 1985). Cements that encase both compacted grains and earlier deformed cements, and themselves are not involved in compaction, surely represent later post-compaction subsurface cements (Figs. 3.8B and 3.9B and C) (Moore and Druckman, 1981; Moore, 1985). Finally, cross-cutting relationships associated with fractures and mineral replacements can clearly establish the relative timing of each cementation and/or replacement event (Moore and Druckman, 1981) (Figs. 3.8C and 3.9D).

While the petrographic relationships outlined above can be used to determine the timing of porosity-modifying events relative to the general paragenesis of a sequence, the actual time, or the depth of burial under which the event took place, cannot be determined. For example, it can be seen that the cement in Fig. 3.9C was precipitated before the rupture of ooids in this specimen, but it is not known how much overburden is necessary to accomplish the failure of the ooids, or to cause the distortion and breakage of the cement itself (Moore, 1985). Some constraints, however, can be placed on the burial depth, temperature, and timing of porosity-related diagenetic events by utilizing trace element and isotope geochemistry, and two-phase fluid inclusions.

Trace element geochemistry of cements and dolomites

Directly precipitated carbonate cements incorporate various trace and minor elements proportionally to the particular element's concentration in the precipitating fluid. It is assumed that the element's incorporation is dominantly by substitution for Ca^{+2}, rather than interstitially between lattice planes, at site defects, as adsorbed cations, or within inclusions (Viezer, 1983). Trace element incorporation is controlled by the distribution coefficient D, as outlined by Kinsman (1969), where m_{Me} and m_{Ca} represent molar concentrations of the trace or minor element of interest and molar concentrations of Ca in the fluid and solid phases.

$$(m_{Me}/m_{Ca})_{Solid} = D\ (m_{Me}/m_{Ca})_{Water}$$

The distribution coefficient of an element is generally determined experimentally by precipitating the solid phase from a solution of known concentration, and measuring the concentration of the element in the solid phase (Kinsman, 1969). Obviously, this process presents a problem in the case of dolomites, which cannot be synthesized under earth-surface temperature and pressure conditions (Viezer, 1983). Behrens and Land (1972) reasoned that the D for Sr and Na in dolomite should be half that for calcite because dolomite contains half as many Ca sites as an equivalent amount of calcite; this reasoning has become common practice (Viezer, 1983). The size of the crystal lattice obviously plays a major role in determining the magnitude of D for the carbonates. The large unit cell of orthorhombic aragonite crystals can accommodate cations larger than Ca (such as Sr, Na, Ba, and U), while the smaller unit cell of the rhombohedral calcite preferentially incorporates smaller cations (such as Mg, Fe, Mn). Hence, the D for Sr in aragonite is much larger than the D for Sr in calcite (Table 3.1).

Initially, the use of trace elements in determining the diagenetic environment of processes such as cementation held great promise (Kinsman, 1969). Sr and Mg are of particular interest because metastable carbonate mineral suites from modern shallow marine environments are dominated by aragonite (Sr rich) and magnesian calcite

Table 3.1. Commonly accepted distribution coefficients for the incorporation of trace elements in calcite, aragonite, and dolomite. The full references to the data sources may be found in Veizer, 1983. Used with permission of SEPM.

Trace element	Reported D	Recommended D	Sources of data
Calcite			
Sr	0.027-0.4 (1.2)	0.13 D.P. 0.05 A→dLMC HMC → dLMC 0.03 LMC →dLMC	Holland et al. (1964), Holland (1966), Kinsman and Holland (1969), Katz et al. (1972), Ichikuni (1973), Lorens (1981), Kitano et al. (1971), Jacobson and Usdowski (1976), Baker et al. (1982)
Na	0.00002-0.00003		Müller et al. (1976), White (1978)
Mg	0.013-0.06 (0.0008-0.12)		Winland (1969), Benson and Matthews (1971), Alexandersson (1972), Richter and Füchtbauer (1978), Kitano et al. (1979b), Katz (1973), Baker et al. (1982)
Fe	$1 \leq x \leq 20$		Veizer (1974), Richter and Füchtbauer (1978)
Mn	5.4-30 (1700)	6 D.P. 15 A →dLMC HMC →dLMC 30 LMC → dLMC	Bodine et al. (1965), Crocket and Winchester (1966), Michard (1968), Ichikuni (1973), Lorens (1981)
Aragonite			
Sr	0.9-1.2 (1.6)		Kitano et al. (1968, 1973), Holland et al. (1964), Kinsman and Holland (1969)
Na	~0.00014 (3-4 x $\underline{D}^{Na}_{calcite}$)		Calculated from figure 1 of White (1977), Kitano et al. (1975)
Mg	~0.0006-0.005		Brand and Veizer (1983)
Mn	0.86 (~½-⅓ $\underline{D}^{Mn}_{Calcite}$)		Raiswell and Brimblecombe (1977), Brand and Veizer (1983)
Dolomite			
Sr	0.025-0.060		Katz and Matthews (1977), Jacobson and Usdowski (1976)
Na	as $\underline{D}^{Na}_{calcite}$) (?)		White (1978)

D.P. = Direct precipitation. dLMC = diagenetic low-Mg calcite.   Modifed from Veizer, 1983

(Mg rich). Stabilization of these shallow marine carbonates to calcite and dolomite involve a major reapportionment of these elements between the new diagenetic carbonates and the diagenetic fluids. Na is of interest because it is a relatively major cation in sea water and brines, and a minor constituent in dilute or meteoric ground waters. Fe and Mn are multivalent, are both sensitive to Eh and pH controls, occur in very low concentrations in sea water, but are present in higher concentrations in ground water and oil field brines (Veizer, 1983). Waters from the major diagenetic environments then, are so different in their trace and minor element compositions (Table 3.2) that the cements precipitated from these diverse waters should easily be recognized by their trace element signatures. As interest in the utilization of trace elements in carbonate studies increased, widespread concern over their validity, and the controls over the magnitude of the most commonly used distribution coefficients surfaced (Katz and others, 1972; Bathurst,

Table 3.2. The mean chemical composition of Jurassic formation waters from the Gulf Coast, U.S.A., north American river waters, and seawater. This table illustrates the wide range of compositional differences seen between meteoric, subsurface, and marine waters. Reprinted with permission of the American Association of Petroleum Geologists.

Composition g/l	Subsurface Smackover Formation[a]		North American River Systems[b]		Seawater[c]
	Range	Mean	Range	Mean	
Cl	31.0-223.0	171.7	0.0008 -1.5	0.04	19.0
Na	5.0- 87.4	67.0	0.002 -0.9	0.04	10.5
Ca	6.8- 55.3	34.5	0.0004 -0.6	0.04	0.4
Mg	0.1- 8.6	3.5	0.0008 -0.2	0.009	1.35
Sr	0.1- 4.7	1.9	0.00001-0.009[d]	0.00067[d]	0.008
Mn		0.03		0.00008[g]	0.00002[g]
Fe	—	0.04	—	0.00040[g]	0.00020[g]
SO$_4$	0 - 4.0	0.4	0.0008 -1.0	0.091	2.7
TDS[e]	51.0-366.0	279.00	0.05 -1.1	0.3	35.0
		Equivalent Ratios			
Mg/Ca	0.05 -0.52	0.17	0.082 - 0.89	0.37	5.3
Na/Cl	0.45 -0.77	0.60	0.72 -61.6	11.65	0.86
Sr/Ca	0.0025-0.085	0.025	0.0009 - 0.0091[d]	0.0035[d]	0.0086[f]
Na/Ca	1.06 -2.41	1.69	0.09 - 5.23	0.70	23.09

[a]Data calculated from Collins, 1974
[b]Data calculated from Livingstone, 1963
[c]Data from Riley and Skirrow, 1965
[d]Data from Skougstad and Horr. 1963
[e]Total dissolved solids.
[f]Value from Kinsman. 1969
[g]Data from Veizer, 1983.

Modified from Moore and Druckman, 1981

1975; Jacobson and Usdowski, 1976; Land, 1980).

Most carbonate geologists realize that distribution coefficients are, for the most part, relatively crude estimates, and can be affected dramatically by temperature, major element composition of the water, and the rate of precipitation of the cement (Katz, 1973; Jacobson and Usdowski, 1976; Veizer, 1983; Mucci and Morse, 1983; Given and Wilkinson, 1985). Table 3.1 outlines the reported range of distribution coefficients for the most commonly used trace and minor elements in the three major carbonate minerals, as compiled by Veizer in 1983. The very wide range of reported D values for most of the trace and minor elements in calcite probably reflects not only the basic complexity of the controls over incorporation of trace and minor elements into the calcite lattice, but also differences in experimental techniques as well. While Veizer (1983) builds a persuasive case for the intelligent use of trace elements in studies leading to the recognition of diagenetic environments, the present uncertainty concerning the magnitude of kinetic controls over distribution coefficients makes their application to ancient rock sequences exceedingly questionable.

Moore (1985) presents a well-documented case history of a Jurassic subsurface calcite cement. Based on stable isotopes, two-phase fluid inclusions, and radiogenic Sr isotopic studies, Moore concluded that these cements precipitated from present oil field brines at elevated temperatures. Table 3.3 presents a comparison of measured trace element composition of the cement to the cement trace element composition that should be expected from the present subsurface fluids. Distribution coefficients that span the range of D's found in Table 3.1 for each of the element's incorporation into calcite were used in construction of the table. It is clear that the trace and minor element composition of these late cements, particularly Sr, does not adequately identify present brines as the precipitational fluid, regardless of the distribution coefficient used. This case history illuminates clearly the problems presently facing the geologist using and trying to interpret trace element compositions of carbonates with the idea of reconstructing original diagenetic environments.

Are trace element studies, then, an exercise in futility? They certainly are at the present stage of understanding, if specific knowledge is to be gained concerning the composition of the diagenetic fluid responsible for a diagenetic event, such as a pore-fill cement. On the other hand, if used intelligently within a solid geologic and petrologic framework, trace elements can provide the geologist with some important information concerning gross characteristics of the diagenetic environments responsible for cementation and dolomitization events.

For example, Banner (1986) suggested using covariation of trace elements and isotopes to construct quantitative models of diagenesis that give pathways on bivariate plots of changing trace element abundances and isotopic compositions in response to fluid-rock interaction. Trace elements used in concert with isotopes can be independent of D values over a wide range, and seem to be diagnostic of the general type of fluid involved in the diagenetic system.

Trace elements may be useful in determining the original mineralogy of rock components, such as ooids, bioclasts, and cements. Sandberg (1983, 1985) and Sandberg and others (1973) used high Sr concentrations in concert with petrography to infer an original aragonite composition for a variety of materials throughout the Phanerozoic. Moore and others (1986) used petrography and the trace elements Sr and Mg to suggest that Upper Jurassic ooids in southern Arkansas were originally both aragonite and magnesian calcite. The inferred aragonite ooids exhibited high Sr concentrations, while

Table 3.3. *Comparison of the measured trace element composition of Jurassic Smackover post-compaction calcite cements to the calculated trace element composition of a calcite precipitated from an average Jurassic formation brine using several commonly used distribution coefficients. Used with permission of SEPM.*

ELEMENT		SMACKOVER BRINE[1]	CALCULATED TRACE ELEMENT COMPOSITION OF A CALCITE PRECIPITATED FROM SMACKOVER BRINE		MEASURED COMPOSITION OF POST-COMPACTION CALCITE (PPM)
STRONTIUM		Equivalent Ratio Sr/Ca	Calculated[2] Sr (ppm) in calcite ppct. from this water with indicated distribution coefficients at 25°C		
			.14 (Kinsman, 1969)	.054 (Katz et al., 1972)	
	Range	.0025-.085	307-10,426	118-4022	
	Mean	.025	3000	1350	<100
MAGNESIUM		Equivalent Ratio Mg/Ca	Calculated[2] Mg (ppm) in calcite ppct. from this water using Katz's (1973) distribution coefficients at 25°C and 70°C		
			(0.573) 25°C	(.0973) 70°C	
	Range	0.05-0.52	696-7,244	1183-12,200	
	Mean	0.17	2368	4021	3160 ± 498
MANGANESE		Equivalent Ratio Mn/Ca	Calculated[2] Mn (ppm) in calcite ppct. from this water using Michard's (1968) distribution coefficient 5.4 at 25°C		
	Range	2.15×10^{-5}-1.07×10^{-3}	64-3174		
	Mean	3.2×10^{-4}	949		498 ± 85
IRON		Equivalent Ratio Fe/Ca	Calculated[2] Fe (ppm) in calcite ppct. from this water using Richter and Fuchtbauer (1978) distribution coefficient of 1 at 25°C		
	Range	2.1×10^{-5} to 4.39×10^{-3}	92-2450		
	Mean	1.42×10^{-3}	793		2014 ± 326

Modified from Moore, 1985

[1]Equivalent ratios from. Table 4.1
[2]Calculated composition of calcite from expression m Sr/Ca Calcite = Distribution Coefficient x m Sr/Ca of Precipitating Fluid (Kinsman, 1969).
[3]Mean composition in ppm determined by electron microprobe analysis is at the L.S.U. Department of Geology, Microanalysis Laboratory by C. H. Moore. Number of Analyses Mg, Fe, Sr. 100; Mn 55.
[4]Equivalent ratios from Meyers, 1974.

the inferred magnesian calcite ooids yielded low Sr, but relatively high magnesium concentrations.

As will be discussed later in Chapters 6 and 7, detailed variations in the Sr and Mg compositions of meteoric cements may reflect the mineralogical composition of the sediments and rocks through which the diagenetic fluids have passed. If the diagenetic fluids pass through, and interact with, metastable mineral suites, such as aragonite and magnesian calcite, increases in the Sr and Mg composition of the cements precipitated from these fluids downflow will reflect these dissolution events (Benson and Matthews, 1971; Benson, 1974).

The pH-Eh sensitivity of Fe and Mn has been used extensively in recent years to infer the environmental conditions under which cements have been precipitated (see Chapters 6 and 7). The oxidizing conditions of the vadose environment favor the Fe^{+3} and Mn^{+4} states and preclude the incorporation of divalent Fe and Mn in the calcite lattice. The general reducing conditions found in the phreatic zone, however, support the incorporation of both in calcite. The influence of Mn and Fe over the cathodoluminescence characteristics of carbonates, therefore, has made cathodoluminescence a useful tool in determining changes, or gradients in oxidation states in ancient hydrologic systems (see Chapter 7).

Finally, trace elements may be used to determine chemical gradients that give useful information concerning direction of paleo-fluid flow in diagenetic terrains (Moore and Druckman, 1981; see Chapter 6, this book), as well as estimates of rock:water ratios (open versus closed diagenetic systems) during diagenesis (Brand and Veizer, 1980; Veizer, 1983).

Stable isotopes

Isotopes are one or more species of the same chemical element (i.e., the same number of protons in the nucleus) differing from one another by having different atomic weights (i.e., different number of neutrons). There are radioactive isotopes, whose precisely determined decay rates are useful for dating geologic samples, and stable, non-radioactive isotopes. While the stable isotopes of an individual element, such as H, O, C, S, and N, have similar geochemical behavior, differences in atomic weight cause significant differences in distribution of the isotopes during natural processes, such as evaporation, condensation, photosynthesis, and phase transformations. These differences form the basis for the utilization of light stable isotopes in diagenetic studies (Anderson and Arthur, 1983).

For carbonate studies, the two most naturally abundant isotopes of oxygen, ^{18}O and ^{16}O, and carbon, ^{13}C and ^{12}C, are commonly used, while sulfur isotopes are used in sequences containing evaporites and sulfide mineralization. The discussion here will be limited to the stable isotopes of oxygen and carbon. Because mass spectrometric analysis can determine isotope ratios more precisely than individual isotope abundances, variations in $^{18}O/^{16}O$ and $^{13}C/^{12}C$ ratios between samples are examined. Variations in these ratios in most natural systems are small, but can be precisely measured. The ∂ notation is useful for expressing these small differences relative to a standard value. ∂-values are reported in parts per thousand, which is equivalent to permil, or $^o/_{oo}$ (Hudson, 1977). For oxygen,

$$\partial^{18}O = (^{18}O/^{16}O_{Sample}) - (^{18}O/^{16}O_{Standard})/(^{18}O/^{16}O_{Standard}) \times 1000$$

The standard commonly used for both oxygen and carbon in carbonates is referred to as PDB (Pee Dee Belemnite, of Cretaceous age). Using the formulation above, the PDB standard would have $\partial^{18}O$ and $\partial^{13}C = 0$. Most marine carbonates are close to this value and hence are small numbers. Oxygen isotopic compositions can also be referred to SMOW (Standard Mean Ocean Water). The conversion of $\partial^{18}O$ values from the PDB to the SMOW scale is approximately (Anderson and Arthur, 1983):

$$\partial^{18}O\ SMOW = 1.03\partial^{18}O\ PDB + 30.86$$

Oxygen and carbon isotopic compositions for the same samples are commonly plotted against each other, making relationships and trends between samples relatively easy to determine. A $\partial^{18}O$ vs $\partial^{13}C$ cross-plot with the extension of the axes through their intersection at 0,0 (see Fig. 3.10) facilitates the comparison of ∂-values as either negative or positive. It should be noted that, today, many isotope geochemists do not use "depleted", or "enriched", to describe stable isotope values, because a ^{18}O enrichment may actually be a ^{16}O depletion. Instead, they advocate "high" or "low" $\partial^{18}O, \partial^{13}C$ values.

Fig. 3.10, taken from Hudson (1977), is such a cross-plot, incorporating along the base of the diagram ranges of the $\partial^{18}O$ (SMOW) composition of natural waters from our three diagenetic realms: meteoric, marine, and subsurface. Isotopic values of marine waters vary several permil about a mean of 0, depending on temperature, evaporation, and dilution from sea ice (Anderson and Arthur, 1983). Under conditions of heavy evaporation, such as is evident in salt ponds and on sabkhas, marine waters can exhibit relatively high $\partial^{18}O$ values. Meteoric waters are relatively low, but show a wide range of oxygen isotopic compositions due to latitude and altitude effects (Hudson, 1977). Subsurface fluids also have a wide range of compositions from $\partial^{18}O = -20$ to $+12$ $^o/_{oo}$ (SMOW) (Land and Prezbendowski, 1981), with most subsurface oil-related brines generally being on the high side of the range (Hudson, 1977). The carbonates precipitated

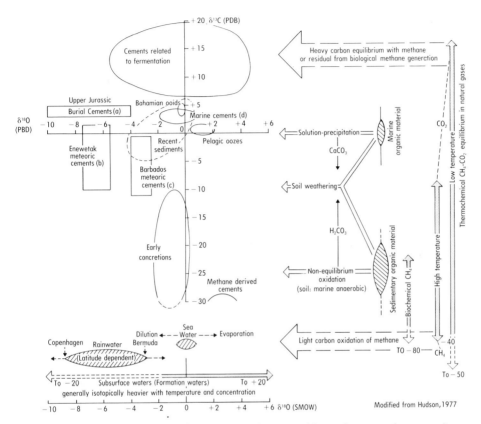

Fig. 3.10. Distribution of carbon and oxygen isotopic compositions of some carbonate sediments, cements, and limestones with some of the factors that control them. Data sources as outlined by Hudson (1977) except as follows: a. Moore, 1985. b. Saller, 1984. c. Matthews, 1974. d. James and Choquette, 1983. Reprinted with permission of the Geological Society of London.

from these waters as cements will generally reflect the oxygen isotopic composition and temperature of the precipitating fluid, as controlled by the temperature-dependent isotopic fractionation between the liquid and mineral phase. Meteoric calcite cements, therefore, will tend to exhibit relatively low oxygen isotopic ratios (Fig. 3.10, Chapters 6 and 7), while shallow-water marine cements will have relatively high oxygen isotopic ratios (Chapters 4 and 5). The oxygen isotopic fractionation between water and calcite is highly temperature-dependent (Friedman and O'Neil, 1977; and many others), with the $\partial^{18}O$ of the calcite becoming increasingly lower with increasing temperature (Fig. 3.10, Chapter 7). Calcites precipitated in the subsurface, even though precipitated from pore fluids with heavy oxygen isotopic composition, can have relatively low $\partial^{18}O$ values that may generally overlap those of meteoric waters (Moore, 1985). Therefore, in any interpretation of the oxygen isotopic composition of a calcite cement, the geologist must have independent evidence of the temperature of formation (i.e., petrography, two-phase fluid inclusions) in order to estimate the original isotopic composition of the water of precipitation and hence the diagenetic environment. Conversely, if the water's isotopic composition is known, or can be assumed, a temperature of precipitation for the calcite can be estimated using the isotopic composition of both the solid and liquid phases and the calcite-water fractionation factor (a), and solving for temperature (using the expression: $10^3 \ln a = 2.78 \times 10^6 T^{-2} (°K) - 2.89$, developed by Friedman and O'Neil, 1977).

Dolomite presents a special problem because the fractionation factor is poorly known and has been, for a number of years, a matter of debate (Land, 1980; Anderson and Arthur, 1983). While extrapolation of high temperature experimental data would give a concentration of ^{18}O in the neighborhood of 6‰ higher than contemporaneous calcite, Land (1980) indicates that a 3-4‰ increase of ^{18}O in dolomite relative to syngenetic calcite is more reasonable.

In addition, as Land (1980, 1985) points out, dolomite commonly originates as a metastable phase (protodolomite) that seems to be susceptible to recrystallization and hence isotopic reequilibration during burial and diagenesis. Reeder (1981, 1983) has shown the presence of complex mineralogical domains in dolomites that surely affect not only the ultimate stability of dolomites, but their isotopic composition as well. These factors seriously impact the interpretation of the isotopic composition of dolomite relative to the nature of the dolomitizing fluids and the diagenetic environment under which dolomitization was accomplished. Owing to the original metastability of many dolomites, the isotopic composition of a dolomite might well reflect the nature of recrystallizing fluids, rather than the composition of the original dolomitizing solution.

The geochemical behavior of carbon is often governed by organic processes, with photosynthesis playing a major role. The range of $\partial^{13}C$ values found in nature is outlined on the right side of Fig. 3.10. Organic carbon exhibits low values of $\partial^{13}C$ (-24 ‰, PDB) relative to the oxidized forms of carbon found as CO_2 (-7 ‰) and marine carbonates (0 to +4 ‰). Extremes in $\partial^{13}C$ values generally involve methane generation either by biochemical fermentation in the near surface, or by thermochemical degradation of organic matter in the subsurface at temperatures greater than 100°C (Hudson, 1977; Anderson and Arthur, 1983). Methane generated from fermentation produces very low $\partial^{13}C$ values (-80 ‰), but the residual organic matter shows high $\partial^{13}C$ values. Oxidation of the methane with subsequent cementation will result in cements with low $\partial^{13}C$ values (Roberts and Whalen, 1975), while oxidation of the residual organic material and subsequent cementation incorporating carbon from this source will lead to cements with high $\partial^{13}C$ (Fig. 3.10) (Curtis and others, 1972). Thermochemically derived methane can indirectly lead to the precipitation of subsurface calcite cements with very low $\partial^{13}C$ (Anderson and Arthur, 1983; Sassen and others, 1987; Heydari and Moore, 1988; see Chapter 9, this book). Soil weathering and carbonate mineral stabilization involving dissolution of marine limestones and sediments and subsequent precipitation of calcite cements in the vadose and the shallow phreatic zones will generally result in cements and limestones with moderately low $\partial^{13}C$ compositions (Fig. 3.10) (Allen and Matthews 1982; James and Choquette, 1984).

Table 3.4 outlines the estimated mass of carbon in the various natural reservoirs found in and adjacent to the earth's crust. It is obvious that most carbon is tied up in sediments and limestones of the lithosphere and that the majority of this carbon is marine in origin and carries a relatively high $\partial^{13}C$ signature (Hudson, 1977; Anderson and Arthur, 1983). Therefore, in any diagenetic process that involves rock-water interaction, such as solution reprecipitation under burial conditions, the relatively heavy rock carbon tends to dominate and to buffer low ^{13}C contributions from biological processes. The lack of appreciable temperature fractionation associated with carbon isotopes (Anderson and Arthur, 1983) supports this buffering tendency.

Cross-plots of $\partial^{18}O$ versus $\partial^{13}C$, as seen in Fig. 3.10, are commonly used to distinguish the diagenetic environment responsible for specific cements. Similar cross-plots detailing the isotopic composition of individual limestone components, such as

Table 3.4. *Major carbon reservoirs, the mass of carbon found in each, and their mean carbon isotopic composition. Mass units, 10^{15} grams carbon. $\partial^{13}C$, permil PDB. Used with permission of SEPM.*

Reservoir	Mass	$\delta^{13}C$‰
1) Atmosphere (pre-1850 CO_2)	610	−6.0-7.0
(~290 ppm)	(range: 560-692)	
2) Oceans—TDC	35,000	0
DOC	1000	−20
POC	3	−22
3) Land Biota (biomass)	(range: 592-976)	−25
	(~10%) may be C_4? plants)	−12
4) Soil Humus	(range: 1050-3000)	−25
5) Sediments (Total)	(range: 500,000-10,000,000)	—
inorganic C	423,670	+1
organic C	86,833	−23
6) Fossil Fuels	>5000	−23

From Anderson and Arthur, 1983

fossils or ooids collected on a regional basis, are useful in determining the pathways of diagenetic change suffered by a limestone, as well as determining the original isotopic composition of the marine components that will give an estimate of the isotopic composition of the paleo-oceanic waters (Lohmann, 1983, 1988).

The two most common stable isotope trends observed on these types of cross-plots are: the J-shaped curve of Lohmann (1988; Fig. 6.24, this book), showing the covariant trend of $\partial^{18}O$ as well as $\partial^{13}C$ toward lower values, during progressive diagenesis under meteoric water conditions; and the dominant trend toward lighter $\partial^{18}O$ values, during burial caused by increasing pore-water temperature and rock buffering of the $\partial^{13}C$ composition of the cement (Moore and Druckman, 1981; Choquette and James, 1987; Fig. 7.17, this book).

Tightly controlled vertical stable isotopic sampling of cores and outcrops has been used to detect geochemical gradients that reflect vadose zones and water table positions (Fig. 3.11). A rapid change in $\partial^{13}C$ composition toward lighter values seems to be characteristic of the vadose zone (Allen and Matthews, 1982). This technique can be

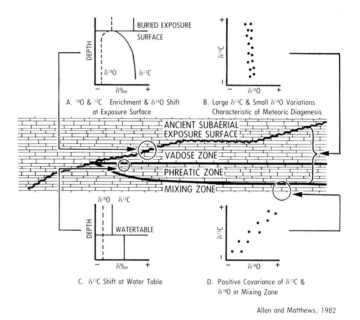

Fig. 3.11. *Schematic diagram showing the anticipated carbon and oxygen isotopic shifts across meteoric diagenetic interfaces such as exposure surfaces, water tables, and meteoric-marine water mixing zones, as compared to the trend expected in the meteoric phreatic zone. Reprinted with permission of the International Association of Sedimentologists.*

useful if applied carefully and is adequately constrained by geologic setting and petrography (see discussion in Chapter 6, this book).

Finally, the $\partial^{18}O$ values of marine limestones and cherts have tended to decrease significantly with increasing age (Fig. 3.12). This temporal variation is a matter of much controversy, leading isotope geochemists to speculate that the isotopic composition of oceans has evolved; oceanic temperatures have decreased (higher in the past); cherts and

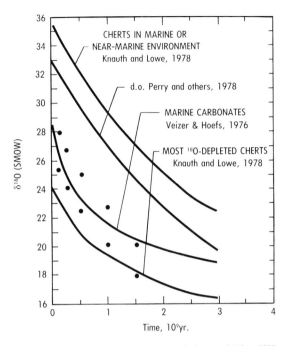

Fig. 3.12. Oxygen isotopic composition of marine carbonates and cherts plotted against time. While the marine carbonates plot near the most depleted ^{18}O chert curve, the shapes of the curves and the range of change are similar. Reprinted with permission of SEPM.

carbonates have continuously reequilibrated during diagenesis; and $\partial^{18}O$ values of ancient limestones and cherts represent diagenetic conditions rather than the temperature and isotopic composition of the sea water from which they originated (Anderson and Arthur, 1983). The uncertainties surrounding the causes of the temporal variation of oxygen isotopic composition of limestones certainly makes the interpretation of stable isotopic data more difficult.

It is obvious from our discussions above, and with a glance at Fig. 3.10, that a calcite cement or a limestone might obtain the same stable isotopic signature under strikingly different diagenetic environments, depending on the temperature of the pore fluid and the rock:water ratio at the time of precipitation. These and other inherent complexities of stable isotopic geochemistry in natural systems require careful use of the technique in carbonate diagenetic studies; constraint of the problem, with a solid geologic and petrographic framework; and, when possible, use of complimentary supporting technology, such as trace elements, radiogenic isotopes, and fluid inclusions (Moore, 1985).

Finally, it should be pointed out, that when dealing with limestones or dolomites that have undergone complex diagenetic histories, each fabric component of the rock (grains, and each diagenetic phase) should be analyzed for stable isotopes separately to avoid whole-rock average values that can be meaningless, or misleading.

Strontium isotopes

The '80s have seen the development of considerable interest in the utilization of strontium isotopes in problems of carbonate diagenesis (Stueber and others, 1984; Moore, 1985; Banner and others, 1988). There are four stable isotopes of strontium: ^{88}Sr, ^{87}Sr, ^{86}Sr, and ^{84}Sr. ^{87}Sr is the daughter product of the radioactive decay of ^{87}Rb, and its abundance has therefore increased throughout geologic time (Burke and others, 1982; Stueber and others, 1984). Since carbonate minerals generally contain negligible rubidium, once ^{87}Sr is incorporated into a carbonate, the relative abundance of that ^{87}Sr will remain constant. However, noncarbonate phases such as clays can have high Rb contents, and small amounts of such contaminants in an ancient carbonate can significantly affect measured ^{87}Sr contents. It is therefore critical for Sr isotopic studies of carbonates to obtain pure samples or to evaluate the effects of impurities that may be present (Banner and others, 1988). The relative abundance of ^{87}Sr is generally expressed as the ratio $^{87}Sr/^{86}Sr$. Finally, isotopic fractionation during natural processes, such as encountered in carbon and oxygen isotope systems in diagenetic studies, is negligible for strontium isotopes (Veizer, 1983).

The $^{87}Sr/^{86}Sr$ signature in sea water is derived from the relative contribution of the various types of rocks that are exposed to chemical weathering on the continents and in

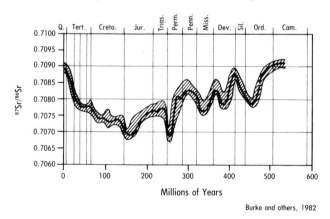

Fig. 3.13. *$^{87}Sr/^{86}Sr$ ratios of marine carbonates plotted against time. The ruled area represents the data envelope. Originally published in Geology, v.10, pp.516-519. Used with permission.*

the ocean basins, and from submarine hydrothermal fluxes. While the $^{87}Sr/^{86}Sr$ is homogeneous in present oceanic waters, it has varied through geologic time in response to patterns of global tectonics (Burke and others, 1982). The strontium present in marine limestones and dolomites, then, should have an isotopic composition that depends on the age of the rocks, reflecting the temporal variations in oceanic $^{87}Sr/^{86}Sr$, rather than representing the *in situ* decay of ^{87}Rb. The latest charts of $^{87}Sr/^{86}Sr$ versus time (see for example, Fig. 3.13) are detailed enough to allow the use of strontium isotopes in a variety of diagenetic problems. If the $^{87}Sr/^{86}Sr$ of a marine carbonate sample differs from the value estimated for seawater at the time of deposition of the sample, then this difference can be attributed to diagenetic alteration by younger seawater or to diagenesis by non-marine fluids.

$^{87}Sr/^{86}Sr$ can be used to date diagenetic events such as dolomitization, if the dolomitization event was accomplished by marine waters and no subsequent recrystallization took place. Saller (1984a and b), using $^{87}Sr/^{86}Sr$, determined that marine dolomites in Eocene sequences beneath Enewetak Atoll started forming in the Miocene, and are probably still forming today. Petrography, trace elements, and the stable isotopes of oxygen and carbon were used to determine the marine origin of the dolomites.

During early meteoric diagenesis, the $^{87}Sr/^{86}Sr$ of the calcites precipitated as the result of the stabilization of aragonite and magnesian calcite will generally be inherited from the dissolving carbonate phases. This rock-buffering is the result of the enormous difference between the high Sr content of marine carbonates relative to the low levels of Sr normally seen in fresh meteoric waters (Moore, 1985; Banner and others, 1989). Sr derived from marine carbonates, then, generally swamps any contribution of Sr from meteoric waters that might exhibit high $^{87}Sr/^{86}Sr$ gained by interaction with terrestrial clays and feldspars. The resulting calcite cements or dolomites will normally exhibit a marine $^{87}Sr/^{86}Sr$ signature compatible with the geologic age of the marine carbonates involved in the diagenetic event. Any enrichment of $^{87}Sr/^{86}Sr$ above the marine value in such a meteoric system would indicate that an enormous volume of meteoric water had to have interacted with the rock to overcome the Sr-buffering of the marine limestones (Moore and others, 1988). This type of diagenetic modeling, using $^{87}Sr/^{86}Sr$ with the support of other stable isotopes and trace elements has been successfully applied to the Mississippian of the Mid-continent region of the U.S. by Banner and others (1989).

Oil field brines and other deep subsurface waters are often highly enriched with Sr relative to that of sea water, presumably because of extensive interaction with the rock pile through geologic time (Collins, 1975). The Sr in these waters is generally enriched with ^{87}Sr, indicating interaction with siliciclastics sometime during their evolutionary journey (Stueber and others, 1984). Because of these considerations, the $^{87}Sr/^{86}Sr$ of calcite cements and dolomites formed in the subsurface can give some insight into the timing of the cementation or dolomitization event. If present brines show elevated $^{87}Sr/^{86}Sr$, and post-compaction cements or dolomites associated with these brines have ratios near the seawater values expected for the age of the enclosing limestones, it can be inferred that the cements or dolomites received most of their Sr from the enclosing limestones, rather than the brines. This indicates that the diagenetic event took place prior to arrival of the waters containing Sr enriched in ^{87}Sr. Phases enriched in ^{87}Sr would presumably have been emplaced in the presence of the present day brine, or a similar pore fluid in the past, and obviously would represent a later event.

Fig. 3.14 illustrates the strontium isotopic composition of rock components, diagenetic phases, and oil field brines from the Lower Cretaceous Edwards Formation, U.S. Gulf Coast. Lower Cretaceous sea water had a $^{87}Sr/^{86}Sr$ of 0.7073. Edwards oil field brines have a strontium isotopic composition of 0.7092, presumably a result of interaction with associated rubidium-bearing shales and sandstones resulting in elevated isotopic ratios. Matrix limestone from the Edwards has a $^{87}Sr/^{86}Sr$ near that of Lower Cretaceous seawater, while post-compaction calcite cements, baroque dolomites, and other diagenetic phases have elevated ratios, indicating that the composition of these phases was influenced by present Edwards brines (Woronick and Land, 1985) (see Chapter 7, this book for other examples of the use of strontium isotopes).

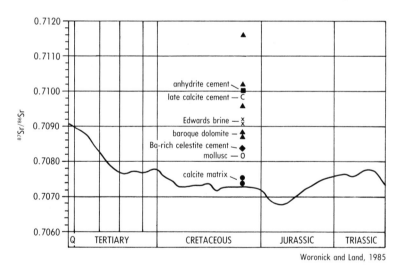

Fig. 3.14. Strontium isotopic composition of limestone constituents and subsurface brines from Lower Cretaceous subsurface sequences in south Texas plotted against the Burke and others (1982) curve. All components except the calcite matrix plots above the Lower Cretaceous marine curve suggesting involvement of basinal waters that had seen siliciclastics before arriving and interacting with these limestones. Used with permission of SEPM.

In summary, strontium isotopic studies, while not definitive when used alone, can provide important constraints on the origin and timing of porosity-modifying diagenetic events as well as the evolution of associated waters. The low concentration levels of Sr found in most ancient calcites and dolomites, combined with the necessity of separating diagenetic phases for analysis, and the need for high analytical precision, can make the use of strontium isotopes both time-consuming and expensive.

Fluid Inclusions

Carbonate cements formed during burial often entrap inclusions of the precipitating fluid at lattice defects or irregularities. These are termed *primary fluid inclusions*. If the inclusions were trapped as a single phase (e.g., one liquid phase) at relatively low temperatures (< 50°C), the inclusions would remain as single-phase fluid inclusions. If, however, the inclusions were trapped as a single phase under elevated temperatures (> 50°C), they would separate into two phases, a liquid and vapor as the inclusion cools, because the trapped liquid contracts more than the enclosing crystal and a vapor bubble nucleates to fill the void. This is a consequence of the different coefficient of thermal expansion for liquids versus solids (Fig. 3.15) (Roedder, 1979; Klosterman, 1981). Secondary inclusions form as a result of fractures or crystal dislocations affecting the crystal after it has finished growing (Roedder, 1979, Klosterman, 1981). The minimum temperature of formation of the crystal containing a two-phase fluid inclusion can be estimated by heating up the sample until the vapor bubble disappears. This is called the *homogenization temperature*.

The following serious constraints must be met before two-phase fluid inclusion homogenization temperatures can be properly used to estimate temperature of formation

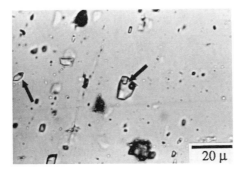

Fig. 3.15. Two-phase fluid inclusions (arrows) in poikilotopic post-compaction calcite cement, Jurassic Smackover, 10170' (3100 m), Walker Creek field, Columbia Co. Arkansas, U.S.A. Plain light.

of the enclosing crystal (Roedder and Bodner, 1980): 1) the inclusion must have been trapped initially as a single phase; 2) only primary inclusions can be utilized; 3) the volume of the cavity cannot change significantly by dissolution or precipitation during burial; 4) the inclusion cannot stretch or leak during burial history, and hence change the volume of the cavity; 5) the fluid in the inclusion must consist of water, or be a water-salt system; and 6) the pressure on the inclusion at the time of entrapment must be known in order to allow pressure correction of the homogenization temperature.

Other information pertinent to the nature of the fluids responsible for precipitation can be developed through the freezing behavior of the fluids within the inclusion. Dissolved salts depress the freezing point of fluids. The salinity of the fluid within an inclusion, and, by extension, the salinity of the precipitating fluid, can be estimated by its freezing temperature (Klosterman, 1981). The gross chemical composition of the fluid within an inclusion can also be estimated by freezing behavior, as well as by Raman spectroscopy (Crawford and others, 1979; Klosterman, 1981).

At first view, the utilization of fluid inclusions in diagenetic studies held great promise, because, potentially, the inclusion is a direct window into the characteristics of the diagenetic fluid responsible for the enclosing crystal. During the '80s a number of diagenetic studies utilized inclusion data to help constrain conditions of formation of subsurface calcite cements and dolomites (Moore and Druckman, 1981; Klosterman, 1981; O'Hearn, 1984; Moore, 1985). During that period, there were a number of significant questions concerning the validity of the results of inclusion studies of carbonate minerals such as calcite and dolomite. Moore and Druckman (1981) reported seemingly primary two-phase fluid inclusions present in early, near-surface, meteoric cements from the Jurassic Smackover. In addition, many reported homogenization temperatures seemed consistently too high relative to the geological setting of the enclosing rocks (Klosterman, 1981; O'Hearn, 1984, Moore, 1985).

Prezbindowski and Larese (1987) present compelling experimental evidence that fluid inclusions in calcite stretch during progressive burial, changing the volume of the inclusion and hence significantly increasing the apparent temperature of formation of the inclusion. These results seem to explain the consistently high homogenization temperatures observed by earlier workers, as well as the presence of two-phase fluid inclusions in cements precipitated at the surface under low temperature conditions, but buried later and exposed to higher burial temperatures (Moore and Druckman, 1981).

Is it futile to use two-phase fluid inclusions in diagenetic studies in the face of the above observations, experimental stretching data, and the other serious constraints on the use of the technique, such as pressure corrections and the problems of secondary inclusions. It is believed that with care and caution, inclusion data can give the worker in diagenesis important information concerning the nature of diagenetic fluids, such as their gross salinity and composition. This information can help interpret oxygen isotopic data of cements relative to formation temperature and assist in correlating diagenetic fluids to cementation and dolomitization events (Moore, 1985; Moore and others, 1988).

SUMMARY

There are three major diagenetic environments in which porosity modification events such as solution, compaction, and cementation are active: the marine, meteoric, and subsurface environments. The meteoric environment with its dilute waters, easy access to CO_2, and wide range of saturation states relative to stable carbonate phases, has a high potential for porosity modification, including both destruction by cementation as well as the generation of secondary porosity by dissolution. Shallow marine environments are particularly susceptible to porosity destruction by cementation because of the typically high levels of supersaturation of marine waters relative to metastable carbonate mineral phases. Changes of carbonate saturation with depth may lead to the precipitation of stable carbonates as cements, as well as the development of secondary porosity by dissolution in oceanic basins and along their margins. The subsurface environment is generally marked destruction of porosity by compaction and related cementation. Later porosity modification by dissolution and cementation is driven by thermal degradation of hydrocarbons and the slow flux of basinal fluids during progressive burial.

The key to the development of viable porosity modification models is the ability to recognize the diagenetic environment in which the porosity modification event occurred. There are a number of tools available to the carbonate geologist today. If used intelligently, they can provide the information necessary to place a diagenetic event in the appropriate environment. Petrography and geologic setting provide a cheap, readily available, relative framework for porosity modification. Trace element and stable isotopic geochemistry of cements and dolomites provide insight into the types of waters involved in the diagenetic event, and hence the event's link to a specific diagenetic environment. Two-phase fluid inclusions can be used to assess the temperature of cement or dolomite formation and the composition of the precipitating or dolomitizing fluid, and hence may provide a tie to a specific diagenetic environment, such as the subsurface.

Geochemically based techniques, however, are relatively expensive, time-consuming, and constrained by significant problems, such as uncertainties in distribution coefficients, temperature fractionation effects, and low concentration values. Two-phase fluid inclusion studies are extremely time-consuming and exhibit significant problems, such as stretching of inclusions during burial, the recognition of primary inclusions, and the necessity for pressure corrections.

None of these tools should be utilized in isolation. Rather, they should be used as complimentary techniques that can provide a set of constraints that allows the carbonate geologist to assess the environment in which a porosity-modifying diagenetic event took place, providing they are used within a valid petrographic and geologic framework.

In the following chapters we shall use many of these tools in our consideration of porosity evolution in each major diagenetic environment.

Chapter 4

NORMAL MARINE DIAGENETIC ENVIRONMENTS

INTRODUCTION

The marine diagenetic realm is an environment of variable diagenetic potential even though the gross chemical composition of marine waters is relatively constant. This variability is fueled by three factors: 1) variations in kinetic energy in surface waters that occur at the sites of carbonate deposition. These variations in energy influence both the flux of diagenetic fluids through the sediments and the rate of CO_2 outgassing; 2) changes in water temperature and pressure when moving from relatively warm surface waters down into the cold regions of the deep oceanic depths; and 3) the intervention of marine plants and animals in the CO_2 cycle through respiration and photosynthesis.

Today, normal marine surface waters between 30° north and south latitude are supersaturated with respect to aragonite, magnesian calcite, and dolomite (Milliman, 1974a; Bathurst, 1975; James and Choquette, 1983). Under these conditions precipitation is the major operational porosity modification process, and variations in rate of water flux through the sediments control the gross distribution of major cementation events.

The saturation state of surface marine waters relative to $CaCO_3$ changes dramatically with increasing depth as a function of changes in partial pressure of CO_2, progressive temperature decrease, and progressive pressure increase with depth (Scholle and others, 1983; James and Choquette, 1983). While pressure is basically a straight line increase with depth in the marine water column, temperature decreases rapidly in relatively shallow waters across the thermocline, immediately below a well oxygenated zone of relatively constant temperature termed the *mixed zone* (Fig. 4.1). Aragonite becomes undersaturated in the thermocline, while the waters remain saturated with respect to calcite and dolomite. Within this zone on the sea floor, or as is most likely along a platform margin, aragonite sediment and rock components dissolve and calcite precipitates as void fill cement (Saller, 1984b). Below 3 km (depth varies between ocean basins), the water becomes undersaturated with respect to calcite, but remains supersaturated with respect to dolomite. In this zone, calcite rock and sediment components dissolve, often to be replaced by dolomite (Saller, 1984a). Little carbonate, other than dolomite, remains below the calcite compensation depth.

The following discussion will cover the detailed diagenetic potential of each of these two normal marine subenvironments—the shallow and deep marine—emphasizing: porosity modification processes active in the environment, and an assessment of their effectiveness; factors controlling the processes; the development of criteria for the recognition of similar diagenetic environments in ancient carbonate rock sequences; and finally, the presentation of several case histories where marine diagenesis has played a major role in the porosity evolution of economically important ancient reef sequences.

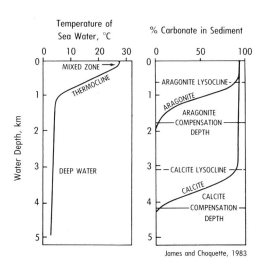

Fig. 4.1. General relationships of temperature and carbonate mineral saturation as a function of depth in oceanic water. Reprinted with permission of the Geological Association of Canada.

SHALLOW WATER, NORMAL MARINE DIAGENETIC ENVIRONMENTS

Introduction to the shallow marine cementation process

While boring algae and fungi are believed to increase the initial intragranular porosity of a sediment during and shortly after deposition (Bathurst, 1966; James and Choquette, 1983), cementation is most certainly the major diagenetic process actively modifying porosity under shallow, normal marine conditions. The cements precipitated are the metastable phases, magnesian calcite and aragonite. Magnesian calcite cements are generally most prevalent in reef-related environments (Land and Goreau, 1970; James and Ginsburg, 1979; Land and Moore, 1980; Purser and Schroeder, 1986), while aragonite cements are most common in intertidal beach rock and hardground environments (Moore, 1973, 1977).

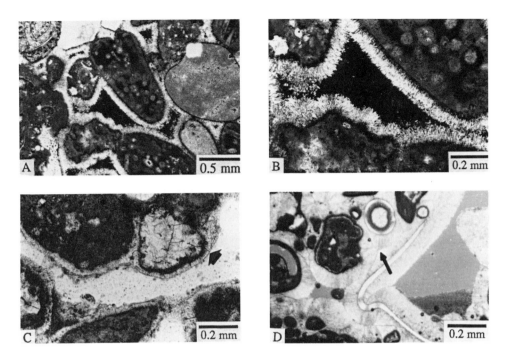

Fig. 4.2. Characteristic marine cements.(A) Aragonite needle cement, modern beach rock St. Croix, U.S.Virgin Islands. Crossed polars. (B) Close view of (A). Crossed polars. (C) Bladed magnesian calcite cement (arrows) (12 mole% MgCO3), modern beach rock St.Croix, U.S.Virgin Islands. Plain light. (D) Fibrous marine magnesian calcite cement, reef-derived debris flow dredged from 900 m (2952') on the island slope of Grand Cayman, British West Indies. Note polygonal suture patterns in cement filled pores (arrows). Plain light.

In general terms, most modern shallow marine cements precipitate as fibrous-to-bladed circumgranular crusts of either aragonite or magnesian calcite mineralogy (Fig. 4.2; see Chapter 3, this book). If the porespace is completely occluded, a polygonal suture pattern that seems to be characteristic of shallow marine cements is developed (Shinn, 1969)(Figs. 4.2D). Magnesian calcite may precipitate as micron-sized crystals, particularly in the reef environment, accumulating as a major component of the internal sediment. This material cannot strictly be classified as a cement, because it does not necessarily bind rock components together. These micrite cements have been reported from beach rocks (Fig. 4.3B) (Moore, 1973, 1977), Bahamian grapestones (Bathurst, 1975), and Bahamian hardgrounds (Dravis, 1979), and seem to be dominantly associated with the activity of algae that bind the grains together. The magnesian calcite appears to precipitate as a result of the living processes of the algae (Figs. 4.3B-D) (Moore, 1973). Aragonite cements, believed to have been precipitated in deeper water (perhaps just above the aragonite lysocline, where aragonite saturation sweeps dramatically toward undersaturation) in association with fore-reef environments, can occur as large botryoidal masses (James and Ginsburg, 1979), as irregular meshes of larger squarely terminated laths (Fig.4.3A), or even as clear, irregular, complex polyhedral crystals reminiscent of calcite spar (Land and Moore, 1980).

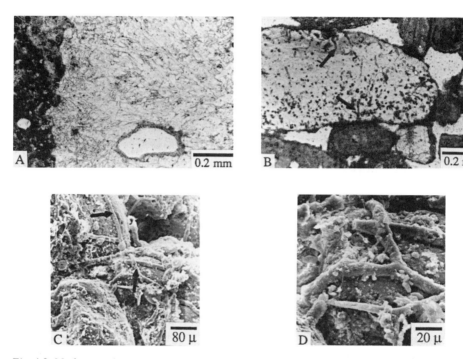

Fig. 4.3. Modern marine cements. (A) Pore fill, aragonite cement forming a mesh work or mosaic of coarse crystals. Sampled from the Jamaica island slope at 212m (696'). Plain light. (B) Micrite magnesian calcite cementing a modern beach rock, St. Croix, U.S.Virgin Islands. Note the dark traces of abundant endolithic algae in the central, large mollusc fragment (arrows). They are particularly abundant at grain contacts.Plain light. (C) SEM photomicrograph of beach rock in (B) showing endolithic algal filaments (arrows) binding the grains together. Thin section of these algal masses appears as pelleted micrite as in (B). (D) Close view of (C).

The stable isotopic composition of modern marine aragonite and magnesian calcite cements as compared to modern ooids, chalks, and bulk reef rock, is shown on Fig.4.4. The modern marine cements generally show high values of ^{18}O, which are in the same compositional range as modern ooids and radiaxial calcite cements from Enewetak, but have considerably lower values than bulk Jamaica reef rock. The $\partial^{13}C$ of modern marine cements are in the same general positive range as other marine precipitates and bulk reef rock from Jamaica. These marine cements are enriched with respect to Mg, Sr, and Na, and depleted relative to Fe and Mn, reflecting the gross stable isotopic and trace element composition of marine waters (Table 3.2). Modern, shallow marine carbonate cements are generally nonluminescent because marine cementation environments commonly exhibit high-energy, oxidizing conditions that preclude the incorporation of Fe and Mn in the carbonate lattice (see Chapter 3).

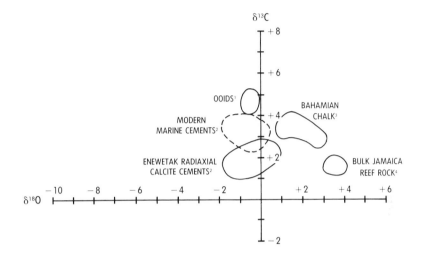

Fig. 4.4. Oxygen, carbon isotope composition of various Holocene marine carbonates. Data sources as follows: 1 and 3, James and Choquette, 1983; 2. Saller, 1984. 4. Land and Goreau, 1970.

Because most surface waters are supersaturated with respect to $CaCO_3$, cementation events are linked directly to the flux of these supersaturated fluids through the sediments. It would require 10,000 pore volumes of saturated marine water to completely fill a pore with carbonate cement, assuming that all available $CaCO_3$ was precipitated from the fluid. With a 10% precipitation efficiency, not unrealistic for a natural system, over 100,000 pore volumes would be required (Scholle and Halley, 1985). Cementation, then, will be favored under the following conditions: in areas where sediments are both porous and permeable; in a high-energy setting, where sediment water flux is maximized and CO_2 degassing can operate; in areas of low or restricted sedimentation rate, or restricted sediment movement, so that the sediment-water interface is exposed for an appropriate length of time to assure adequate pore-water exchange; in locales of high organic activity where framework organisms can stabilize substrates and erect large pore systems, and organic activity can impact the partial pressure of CO_2; and finally, in regions where surface waters exhibit higher than normal saturation values for $CaCO_3$.

The kinetic factors necessary for cementation obviously limit the volume and distribution of cementation under shallow normal marine conditions. Fig. 4.5 outlines the distribution of shallow marine cementation that can be expected and that has been observed in modern marine environments. Significant cementation that seriously impacts porosity is concentrated at the shelf margin in conjunction with reef development. The high-energy intertidal zone, in association with beach sedimentation, and hardgrounds associated with ooid shoals and areas of nondeposition are other, less important zones of shallow marine cementation. Each of these cementation environments will be discussed in subsequent sections of this chapter.

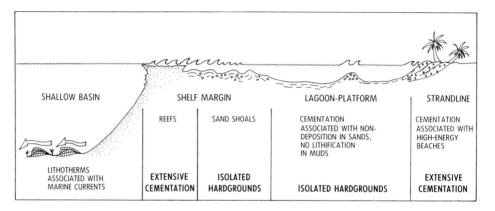

Fig. 4.5. Diagenetic processes associated with shallow marine depositional environments.

Recognition of ancient shallow marine cements

The major problem in recognizing ancient marine cements rests in the fact that all modern marine cements are metastable phases that will ultimately stabilize to calcite and dolomite through some process of dissolution and reprecipitation. How much of the unique textural and geochemical information carried by the original marine cement is lost during this transformation to its more stable phase? Magnesian calcite generally transforms to low magnesian calcite with little loss of textural detail (James and Choquette, 1984). Ancient shallow inferred marine magnesian calcite cements, then, should exhibit the characteristic bladed-to-fibrous textures enjoyed by their modern counterparts (Compare Figs. 4.2D to 4.6A and B). During this transformation, however, some geochemical information is invariably lost. If stabilization takes place in an open system in contact with meteoric waters, much of the original trace element and isotopic signal contained in the original marine cement may be lost. If, however, stabilization is accomplished in a closed system in contact with marine waters, much of the original marine geochemical information may be retained. Therefore, inferred magnesian calcite cements should be expected to exhibit relatively heavy stable isotopic compositions and to be enriched relative to Sr and Na (James and Choquette, 1983). Lohmann and Meyers

(1977) documented the correlation of the presence of microdolomite to original magnesian calcite mineralogy in Paleozoic crinoid plates and associated fibrous cements.

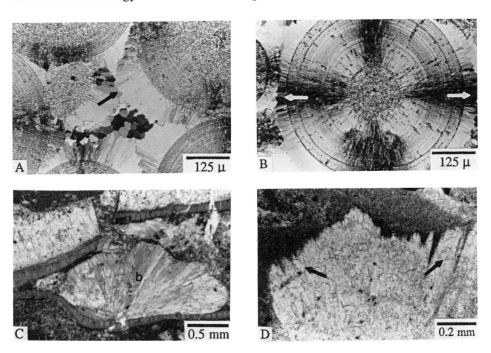

Fig. 4.6. Ancient marine cements. (A) Jurassic fibrous marine calcite cement inferred to be originally magnesian calcite (Moore and others, 1986). Note the polygonal suture pattern where these cements come together in the center of the pore (arrows). Jurassic Smackover Formation, subsurface, south Arkansas, U.S.A. Crossed polars. (B) Jurassic fibrous marine calcite cement precipitated in optical continuity with a radial ooid. It is thought that both the cement and ooid were originally magnesian calcite. Same sample as (A). Crossed polars. (C) Preserved Pennsylvanian botryoidal aragonite (b), Sacramento Mountains, New Mexico. Plain light. (D) Upper Permian to Lower Pennsylvanian (Laborocita Formation) botryoidal aragonite (?) that has been recrystallized to calcite. Note the fine crystalline calcite mosaic (arrows) through which the individual botryoidal rays are seen as ghosts. Compare to preserved aragonite in (C). Plain light.

Aragonite cement, however, shows significantly more textural loss than does magnesian calcite during stabilization, and indeed in some cases where fresh waters are involved in stabilization, all textural information may be destroyed by dissolution before calcite is reprecipitated in the resultant void (James and Choquette, 1984). In the case of total dissolution, the presence of a precursor aragonite marine cement is impossible to determine. When stabilization takes place in marine pore fluids or in waters nearly saturated with respect to $CaCO_3$, however, ghosts of the original aragonite cement texture and some of the original geochemical information may be retained within the replacive calcite mosaic. Sandberg (1983) developed a set of criteria useful in determining original aragonite mineralogy of grains and/or cements (Table 4.1). One of the cornerstones of this set of criteria is the common occurrence of original aragonite as relicts in the enclosing coarse calcite mosaics. These calcite-after-aragonite marine cements often exhibit elevated Sr composition and often show high ^{18}O values relative to meteoric-derived cements (James and Choquette, 1983). Massive botryoids of mosaic calcite associated with relict aragonite, and high Sr compositions are believed to be examples of seafloor cements that were once composed of aragonite (Figs. 4.6C and D) (Assereto and Folk, 1980; Mazzullo, 1980; Bathurst, 1982; Sandberg, 1985). Similar botryoids occur in Mississippian to, possibly, Lower Jurassic rocks, and have been reported from the Lower Cambrian and the Middle Precambrian (Sandberg, 1985).

Table 4.1. Criteria for the recognition of ancient aragonites. (Reprinted by permission from Nature, Vol. 305, pp 19-22. Copyright (C) 1983, Macmillan Journals Limited)

1. *Still aragonite.* Encountered in quite old rocks.
2. *Mosaic of generally irregular calcite spar containing oriented aragonite relics.* Crystal boundaries commonly cross-cut original structure defined by organic or other inclusions. Replacement crystals normally $10-10^3$ times larger than replaced aragonite crystals.
3. *Calcite mosaic as for Criterion 2, but without aragonite relics.* Original aragonite mineralogy supported by elevated Sr^{2+} content, relative to levels reasonably expected in dLMC resulting from alteration of originally calcitic constituents.
4. *Calcite mosaic as for Criterion 3, but Sr^{2+} not elevated, or not measured.* Differing Sr^{2+} content in the resulting calcites can reflect differing degrees of openness of the diagenetic system.
5. *Mold or subsequently calcite-filled mold.* Common for aragonitic constituents, but equivocal because observed for non-aragonitic constituents in undolomitized limestone.

¹Arranged in order of decreasing reliability, from Sandberg (1983).

Can the concept of uniformitarianism be used in studies of carbonate porosity to assume that marine cements have always been metastable magnesian calcite and aragonite? There is increasing evidence led by the work of Mackenzie and Pigott(1981), Sandberg (1983, 1985), and Given and Wilkinson (1985) suggesting that the nature of abiotic marine precipitation, such as ooids and marine cements, has changed in a cyclical fashion through the Phanerozoic. These cycles apparently reflect oscillations in Pco_2 driven by global tectonics and result in relatively long periods of time during which abiotic precipitation is dominated by calcite, followed by periods such as are found in today's ocean, where aragonite is the dominate marine precipitate (Fig. 4.7). In the Lower Paleozoic and in the Mesozoic periods, which are believed to have been marked by calcite seas, marine cements should dominantly have been calcite, and should exhibit well-preserved textural and geochemical information. In the Upper Paleozoic and the Tertiary, aragonite and, perhaps, magnesian calcite, marine cements should dominate, and recognition of these metastable precursors may be more difficult.

Finally, as indicated in Chapter 3, there is also some evidence that the stable isotopic composition of marine waters has systematically changed through time, as reflected in the cross-plot of $\partial^{18}O$ and $\partial^{13}C$ compositions of marine cements from lower

Cambrian to Holocene (Fig. 4.7). These changes parallel the results of Veizer and Hoefs (1976) (Fig. 3.12), and must be taken into account when utilizing stable isotopic compositions of cements to help determine precipitational environment.

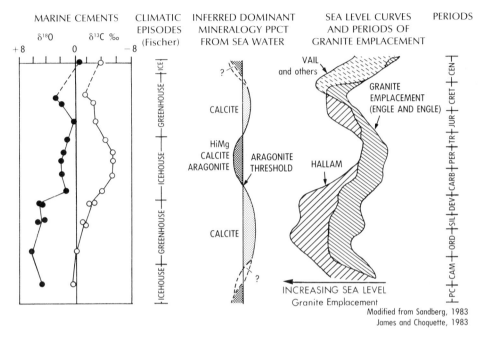

Fig. 4.7. (Left) Variations in oxygen and carbon isotopic composition of marine cements through time. Data sources may be found in James and Choquette, 1983. (Right) The possible distribution of calcite and aragonite seas through time as a function of climatic episodes, sea level, and periods of granite emplacement. (Left) Reprinted with permission of Geological Association of Canada. (Right) Reprinted by permission from Nature, Vol. 305, pp. 19-22. Copyright (C) 1983, Macmillan Journals Limited.

Diagenetic setting in the intertidal zone

The beach setting is ideal for the precipitation of marine cements. High-energy conditions, including both wave and tidal activity, as well as the presence of relatively coarse, highly porous, and permeable sediments, ensure that adequate volumes of supersaturated water are able to move through the pore system to accomplish the cementation. While beach rock cementation often occurs in zones of mixing between meteoric and marine waters (Moore, 1973, 1977), it is believed that CO_2 degassing, rather than the mixing phenomenon itself, is the prime cause for precipitation (Hanor, 1978). Tidal pumping and wave activity provide the most logical mechanism for the CO_2

outgassing in the beach shoreface zone. As was mentioned earlier, beach rock cementation may also be accomplished by stabilizing a beach shoreface by algal binding, accompanied by cementation related to the living processes of the algae (Figs. 4.3B-D).

Active beach progradation will generally not allow beach rock formation because the sediment-water interface is not exposed long enough for significant cementation to occur. Cemented zones are typically thin (< 1 m in the Caribbean, but can be thicker in areas of high tidal range such as the Pacific), and occur in the upper shoreface, with an upper limit controlled by mean high tide (Fig 4.8A). While most cementation in the beach shoreface takes place in the marine phreatic zone, in areas of relatively high tidal range, marine vadose cements often occur in beach rock (Figs. 4.8C and D). Cementation in a

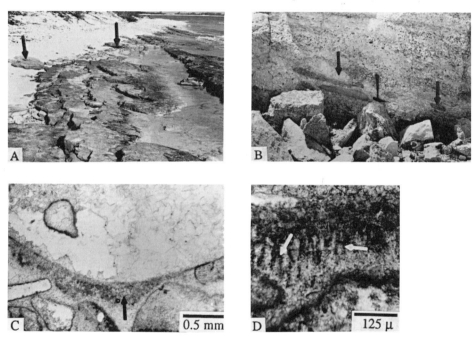

Fig. 4.8. (A) Beach rock, Grand Cayman, British West Indies. Note the blocks (arrows) that have been torn up and are being reincorporated into the beach. (B) Beach rock clasts incorporated in a Lower Cretaceous beach sequence (arrows), northcentral Texas, U.S.A. (C) Microstalactitic cement (arrow) occurring in beach rock clasts from outcrop shown in (B). (D) Close up of (C) showing the fibrous ghosts (arrows) occurring in microstalactitic calcite mosaic suggesting that the beach rock cement was aragonite. Plain light.

high-energy zone, like a beach, is often followed by the destruction of the beach rock bands during storms and in the formation of boulder and breccia units associated with the shoreface environment (Figs. 4.8A and B).

Beach rock cements are generally fibrous-to-bladed crusts that may completely occlude available pore space (Fig. 4.2). Today, directly precipitated beach rock cements are both aragonite and magnesian calcite, with aragonite generally dominant. Microcrystalline cements in beach rock, associated with algae, are commonly magnesian calcite (Fig. 4.3) (Moore, 1973, 1977).

Beach rock cementation is common in ancient carbonate shoreface sequences (Inden and Moore, 1983) and is relatively easy to recognize in outcrop and core because of its association with suites of sedimentary structures characteristic of shoreface sedimentation. In addition, the presence of large clasts with fibrous-to-bladed cements similar to present shallow marine cements (Fig. 4.8B), and fibrous pendant cements (Figs. 4.8C and D), indicating early cementation in the intertidal zone, are characteristic of early marine shoreface cementation (Inden and Moore, 1983; James and Choquette, 1983).

Table 4.2. *Porosity %, depositional and diagenetic fabrics for lithologies characteristic of depositional environments associated with a Lower Cretaceous beach complex. Reprinted with permission of 1972 International Geological Congress.*

Summary table of depositional and diagenetic fabrics and features averaged for individual depositional environment										
	Average Allochems	Average Cement[1]	Average Micrite Matrix[1]	Average Dolomite %[2]	Average Silica Total %[2]	Average Total Leached %	Measured Porosity (%) Average	Measured Permeability (md) Average[3]	Dominant Pore Types[4]	Secondary Pore Types[4]
	Total %	Total %	Total %							
Back-shore	—	—	—	90%	<5%	?	33%	18.3	BC	sxS-MsMO
Upper Foreshore	65	35	0	<5%	28%	7	14.4	32.8	crS-msMO	crS-msVUG crP-msBP
Lower Foreshore	68	25	7	<5%	10%	4	24	134	sxS-msMO	sxS-msVUG
Offshore	44	15	41	<5%	<5%	2	30	39	sxS-msMO	crS-msVUG

1. Average % data on allochems, cement and matrix taken from actual point counts on a vertical sample grid (15 total) through the beach sequence, averaged by depositional environment.

2. Average % data on dolomite and silica based on X-ray diffraction analysis of same 15 samples from 1 above, and averaged by depositional environment.

3. Measured porosity and permeability averages from an adjacent vertical sample grid (15 total) through the beach sequence, averaged by depositional environment.

4. Porosity types follow scheme of Choquette and Pray, 1970, and are based on visual observation of thin sections from the vertical sample grids of 1 and 3 above.

Modified from Moore and others, 1972

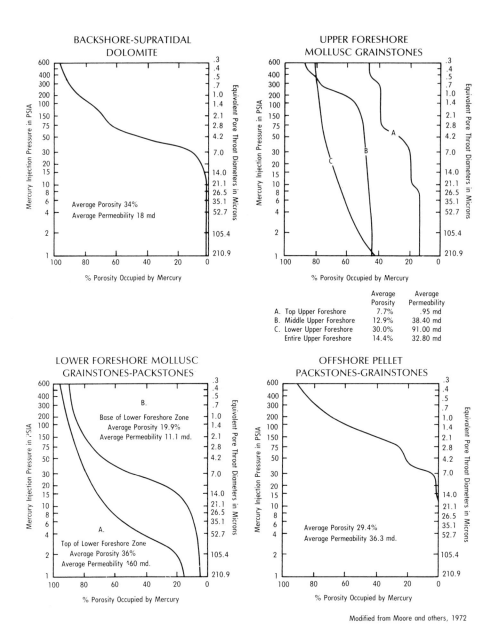

Fig. 4.9. Mercury injection curves and porosity-permeability of the major lithofacies characteristic of the depositional environments associated with a Lower Cretaceous carbonate beach sequence, central Texas, U.S.A. Reprinted with permission of 1972 International Geological Congress.

Porosity modification in the beach intertidal zone by cementation is generally limited to the upper part of the upper foreshore. Moore and others (1972) described the pore system evolution of a Cretaceous carbonate beach sequence from central Texas. Table 4.2 details the distribution of grains, cement, micrite matrix, porosity-permeability, and pore types as a function of depositional environment. Based on the distribution of micrite matrix and the percentage of grains, the upper foreshore should have had an original porosity of approximately 35%. Early fibrous and bladed cements total exactly 35% in the upper foreshore zone, indicating that shortly after deposition, all available intergranular pore space in the upper foreshore was occluded by marine cements, as beach rock. The intensity of early marine cementation decreases dramatically toward the base of the beach. Today, all porosity in this Cretaceous sequence is secondary, and its distribution in the beach sequence (with the greatest porosity in the offshore and the poorest porosity in the upper foreshore) is opposite the original porosity distribution expected from such a sequence. Early marine cementation sealed off the porous upper foreshore and later diagenetic waters were preferentially directed through the lower foreshore and offshore zones, creating significantly better quality reservoir rock by selective dissolution (Fig. 4.9). The upper .5 m of the beach sequence is tight and could form a reservoir seal. The backshore, resting immediately above the beach, is a dolomitized supratidal sequence with perhaps the best porosity development in the entire beach complex (Fig.4.9). This sequence, no doubt, was originally mud-dominated, and seems to have been flushed by meteoric water after dolomitization, which contributed significantly to favorable porosity development (see Chapter 5).

Beach rock cementation then, has a tendency to vertically compartmentalize reservoirs. Perhaps the best example of this type of modification in a reservoir setting can be found in the Upper Jurassic Smackover Formation at Walker Creek, in southern Arkansas (Brock and Moore, 1981). In this field, the reservoir rock is an ooid grainstone with preserved primary intergranular porosity. The field is developed over subtle salt structures that were active during deposition of the Smackover ooid sands. Well developed fibrous-to-bladed marine cements are only present in association with early structure, and presumably formed as beach rock during ephemeral island development over the crests of the salt structures (Figs. 4.6A and B). The cemented zones are thin (less than a meter), but usually the cement completely occludes the pore space and helps compartmentalize the reservoir into multiple-producing horizons (Brock and Moore, 1981; see discussion of the Walker Creek field in Chapter 7).

Modern shallow water submarine hardgrounds

Thin, discontinuous, submarine-cemented crusts are common in certain modern carbonate environments such as the Bahama Platform, Shark Bay in western Australia, and along the margins of the Persian Gulf (Taft and others, 1968; Shinn, 1969; and Dravis, 1979). In the Bahamas, submarine hardgrounds are most common along the margins of the platform, associated with relatively high-energy conditions such as ooid or grapestone sand shoals. The frequency of hardground development seems to decrease into the platform interior away from high-energy bank margins (Dravis, 1979). In the Persian Gulf, hardground development seems to be much more widespread, and may even involve siliciclastics (Shinn, 1969). In both the Persian Gulf and the Bahamas, the dominant cement mineralogy and morphology is fibrous aragonite, with subordinate magnesian calcite. Modern hardgrounds invariably show cementation gradients decreasing away from the sediment-water interface, and with the exception of the algal-related magnesian calcite, form directly by precipitation from supersaturated marine waters. It would seem that sedimentation rate and/or substrate stabilization play the major role in hardground formation, with hardgrounds forming where sediments are either not being deposited, or where a substrate has momentarily been stabilized by subtidal algal or bacterial mats (Shinn, 1969; Dravis, 1979). Cessation of sedimentation or substrate stabilization need not involve a long period of time before the development of a hardground by cementation. Both Dravis (1979) and Shinn (1969) indicated that hardgrounds could form in the space of months.

The regional distribution of hardgrounds is probably related to surface water saturation relative to $CaCO_3$. The margin of the Bahama Platform is an area of higher $CaCO_3$ saturation caused principally by CO_2 degassing driven by high wave and tidal energy, progressive heating of colder marine water, and photosynthetic activity of shallow-water algae. Hence, hardground development would be expected to die out toward the platform interior, as reported by Dravis (1979). Development of hardgrounds in the Persian Gulf is more widespread than in the Bahama Platform. This is apparently because of the very high state of supersaturation of Persian Gulf surface water with $CaCO_3$, resulting from the progressive development of hypersalinity along the margins of the Gulf (Shinn, 1969).

Recognition and significance of ancient hardgrounds

Shallow marine hardgrounds are common in the geologic record (Purser, 1969; Bromley, 1978; Wilkinson and others, 1982; James and Choquette, 1983; Halley and others, 1983). They occur as an integral part of ooid-bearing ancient platform margin sequences (Halley and others, 1983), within coarse grainstones in normal marine cratonic sequences (Wilkinson and others, 1982), or as the terminal phase of shallowing upward sequences (Purser, 1969; James, 1979; see Chapter 1, this book).

Hardgrounds may be recognized in core by the presence of an abraded, sometimes irregular bored surface, exhibiting the holdfasts of sessile organisms and an encrusting fauna on the surface with intraclasts derived from the hardground incorporated in the sedimentation unit above (Rassmann-McLaurin, 1983; James and Choquette, 1983). The surface may be stained with manganese, and glauconite is often found associated with the hardground, perhaps marking a decrease in sedimentation rate. Petrographically, hardgrounds commonly show typical marine cement textures that may be penetrated by marine endolithic algae and that may be associated with marine internal sediments (Fig. 4.10). Compaction effects that involve adjacent sedimentation units are commonly

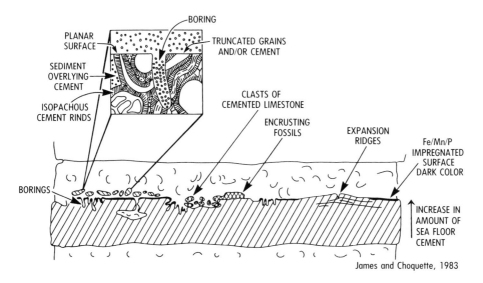

Fig. 4.10. Criteria for the recognition of marine hardgrounds. Reprinted with permission of the Geological Association of Canada.

absent within hardgrounds (Rassmann-McLaurin, 1983).

While the actual volume of marine cement associated with hardgrounds is small, and the effect on the total porosity of a sedimentary sequence is minimal, they can, like beach rock, act as a reservoir seal, effectively compartmentalizing a reservoir and slowing or stopping vertical fluid migration.

Diagenetic setting in the modern reef environment

The locus of marine cementation in shallow marine carbonate environments is the shelf-margin reef. The framework reef provides a remarkably favorable environment for the precipitation of marine cement. The reef itself furnishes a porous stabilized substrate that allows the uninhibited movement of supersaturated marine water through its pores, allowing it to interact with a variety of carbonate material. Reefs are generally sited along shelf margins, in areas of high wave and tidal activity, ensuring the constant movement of vast quantities of marine waters through the reef framework. The shelf-margin site supplies the reef with unlimited volumes of fresh, cool, marine water in the process of being degassed relative to CO_2 by agitation, warming, and organic activity, leading to further saturation with respect to $CaCO_3$, and ultimately to massive cementation within the reef frame. This process has been well documented by numerous workers in modern reefs at sites around the world including: Jamaica (Land and Goreau, 1970; Land and Moore, 1980); Bermuda (Ginsburg and Schroeder, 1973); Red Sea (Friedman and others, 1974; Dullo, 1986); Belize (James and others, 1976; James and Ginsburg, 1979); Panama (Macintyre, 1977); Great Barrier Reef (Marshall and Davies, 1981; Marshall, 1986); Mururoa Atoll (Aissaoui and others, 1986); south Florida (Lighty, 1985); and many other sites and authors.

While both aragonite and magnesian calcite cements are associated with modern marine reef lithification, magnesian calcite is today the dominant marine cement mineralogy (Macintyre, 1985; Purser and Schroeder, 1986). Fibrous aragonite commonly occurs as epitaxial overgrowths on aragonite substrates such as coral septa (James and Ginsburg, 1979). Aragonite may grow into larger cavities in a spherulitic habit that can develop into spectacular large botryoidal masses, such as were described by James and Ginsburg (1979) from the deeper fore reef (>65m) of Belize.

Modern magnesian calcite reef-related cements occur as two distinct types: peloidal aggregates of micrite-sized crystals bearing a radiating halo of coarser scalenohedral crystals (Fig. 4.11A), and thick fibrous-to-bladed crusts lining larger voids (Fig. 4.11B). The peloidal magnesian calcite was believed by Land and Goreau (1970) to be precipitational rather than sedimentological in origin, based on the lack of aragonite in the peloids. The pelleted nature of this most common feature of submarine lithification of reefs is generally believed to be the result of repeated nucleation around centers of growth in a highly supersaturated fluid, although the pelletizing action of cryptic organisms and calcification of algal filaments may also play a role in the development of the peloidal habit (Land and Moore, 1980; Macintyre, 1985).

The stable isotopic composition for both magnesian calcite and aragonite submarine, reef-related cements show high ^{18}O values and, relative to cements precipitated from normal marine waters, such as beach rocks, exhibit higher ^{13}C values, perhaps indicating some organic intervention in their precipitation (Land and Goreau; 1970, Land and Moore, 1980) (Fig. 4.4, this book). Aragonite cements, including the massive botryoidal cements of Belize, have an average Sr composition of 8300 ppm, and an average Na composition of 2000 ppm, both within the range expected from aragonites precipitated from marine waters (James and Ginsburg, 1979).

Cement distribution in modern reefs is strongly skewed toward the seaward margins of the reef, and indeed the leeward edges are frequently completely uncemented

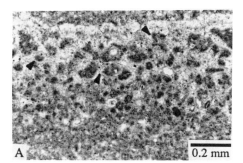
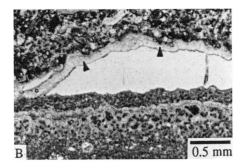

Fig. 4.11. (A) Pelleted, magnesian calcite internal sediment filling a cavity in reef frame, Jamaica. Each pellet is surrounded by a halo of magnesian calcite cement (arrows). Sample taken in 212 m (695') of water. Plain light. (B) Large view of cavity shown in (A) showing a thick crust of magnesian calcite cement (arrow) across the roof of the cavity. Plain light.

(James and others, 1976; Macintyre, 1977; Marshall and Davies, 1981; Marshall, 1986). This distribution seems to be in response to higher energy conditions and cooler oceanic water temperatures along the seaward edges of the reef. The large volume of water moved through the reef framework by waves and tides, and CO_2 degassing by warming and agitating the cold marine waters would result in the intense cementation observed along the seaward margins of modern reefs. Reef cementation often seems to be concentrated at the reef-water interface, and may die out toward the interior of the reef, sheltering zones of high porosity, particularly if framework accretion is rapid (Macintyre, 1977; Lighty, 1985). In areas with slower reef accretion rates, such as in Jamaica, or in the deeper fore-reef zones of Belize, the entire seaward margin of the reef can be solidly lithified, destroying most porosity (Land and Goreau, 1970; James and Ginsburg, 1979; Land and Moore, 1980). During drilling operations on the Belize barrier reef at Carrie Bow Key, Shinn and others (1982) encountered pervasively cemented framework on the seaward margin of the reef, and uncemented framework less than 600 m landward of the reef crest.

The actual reef lithification process, and by extension, early porosity evolution, is much more complex than simple, passive, marine cementation. Three distinct processes, operating simultaneously, affect early lithification and porosity modification in the reef environment: bioerosion, internal sedimentation, and marine cementation. During the initial development of reef framework, porosity potential is practically unlimited because of the following factors: common formation of large shelter voids during the development of the framework of the reef (some of these voids are so large that a person can actually swim through them, as shown in Fig. 2.8A); the presence of copious intragranular porosity within the framework organisms, such as corals (see Chapter 2, this book); the presence of significant intergranular porosity within associated sand-dominated sediments; and, finally, the development of extensive secondary porosity within the reef organic framework, as a result of the activity of boring organisms such as pelecypods, sponges, algae, and fungi (Chapter 2, this book) (Fig. 4.12).

Reef-associated marine waters contain suspended sediments derived from the physical and biological breakdown of the reef framework in high-energy environments, as well as calcareous phytoplankton and zooplankton (Moore and Shedd, 1977). This suspended material is forced into, and is ultimately deposited within the framework of shelf margin reefs as internal sediments by the flow of immense volumes of marine water through porous reef framework driven by tides and wave activity (Ginsburg and others, 1971). These sediments consist of unidentifiable silt- and clay-sized aragonite and magnesian calcite, as well as coccoliths, planktonic foraminifera, planktonic molluscs, and silt-sized chips derived from the activity of clionid sponges (Moore and others, 1976).

These internal sediments are cemented, dominantly by fine crystalline magnesian calcite, very shortly after they are deposited within the voids developed in the framework. Deposition seems to be episodic, and cements become much coarser at times of nondeposition along the sediment water interface in larger cavities. The final fabric is one of fine, intergranular cements interrupted periodically by horizontal, geopetal crusts of coarser cement that can be traced up the walls and across the roof of the remaining cavity (Fig. 4.11) (Land and Moore, 1980; James and Choquette, 1983). Much of the internal sediment is pelleted, and often is monomineralic (totally magnesian calcite) with no marine planktonic fossils. This material is believed to have been precipitated directly from marine waters, and is central to the controversy concerning the mechanism of pelletization, as mentioned above and discussed by MacIntyre in 1985. Regardless of whether the material was pelleted by organic processes, then overgrown by coarser cements, or was aggregated by repeated nucleation as espoused by MacIntyre, it fills porespace, is cemented, and is an important component of early reef lithification and porosity modification.

Reef framework, cemented internal sediments, and marine cements are penetrated and destroyed by the galleries and borings of various marine organisms, including lithophagus pelycopods and clionid sponges. These galleries and borings can displace more than 50% of the lithified framework of the reef with relatively large organism-filled voids at any one time (Moore and Shedd, 1977) (Fig. 4.12A). Upon death of the organism, an empty void that immediately begins to be filled with internal sediment that is quickly cemented is produced (Figs. 4 12C and D). In reefs where framework accretion is not exceptionally fast, the process is repeated innumerable times until an original porous, recognizable reef framework is reduced to a relatively nonporous pelleted lime wackestone within which little reef framework is preserved (Fig. 4.12A) (Schroeder and Zankl, 1974; Land and Moore, 1980).

The repetitive application of this tripart set of concurrent destructive-constructive

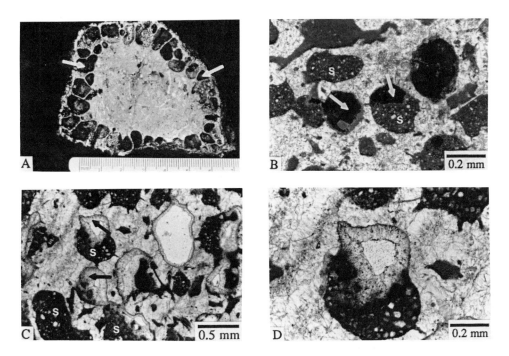

Fig. 4.12. (A) Modern reef-related rock taken from the Jamaica island slope at 212 m (695'). Entire periphery of rock has been penetrated by Clionid sponges, forming open galleries when the sponges die (arrows). (B) Clionid galleries (arrows) penetrating modern reef frame. Galleries being filled with pelleted internal sediment (s). Jamaica island slope at 181 m (594'). Plain light. (C) Clionid galleries showing extensive cementation (arrows) by magnesian calcite cement in association with internal sediments (s). Same sample as (B). Plain light.. (D) Close view of (C). Plain light.

processes is not only the cause for massive early porosity reduction in reef sequences, but also the difficulty geologists frequently encounter when trying to recognize framework reefing in cored subsurface material. Because the shelf margin reef is the one environment where these three processes are active over relatively long periods of time, the recognition of the activity of concurrent internal sedimentation, marine cementation, and bioerosion can be useful criteria for the recognition of the reef environment in cored subsurface sequences (Schroeder and Zankl, 1974; Land and Moore, 1980).

Recognition of reef-related marine diagenesis in the ancient record

Detailed studies of ancient reef sequences indicate that many of the same textural and porosity modification patterns seen in modern reefs are mirrored in their ancient counterparts, indicating that concurrent internal sedimentation, marine cementation, and bioerosion have long been the most important factors controlling the nature of early marine lithification within the reef environment (Davies, 1977; Petta, 1977; Walls and others, 1979; Mazzullo, 1980; James and Klappa, 1983; James and Choquette, 1983; Walls and Burrowes, 1985; Schroeder and Purser, 1986; Kerans and others, 1986; Rabat and others, 1986). Clionid borings have been observed as far back as the Devonian (Moore and others, 1976), and have played a major role in textural modification during early marine lithification in Jurassic coralgal reefs across the Gulf of Mexico (Crevello and others, 1985), Cretaceous rudist reefs of central Texas (Petta, 1977), and Upper Cretaceous of Tunisia (Rabat and others, 1986). The recognition of clionid borings and their concurrent penetration of framework, internal sediments, and cements, then, may be a prime criterion for the recognition of early marine lithification of ancient reef sequences through most of the Phanerozoic.

Pervasive early cements, believed to have been precipitated from marine waters, have been described from reef sequences throughout the Phanerozoic (Schroeder and Purser, 1986). In ancient rock sequences these inferred marine cements associated with reef environments exhibit two basic fabrics: 1) isopachous radiaxial fibrous calcite with

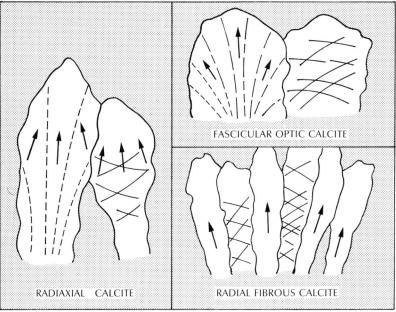

From Kendall, 1985

Fig. 4.13. Schematic diagrams illustrating the differences between radiaxial, fascicular optic, and radial fibrous calcites. For each diagram arrows record fast vibration directions, narrow continuous lines represent twin-planes, and discontinuous lines represent subcrystal boundaries. Used with permission of SEPM.

perfectly preserved, but sometimes complex fibrous crystal structure, usually with undulose extinction (Fig. 4.14); and 2) spherulitic fibrous isopachous fringes and botryoidal masses, recrystallized into a fine-to-coarse calcite mosaic with the original fibrous crystal morphology preserved as ghosts outlined by fluid and solid inclusions (Figs. 4.6C and D) (Sandberg, 1985).

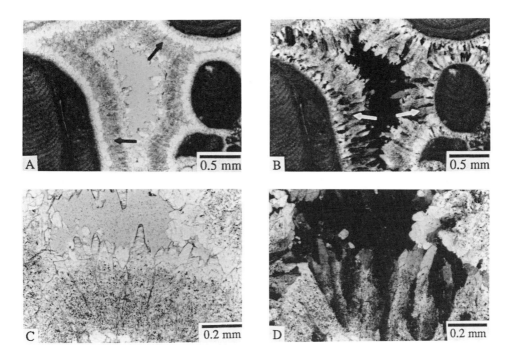

Fig. 4.14. Radiaxial calcite cement, at 640 m (2100') F-1 well Enewetak Atoll. Rocks are Miocene in age. Note inclusion rich zone (arrows). Plain light. (B) Same as (A) Note irregular extinction patterns (arrows) characteristic of radiaxial calcite under crossed polars. (C) Close view of (A) showing the inclusion rich zone. Plain light. (D) Same as (C), with crossed polars.

Radiaxial fibrous calcites show a range of textures, as schematically presented in Fig. 4.13. Radiaxial fibrous calcite proper (as originally defined by Bathurst in 1959) is identified by a pattern of subcrystals within each crystal that diverge away from the substrate in an opposing pattern of distally-convergent optic axes, giving a corresponding curvature to cleavage and twin lamellae (Kendall, 1985). Fascicular-optic calcite (Kendall, 1977a) is a less common type distinguished by the presence of a divergent pattern of optic axes coinciding with that of the subcrystals. Radial fibrous calcites (Mazullo, 1980) do not have undulose extinction but are formed of turbid crystals. These radiaxial fibrous calcites were originally believed to represent the recrystallization fabric of an acicular marine aragonite precursor (Kendall and Tucker, 1973; Bathurst, 1975; Davies, 1977; Bathurst, 1982). Sandberg (1985) and Kendall (1985), basing their interpretations on petrographic considerations, suggest that these cements were originally precipitated as a calcite cement that may have had an original high magnesian calcite mineralogy because of the common occurrence of microdolomite inclusions. Saller (1986) describes radiaxial fibrous calcite cements of low magnesian calcite mineralogy precipitated into aragonite dissolution voids in Miocene limestones from cores taken along the margin of Enewetak Atoll (Figs. 4.14 and 4.22). Saller develops compelling isotopic (Fig. 4.4) and petrographic evidence to indicate that these cements were precipitated directly from deep, cool marine waters located beneath the aragonite compensation depth and above the calcite compensation depth.

The second common group of reef-related cement fabrics, the spherulitic fibrous fringes and botryoidal masses described by Mazullo (1980), James and Klappa (1983), and others, are generally thought to have originally been aragonite marine cement subsequently recrystallized to calcite (Fig. 4.6) (Sandberg, 1985). These inferred aragonite marine cements are only common during certain spans of geologic time: Middle Pre-Cambrian (Grotzinger and Read, 1983); Early Cambrian (James and Klappa, 1983); Late Mississippian (Davies, 1977; Mazzulo, 1980) to Early Jurassic (Burri and others, 1973); and mid-to-late Cenozoic (Sandberg, 1985) (Fig. 4.6, this book).

This distinctly clumped temporal distribution pattern of aragonite cements strongly parallels the oscillatory temporal trends in abiotic carbonate mineralogy established by Mackenzie and Pigott (1981) and Sandberg (1983) (see introduction to this chapter; Fig. 4.7). These trends were based primarily on the interpretation of the original mineralogy of Phanerozoic ooids. The periods when inferred aragonitic ooids are common coincide with the documented occurrences of botryoidal aragonite (Fig. 4.7). During these intervals of apparent aragonite inhibition, calcite seems to have been the dominant precipitational phase in marine waters, and hence radiaxial calcite cements dominated the shallow and deep reef environments. To date, modern radiaxial low-magnesian calcite cements have only been found within shelf-margin reef sequences bathed with cold marine waters below the aragonite compensation depth (Saller, 1986).

In the following pages the effects of shallow marine diagenetic processes on porosity evolution will be explored. Two economically important reef trends will be presented as case histories: the Lower-Middle Cretaceous reef-bound shelf margin of the Gulf of Mexico, anchored on the south by the Golden Lane of Mexico; and the middle Devonian reef trends of western Canada epitomized by the Leduc, Rainbow, and Swan Hills reefs.

Porosity evolution of reef-related Lower-Middle Cretaceous shelf margins: the Golden Lane of Mexico and the Stuart City of south Texas

The Lower-Middle Cretaceous shelf margin is one of the most dominant geologic features of the Gulf of Mexico, forming a 2500-km-long structural and sedimentological boundary around the Gulf of Mexico basin (Meyerhoff, 1967; Winker and Buffler, 1988) (Fig.4.15). The southwestern segment is a shallow marine carbonate shelf that can be traced in outcrop from southeastern Mexico to central Texas. The shelf margin is exposed at the surface in Mexico as a result of Laramide tectonism, but rapidly dips into the subsurface to the north and south.

The Lower Cretaceous shelf margin sequence of south Texas and northeastern Mexico consists of two distinct progradational reef-dominated shelf margin sequences called the *Stuart City* (El Abra in Mexico) and *Sligo* (Cupido in Mexico), separated by a regional drowning event marked by the Pearsall Shale (Fig. 4.15). In Texas, the Stuart City is characterized by the development of persistent, thick (up to 800 m), rudist-dominated reefs and reef-related facies buried some 5000 to 6000 m in the subsurface (Winter, 1962; Bebout and Loucks, 1974; Achauer, 1977). Seaward of the shelf margin, the sequence passes rapidly into dark, pelagic, foraminiferal limestones, termed the Atascosa Formation, or Group (Winter, 1962). Shelfward, the Stuart City changes to the relatively low-energy shelf sequences of the Glen Rose Formation and the Fredericksburg Group (Fig. 4.15). These units ultimately crop out across a wide region of central and west Texas. The subsurface Stuart City extends west and south in the subsurface into Mexico where it is termed the El Abra Formation (Rose, 1963).

In southeastern Mexico, reef-dominated shelf margin development was continuous from the Lower through Middle Cretaceous, and perhaps extended up into the Upper Cretaceous. These shelf margins were developed in the El Abra Formation on a series of large, structurally-controlled, epicontinental platforms called the *Valles-San Luis Potosi* and *Tuxpan* (Golden Lane) platforms. The El Abra margin was exposed by Laramide tectonism along the Valles-San Luis Potosi Platform (Fig. 4.15) (Winker and Buffler, 1988).

The subsurface Stuart City trend in Texas was extensively penetrated during hydrocarbon exploration in the late '50s and early '60s in the search for a south Texas "Golden Lane." While the depositional setting and facies were believed to be similar to those encountered in the El Abra Formation of the Golden Lane in northeastern Mexico (Rose, 1963) (Fig. 4.18), the south Texas Stuart City trend has been an economic disappointment. It has produced only small quantities of gas to date, in contrast to the billions of barrels of oil recovered from the El Abra (Boyd, 1963; Coogan and others, 1972; Bebout and Loucks, 1974).

The Stuart City is a marginal hydrocarbon trend because of a general lack of

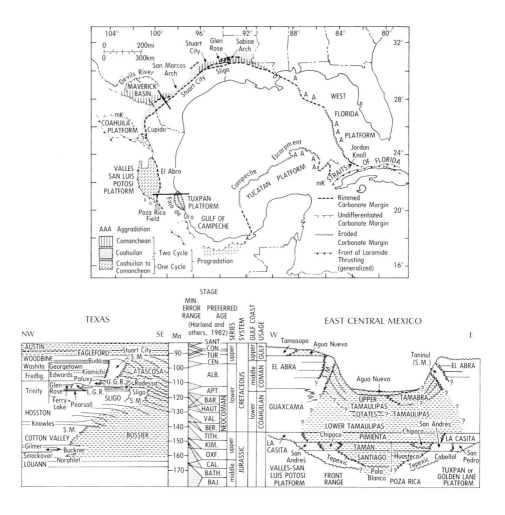

Fig. 4.15. (Top) Lower and middle Cretaceous carbonate shelf margins of the circum-Gulf province (Winker and Buffler, 1988). Location of two cross sections indicated by solid black lines. (Lower left) Stratigraphic relationships of the Mesozoic, south Texas showing the two-cycle shelf margin. (Winker and Buffler, 1988). (Lower right) Stratigraphic relationships of the Mesozoic in east-central Mexico and their correlation with south Texas (Winker and Buffler, 1988). Reprinted by permission of the American Association of Petroleum Geologists.

porosity and permeability in its potential reservoir facies. The Stuart City reef and reef-related sequences were formed, as similar sequences are today, with extensive primary growth porosity, as well as intergranular and intragranular porosity. Petta (1977), in a study of an outcrop Glen Rose rudist reef complex equivalent to the subsurface Stuart City, found that internal sedimentation, marine cementation, and clionid bioerosion developed fabrics identical to those encountered in modern framework reefs, and were the dominant processes causing early porosity loss in this Cretaceous reef sequence. Achauer (1977) and Prezbindowski (1985) documented massive early porosity loss in the Stuart City shelf margin rudist reef sequences by cementation. Achauer (1977) noted that most of the porosity loss was the result of marine cementation by radiaxial calcite, and indicated that these cements were primarily encountered along the shelf margin. Prezbindowski (1985) estimated that some 21% of the Stuart City's primary porosity was lost by cementation (out of a total of 38%) before the sequence was buried 100 m (Fig. 4.16). While some of this early porosity loss seems to have been by cementation under meteoric phreatic conditions, the majority of the porosity loss is by marine cementation, since there is little evidence for substantial subaerial exposure of the Stuart City (Prezbindowski, 1985). Only 9% of the remaining porosity was lost during the next 4000 m of burial, leaving an average porosity encountered in the subsurface of some 8% (Fig. 4.16). This porosity is generally unconnected, occurring as intragranular porosity associated with either the living cavities of rudists, or the tests of foraminifera, leading to the very low average permeability values encountered in the south Texas Stuart City (Bebout and Loucks, 1974).

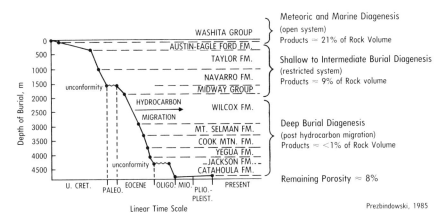

Fig. 4.16. (Left) Approximate burial curve for the Lower Cretaceous Stuart City reef trend, south Texas. (Right) Diagenetic environments and porosity evolution with burial. Used with permission of SEPM.

The Golden Lane is a trend of highly productive oil fields developed in what is generally considered to be reefal facies of the El Abra Formation, along the margins of the Tuxpan Platform, located some 75 km east of the main Lower-Middle Cretaceous shelf margin (Boyd, 1963) (Fig. 4.17). While this trend has been producing since 1908, little is actually known about the nature of the reservoir rock because most wells are completed in the uppermost El Abra, and little core material is available (Coogan and others, 1972). The correlation of environment and facies with the south Texas Stuart City, is based on outcrop studies of exposures of El Abra platform and related facies north and

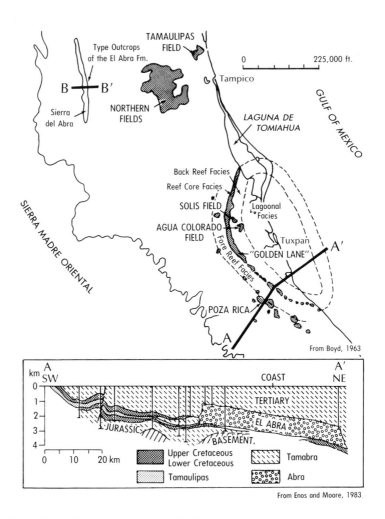

Fig. 4.17. (Top) General location map of the Golden Lane area. Major oil field shown in stippled pattern. Line A-A" is location of cross section below. Line B-B' is location of cross section across the type outcrops of the El Abra Formation. (Bottom) Subsurface cross section across the Golden Lane to the Sierra Madre Oriental. (Top) Used with permission of Corpus Christi Geological Society. (Bottom) Reprinted by permission of the American Association of Petroleum Geologists.

west of the productive trend along the margins of the Valles-San Luis Potosi Platform (Rose, 1963; Enos, 1974; Carrasco-V., 1977; Enos, 1986). At this site the platform-to-basin facies tracts can be traversed, and shelf-margin rudist reefs observed, giving way to fore reef talus and ultimately to the dark, laminated, basinal limestones of the Tamaulipas Formation, a facies tract that most researchers believe parallels that encountered in

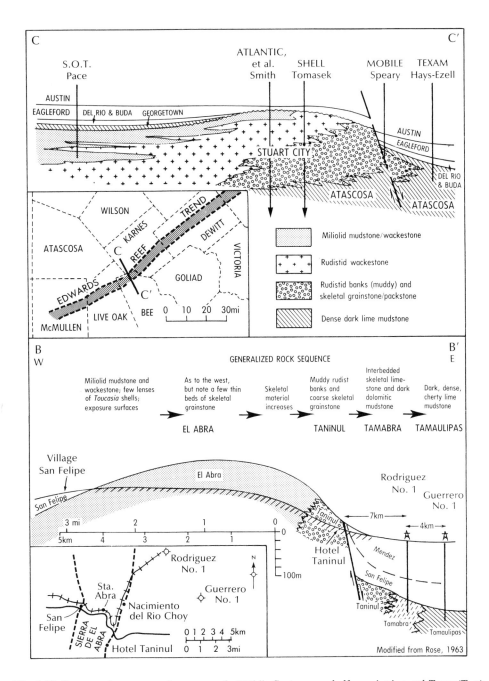

Fig. 4.18. Comparative cross sections across the Middle Cretaceous shelf margins in south Texas (Top) and south-east Mexico at El Abra (Bottom). Locations shown in inset maps, lithologic patterns shown in top figure. Used with permission of Corpus Christi Geological Society.

the shelf-margin wells penetrating the Stuart City in south Texas (Rose, 1963; Bebout and Loucks, 1974) (Fig. 4.18).

The syndepositional diagenetic history of the El Abra shelf margin also seems to parallel that of the south Texas Stuart City, with massive marine cementation, dominated by radiaxial calcite, which forms an early porosity destructive stage (Shinn and others, 1974). Shortly after deposition, however, the subsequent geologic history of these analogous sequences dramatically diverges. The Golden Lane platform, only 70 km east of the Sierra Madre Oriental, experienced intense deformation associated with Larimide tectonics shortly after deposition of the El Abra. The platform was uplifted and tilted toward the east, causing subaerial exposure and intense karsting to the western margin of the platform, in turn resulting in the development of extensive cavernous porosity in the El Abra shelf-margin facies (Coogan and others, 1972) (Fig. 4.17).

In contrast, the south Texas Stuart City shelf margin trend is located some 500 km from the locus of Laramide tectonism, and hence was rapidly buried into the deep subsurface, with no opportunity for significant subaerial exposure nor the occasion to develop the extraordinary secondary cavernous porosity that is the key to the phenomenal hydrocarbon production seen in the Golden Lane for the past 90 years (Fig. 4.16) (Rose, 1963; Bebout and Loucks, 1974; Prezbindowski, 1985).

Porosity evolution of Middle Devonian reef complexes: Leduc, Rainbow, "Presqu'ile," and Swan Hills reefs, Western Canadian Sedimentary Basin

The western Canadian Sedimentary Basin extends over 1900 km (1200 miles) north to south and is over 550 km (350 miles) wide. The Precambrian Shield forms the eastern margin, and the Cordilleran tectonic front outlines the western border (Klovan, 1974) (Fig. 4.19). The Middle and Upper Devonian sedimentary sequences deposited within and along the margins of this enormous intracratonic basin harbor 62% of Alberta's recoverable, conventional crude oil in reef-related limestones and dolomites (Davies, 1975). The Middle and Upper Devonian Givetian and Frasnian series represent a basic north-to-south transgression of the basin with progressive onlap of reef-bound shelves through time (Fig. 4.19). There are three major intervals of reef development : 1) the Givetian "Presqu'ile" shelf-margin reef complex fronting the Hare Indian Shale Basin, with a broad shelf lagoon extending to the east and south upon which the high-relief Rainbow reef complexes developed. This shelf lagoon was highly restricted, and became evaporitic. The Rainbow reefs were ultimately buried by massive Muskeg evaporites. 2) Frasnian age, low-relief Swan Hills reefs were developed on the Slave Lake carbonate shelf along the margins of the Peace River Arch. 3) the Frasnian age Leduc reefs were developed as moderate relief, linear trends, on the Cooking Lake platform. The Leduc reefs were ultimately drowned by Ireton shales and associated Duperow evaporites (Fig. 4.19).

Devonian reefs of western Canada consist of a bank margin of stromatoporoid-coral framework and fore-reef debris, with bank interior facies consisting of low-energy lagoonal wackestones and mudstones, mobile sand flats, and some low-energy supratidal sequences (McCrosson and Glaister, 1964; Klovan, 1964; McGillivray and Mountjoy, 1975; Walls 1977; Walls and Burrowes, 1985). Buildup growth is often interrupted by

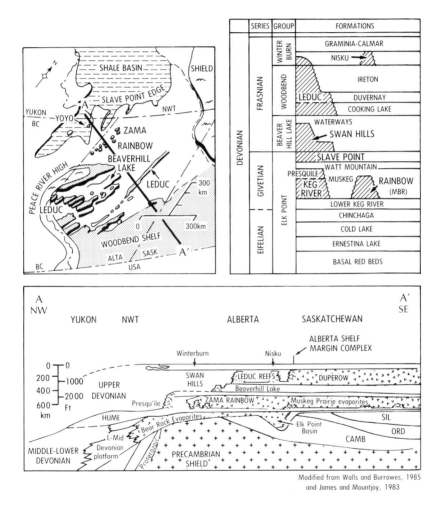

Fig. 4.19. (Top left) Distribution and paleogeography of Middle and Upper Devonian reef carbonates in western Canada. (Top right) Devonian stratigraphy in western Canada illustrating stratigraphic position of reef carbonates. (Bottom) Subsurface schematic cross section along line A-A' (shown in top left) through the western Canadian sedimentary basin illustrating the stepped onlap mode during Late Devonian time. Used with permission of SEPM.

subaerial exposure and submarine hardgrounds. Reef growth is generally initiated on submarine hardgrounds (Fischbuch, 1968; Wendte, 1974; Harvard and Oldershaw, 1976).

The diagenetic history of Devonian reefs has been the subject of numerous studies emphasizing various aspects of the diagenetic history of isolated reef complexes, generally without regard to ultimate porosity evolution or any attempt to discover unifying principals that might be useful as predictive tools (Walls and Burrowes, 1985). The synthesis of Walls and Burrowes (1985) provides a comprehensive view of the diagenetic history of four of the most economically important western Canadian Middle-Upper Devonian buildups. Their work reveals that the margin of each bank has quite a different diagenetic history, and hence porosity evolution, than the adjacent bank interior (Walls and Burrowes, 1985). These differences are illustrated in Figs. 4.20 and 4.21, where porosity evolution is traced from deposition to the present and cementation

Table 4.3. Comparison of the reservoir quality and characteristics of various Devonian reef margins and interiors. Used with permission of SEPM.

REEF COMPLEX		ϕ MEAN (%)	K MEAN (md)	RESERVOIR CHARACTER
LEDUC (GOLDEN SPIKE)	M	7	25	POOR PERM.; VERT. & HORIZ. PERM. BARRIERS
	I	12	60	"STRATIFIED"; HORIZ. PERM. BARRIERS
SWAN HILLS	M	15	170	GOOD PERM.; FEW PERM. BARRIERS
	I	9	25	"STRATIFIED"; HORIZONTAL PERM. BARRIERS
RAINBOW MBR	M	-8—LS 6-DOL.	35 250	VARIABLE PERM. LENSOID RESERVOIRS
	I	13-LS 7-DOL.	90 310	VARIABLE PERM. LIMITED PERM. BARRIERS
SLAVE PT.— "PRESQU'ILE" (DOLOMITE)	M	10	75	GOOD VERTICAL PERM.; LENSOID RESERVOIRS
	I	7	50	LIMITED PERM.

M—REEF MARGIN   I—REEF INTERIOR   From Wall and Burrows, 1985

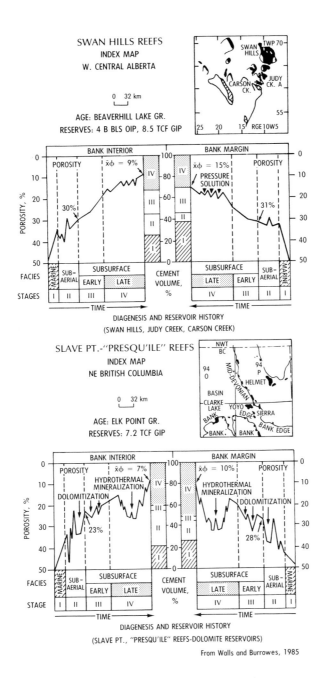

Fig. 4.20. Summary of diagenesis and reservoir history for the Swan Hills reefs (Top) and the Slave Pt. "Presqu'ile" reefs (Bottom). Bank interiors and bank margins are compared. Porosity evolution through time, and cement volume in % are estimated. Used with permission of SEPM.

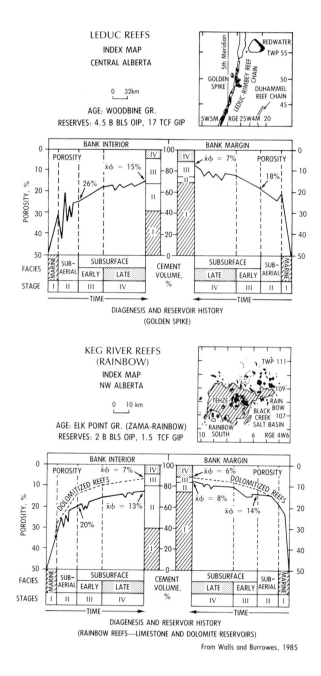

Fig. 4.21. Summary of diagenesis and reservoir history for the Leduc reefs (Top) and the Keg River (Rainbow) reefs (Bottom). Bank interiors and bank margins are compared. Porosity evolution through time, diagenetic environment, and cement volume in % are estimated. Used with permission of SEPM.

environment is inferred. A direct comparison of reservoir quality and characteristics of the four reef complexes is compiled in Table 4.3.

Walls and Burrowes (1985) conclude that marine cementation dominated by radiaxial calcite is the single most important diagenetic process affecting porosity in Middle and Upper Devonian reefs of western Canada. In the Leduc and Rainbow reef complexes, marine cementation almost completely occludes depositional porosity along the reef margins, while the volume of marine cement in the bank interiors is less than half that encountered along the margins. The lack of massive marine cementation and the presence of secondary porosity developed during intermittent subaerial exposure of the bank interiors, give Leduc and Rainbow bank interiors better reservoir quality than their adjacent bank margins (Fig. 4.20; Table 4.3). However, the bank interiors of Leduc and Rainbow reefs generally do show more subsurface cements associated with pressure solution, than do the bank-margin sequences. Pressure solution, cementation, and multiple exposure surfaces tend to horizontally compartmentalize the favorable porosity-permeability trends of these bank interiors (Table 4.3).

The "Presqu'ile" and Swan Hills reef complexes contrast sharply with Leduc and Rainbow (Figs. 4.20 and 4.21; Table 4.3). The volume of marine cements found in the margins of these two reef complexes is generally half that found at Leduc and Rainbow. As a result, subsurface cements and dolomitization are much more important porosity modification processes in the Presquile and Swan Hills sequences. Finally, because of preserved primary porosity and dolomitization, the best quality reservoir rock is found in "Presqu'ile" and Swan Hills bank margins, rather than in their bank interiors, as seen in Leduc and Rainbow.

Why do the Leduc and Rainbow reefs exhibit so much more intense marine cementation than "Presqu'ile" and Swan Hills? Two factors seem important. Both Leduc and Rainbow are closely associated with evaporitic conditions (Duperow and Muskeg evaporites), possibly implying that marine waters present during deposition and syngenetic diagenesis were tending toward hypersalinity, and hence could have been more saturated with respect to $CaCO_3$. Both Leduc and Rainbow show moderate-to-high relief relative to the seafloor, and both face broad open-water shelves that might result in higher-energy conditions, with high marine water flux, and a propensity to cementation. "Presqu'ile" and Swan Hills seem to be developed under lower-energy situations because of bank-margin geometry and relative relief above the seafloor; therefore, they are less likely to exhibit massive marine cementation. While these contentions have yet to be tested by a comprehensive geochemical, petrographic, and biofacies assay, it is obvious that the ability to predict these types of diagenetic trends on a regional basis is extremely important to future exploitation and exploration strategy in the western canadian sedimentary basin, as well as similar basins in the geologic record.

These examples from the Gulf of Mexico and Canada illustrate clearly the importance of syngenetic marine diagenetic processes on porosity evolution in shallow marine reef, and reef-related sequences. The message to the explorationist is clear. While marine cementation often destroys the original depositional porosity of reef sequences, some reef trends have been found to be enormously productive. These productive reef trends invariably have seen massive diagenetic overprints that have rejuvinated original depositional porosity by dissolution, dolomitization, and fracturing. Fortunately, the original aragonitic mineralogy of many reef-building organisms enhances the effectiveness of secondary porosity generation by meteoric-related dissolution. The original magnesian calcite mineralogy of other reef-related organisms may well enhance the dolomitization potential of reef sequences. The common location of reefs at the terminal phase of shallowing upward sequences improves the probability of meteoric diagenetic enhancement of porosity. Finally, the shelf margin position of reefs is favorable for the movement of later, basinal-derived fluids through and around the reefs. These fluids are often capable of dolomitization and dissolution, and are commonly the carriers of hydrocarbons formed in basinal shales. Reefs then, are attractive exploration targets, but require, perhaps more than any other depositional setting, a total rock history approach incorporating diagenetic models, before success is assured.

The discussion will now move from warm, surface marine waters, and focus on the cold, dark, world of the deep marine environment, and examine the porosity modification processes operating there.

DEEP MARINE DIAGENETIC ENVIRONMENTS

Introduction to diagenesis in the deep marine environment

Diagenesis in those waters below the surface mixing zone, within and below the thermocline, is marked by the incremental dissolution, or recrystallization, of certain carbonate phases that are out of equilibrium with adjacent, progressively colder marine water, as well as, concurrent precipitation of carbonate phases in equilibrium with this colder marine water (Fig. 4.1) (introductory remarks to this chapter, Scholle and others, 1983).

Open marine waters, distant from carbonate shelves and platforms, support a dominantly calcite planktonic community of coccoliths and planktonic foraminifera with a minor aragonite component consisting of pteropods. The abyssal plains, therefore, are generally covered with calcite-dominated pelagic sediments, and magnesian calcite and aragonite are not a factor until a shelf or platform margin is encountered (Scholle and others, 1983). Fine-grained, metastable carbonates originating in the shallow waters of the platform are commonly swept off the platform and deposited as a peri-platform ooze, where they come into contact with cold waters undersaturated with respect to aragonite and presumably high magnesian calcite (Berner and others, 1976). Finally, major constructional carbonate platform margins, including atolls, can be the site of interaction of cold, undersaturated (with respect to aragonite, or even calcite) marine waters with coarse, porous, mineralogically metastable, platform margin sequences.

The following discussion is keyed to the processes operating on mineralogically stable pelagic, as well as metastable platform-derived materials, within two oceanographic zones: the zone from the aragonite lysocline to the calcite compensation depth; and the zone below the calcite compensation depth.

Diagenesis within the zone of aragonite dissolution

While sea-floor lithification, as thin, two-dimensional hardgrounds, has been known since the nineteenth century (Fischer and Garrison, 1967), reports of major deep water lithification only began to surface in the mid '70s. Neumann and others (1977) described numerous, large, cemented carbonate mounds, which they termed *lithoherms*, in the straits of Florida. These lithoherms were encountered between 600 and 700 m in an area of strong bottom currents. Cementation consists of finely crystalline magnesian calcite containing 14 mole % $MgCO_3$.

Perhaps a more significant discovery was the report of Schlager and James (1978) detailing the relatively rapid lithification of Bahamian peri-platform oozes in the Tongue of the Ocean, apparently by dissolution of aragonite and precipitation of low-magnesian calcite between 700 and 1960 m water depth. Lithification has taken place within 100,000 years and is accompanied by unquestioned evidence of aragonite solution, and loss of magnesium from the magnesian calcite allochems such as coralline algae and foraminifera. The starting material is a peri-platform ooze consisting of aragonite and magnesian calcite. Lithification is by 2-4 micron polyhedral calcite containing 3.5 to 5 mole% $MgCO_3$. This material is isotopically in equilibrium with the cold waters (3 - 5°C) found at the sites of lithification (Fig. 4.4) (Schlager and James, 1978). The final product is a low-magnesian calcite micrite. Cementation occurs along the margins of gullies cut through unconsolidated ooze. Lithification extends only a meter or so below the surface and is always confined to the steep-to-vertical surfaces that do not receive sedimentation. The porosity of unconsolidated ooze is near 75%, while the porosity of adjacent strongly-cemented, chalky limestone is 36%, a loss of nearly 40% porosity during the seafloor lithification process.

The patterns of lithification seen in the Tongue of the Ocean would seem to parallel the mechanisms thought responsible for pelagic hardgrounds reported in widely

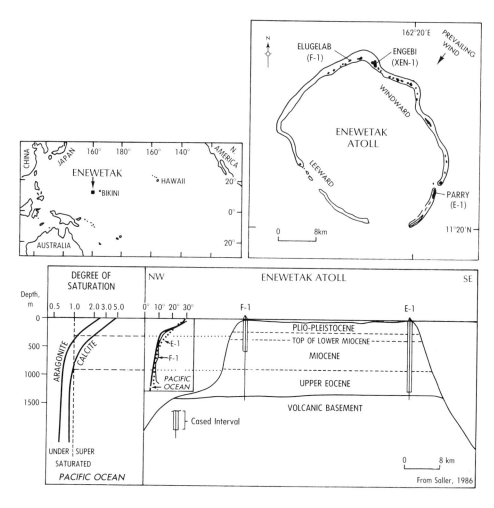

Fig. 4.22. (Top left) Location map for Enewetak Atoll, western Pacific. (Top right) Enewetak Atoll showing location of the major deep drill holes on the atoll. (Lower left) Saturation state of aragonite and calcite as a function of depth in the Pacific Ocean (Scholle and others, 1983). (Lower right) Subsurface stratigraphy of Enewetak Atoll based on deep drill holes. Temperature - depth relationships found in the two deep drill holes are compared to adjacent marine waters on the left side of the lower right diagram. Used with permission of SEPM.

separated sites around the world, and indicate that the process may be much more common today than was previously believed (Schlager and James, 1978).

Saller's work on Enewetak (1984a; 1984b; 1986) provided the first clear view of the effects of deep marine waters on the coarse, mineralogically metastable carbonate sequences exposed on the steep constructional margins of platforms and oceanic atolls.

A series of observational drill holes penetrate the entire carbonate section of Enewetak Atoll (some 1400 m) through the Upper Eocene, which lies directly on volcanic basement (Fig. 4.22). The facies encountered in cores from these drill holes mirror present Enewetak Atoll margin facies, including lagoon margin, backreef, reef crest, and fore-reef facies. This discovery indicates that the atoll margin has retained similar facies, and has been located near its present site since initiation of the atoll, atop a marine volcano, in the Eocene. Lithologies range from boundstones to grainstones and packstones and were originally composed of magnesian calcite and aragonite (Saller, 1984b; 1986).

The upper 60-100 m of the carbonate sequence on Enewetak representing the Holocene and Pleistocene show the effects of repeated exposure to subaerial conditions during sequential glacial-related sea level lowstands. There is some evidence of subaerial exposure during the Middle-to-Upper Miocene, but no evidence of major unconformities in the section below 200 m. The carbonate section of the Lower Miocene and Upper Eocene, then, has been continuously under the influence of progressively deeper marine water of the western Pacific Basin since the Upper Eocene (Saller, 1984b; 1986).

One of the deep drill holes on Enewetak, though located up to 3 km from the adjacent ocean, and cased to below 600 m, exhibits tidal fluctuations in phase with marine tides and displays similar amplitudes. These observations indicate free marine circulation with adjacent ocean waters through deep carbonate units. In addition, the thermal profile of this well tracks that of adjacent ocean water, reinforcing the conclusion that deep carbonate sequences on the margin of the atoll are in open communication with adjacent cold marine waters below the aragonite compensation depth (375 m, as shown on Fig. 4.22) (Saller, 1984b).

In cores from these wells, taken at depths between the present aragonite compensation depth (375 m) and the present calcite compensation depth (1000 m), Saller (1984b; 1986) documents the pervasive dissolution of aragonite, and the precipitation of radiaxial and bladed calcite cements containing an average of 3.2 mole% $MgCO_3$ (Fig. 4.14). These cements are often precipitated into bioclast molds formed by dissolution of *Halimeda* segments and other aragonite grains, and are isotopically in equilibrium with present cold, deep, marine waters (Fig. 4.4). Saller (1986) also documents the progressive loss of magnesium from magnesian calcites through this zone, confirming the observations of Schlager and James (1978) concerning mineralogical changes associated with the lithification of peri-platform oozes below the aragonite compensation depth along the margins of Tongue of the Ocean.

Fig. 4.23 illustrates Saller's (1984b) estimates for the percentage of diagenetic alteration, as cementation, dissolution, and dolomitization, seen in one of the two, deep, Enewetak core holes. In the upper section, where subaerial exposure was common, and where the section is presently above the aragonite compensation depth, dissolution was the major active diagenetic process. Between the aragonite and calcite compensation

112   NORMAL MARINE DIAGENETIC ENVIRONMENTS

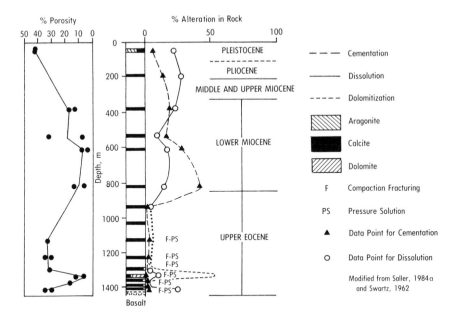

Fig. 4.23. Porosity (Left) and patterns of diagenetic alteration (Right) exhibited by the F-1 deep drill hole at Enewetak Atoll. Used with permission of authors.

depths, cementation was the dominant process, while below the calcite compensation depth, dolomitization became dominant. Porosity data from this well (Swartz, 1962), as plotted on Fig. 4.23, seem to indicate that radiaxial calcite cementation, in relatively deep water along platform margins, is important, and can have a significant impact on porosity development. As indicated earlier in this chapter, ancient radiaxial calcite cements often occur in sequences that were, or could have been, in contact with circulating marine waters during or shortly after deposition in settings similar to the Enewetak situation. These sequences include: Late Paleozoic bioherms (Wilson, 1977); the Devonian Miette complex in Alberta (Mattes and Mountjoy, 1980); the Devonian reef complex of western Australia (Kendall, 1985; Kerans and others, 1986); and finally, the Lower Cretaceous shelf margin, south Texas and Mexico (Bebout and Loucks, 1974; Achauer, 1977; Prezbindowski, 1985; Enos, 1986).

Dolomitization below the calcite compensation depth

Immediately below the calcite compensation depth at Enewetak, cementation and dissolution seem to cease, and porosity, relative to the radiaxial calcite zone above, increases dramatically, presumably because of lack of pore-fill cement rather than an increase in dissolution (Fig. 4.23). Calcite dissolution and pervasive dolomitization begin to affect the section in the Enewetak cores at about 1200 m, fairly close to the base of the carbonate sequence (Fig. 4.24). Saller (1984a) explains these dolomites as the result of dolomitization by cold, deep, marine waters, undersaturated with respect to

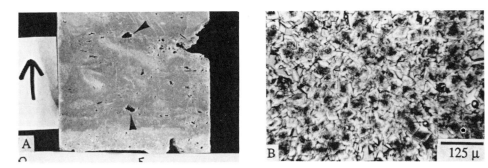

Fig. 4.24. (A) Slab photo showing pervasive dolomitization of Upper Eocene sequence in well F-1 at 1350 m (4429'). Note solution vugs (arrows). (B) Thin section photomicrograph of (A) showing cloudy centers and clear rims in each dolomite rhomb. Note that significant porosity (p) is still present. Photographs furnished by Art Saller.

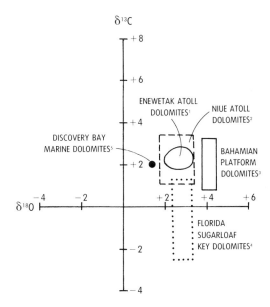

Fig. 4.25. Oxygen and carbon isotopic composition of Enewetak Atoll dolomites compared to other reported marine dolomites. Data from: 1. Saller, 1984; 2. Aharon and others, 1987; 3. Vahrenkamp and Swart, 1987; 4. Carbello and others, 1987; 5. Mitchell and others, 1987.

calcite, but still saturated with respect to dolomite, flushing through porous reef-related carbonate sequences. Stable isotope (Fig. 4.25) and trace element compositions of the dolomites are compatible with precipitation from normal marine waters. Strontium isotope composition of the dolomites contained in Upper Eocene strata indicates that dolomitization commenced in the Miocene, and probably continues on to the present (Saller, 1984a).

Numerous occurrences of dolomite (usually as small quantities of dolomite rhombs) have been reported from deep marine sites, far removed from shallow platforms and shelves, in the Pacific, Atlantic, and the Indian oceans. The little isotopic data available certainly indicate a marine origin under cold temperature conditions, suggesting formation near, or within, the calcite lysocline (Milliman, 1974a). Dolomites formed under these conditions, perhaps in association with marine manganous hardgrounds, should be relatively rare in the geologic record, and of little significance to questions of porosity evolution.

The thermal convection model of marine water dolomitization

Simms (1984) applied a marine-driven Kohout (thermal) convection model to the Bahamas Platform to explain the pervasive dolomitization of Tertiary-to-Pleistocene carbonates within the core of the platform. The Kohout model is an open-cycle thermal convection which develops because of a strong horizontal density gradient between the cold ocean water around the platform and the warmer interstitial waters heated by the geothermal gradient within the platform (see Fig. 9.1) (Kohout, 1965). These dolomites were previously believed to have been the result of fresh-marine water mixing. Simms (1984) makes a persuasive case for the importance of Kohout convection-related dolomitization along the steep seaward margins and into the cores of major carbonate platforms and shelves. Vahrenkamp and Swart (1987b) noted that the stable isotopic composition of the Bahamaian dolomites, mentioned by Simms, were decidedly high relative to ^{18}O (+4 $^o/_{oo}$, PDB) and ^{13}C (+3.2 to +1.5 $^o/_{oo}$, PDB), suggesting a marine-related dolomitizing fluid (Fig. 4.25). The trace element composition (Vahrenkamp and Swart, 1987a), as well as the $\partial^{13}C$ composition of these dolomites, exhibits a well-defined gradient from platform-margin-to-platform-interior, suggesting that the marine dolomitizing fluids were under the influence of a powerful margin-to-interior flow system similar to Simms' Kohout convection model. Hardie (1987) presents some compelling arguments in favor of extensive dolomitization by normal marine waters, such as would be encountered within a Simms-style thermal convection system. Hardie's arguments for marine dolomitization and the marine thermal convective model are certainly supported by the recent discovery of apparent normal marine dolomites at Sugarloaf Key in Florida (Carballo and others, 1987) and within a marine hardground on a fringing reef in Jamaica

(Mitchell and others, 1987) (Isotopic composition shown in Fig. 4.25).

Saller (1984a and b) used thermal convection as a driving mechanism for moving the large volumes of marine water through atoll margin carbonate sequences needed to accomplish the level of dolomitization observed at Enewetak. Earlier, Swartz (1958) had postulated the presence of thermal convection systems within western Pacific atolls based on an extensive geothermal study of Enewetak and Bikini atolls. Aharon and others (1987) attributed the extensive dolomitization of the south Pacific uplifted Niue Atoll to thermal convection of marine waters through the core of the atoll. As at Enewetak, the dolomites seemed to be in isotopic equilibrium with surrounding marine waters (Fig. 4.25). Trace element gradients within the atoll carbonates suggest that the dolomitizing fluids interacted with the volcanic foundation of the atoll, indicating a major convective flow system of marine waters through the volcanic pedestal into overlying carbonates.

The dolomites at Niue Atoll and those within the core of other atolls in the south and western Pacific, including Enewetak, were originally explained by either fresh-marine water mixing (Bourrouilh, 1972; Rodgers and others, 1982; Major, 1984; Aissaoui and others, 1986), or evaporative brine reflux (Schlanger, 1965; Berner, 1965; Schofield and Nelson, 1978). Saller (1984b) and Aharon and others (1987) have established a compelling case based on isotopes (Fig. 4.25), trace elements, stratigraphy, and hydrology for atoll dolomitization by marine waters driven by Kohout (thermal) convection. Is this convective model of dolomitization vertically restricted because of its tie to the calcite lysocline, and hence deep, cold waters undersaturated with respect to calcite; or can it be applied to the shallower portions of the atolls as well? It would seem that the most important feature of the model is the ability to move a significant volume of marine, Mg-bearing water into and through the porous limestones of the atoll core. Above the calcite lysocline, as at Enewetak, the result of convective movement of marine waters through the atoll margin may be massive calcite cementation. As marine waters move further into the interior of the atoll by convective displacement, the Mg/Ca ratio of the fluid should increase because of the previous calcite precipitation, calcite saturation should drop below equilibrium, and pervasive dolomitization of the atoll core should result. This system could be operative to the point where convection is broken up by multiple-cemented zones associated with unconformities related to Pleistocene sea level lowstands (starting at 230 m below sea level on Enewetak, as described by Goter in 1979). Unfortunately, at Enewetak, and other Pacific atolls that have not been uplifted, the available deep drill holes are all located around the margin of the atoll, and the model has not been fully tested.

The Kohout convection model is useful in explaining pervasive platform and platform margin dolomitization that cannot be related stratigraphically to evaporative sequences, and for which a marine-meteoric water mixing model is not viable because of geochemical and geohydrological constraints. The Middle Cretaceous Tamabra

Limestone at Poza Rica, was thought by Enos (1977) to represent deep-water, mass-flow deposits from the El Abra-Golden Lane platform margin to the east. These deep-water sequences are commonly dolomitized, and Enos (1988) favors dolomitization as the result of deep circulation of meteoric waters from the surface of the adjacent, exposed platform. As indicated by Enos (1988), however, the geologic setting, patterns of diagenetic modification, and stable isotopic composition of the dolomites are all compatible with dolomitization by marine waters driven through the platform margin by Kohout convection. The Poza Rica trend, then, could well represent an economically significant example of porosity enhancement by deep marine dolomitization driven by thermal convection.

SUMMARY

Porosity modification in the marine diagenetic environment is centered on cementation and dolomitization. There are two major subenvironments where these processes are operative: the shallow water, normal marine diagenetic environment within the warm, well mixed, oceanic surface zone; and the deep marine, diagenetic environment, in the density-temperature structured oceanic water column below the surface mixing zone.

Porosity evolution in the shallow water, normal marine environment is dominated by porosity loss through cementation. The two major sites of marine cementation are the high-energy intertidal zone and the shelf-margin framework reef, where high-energy, stable substrates, organic activity, and high porosities-permeabilities ensure a high rate of fluid flux and the elevated saturation states necessary for significant cementation. While intertidal cementation (as well as cementation associated with hardgrounds) tends to be vertically and laterally restricted, these cemented zones can serve as reservoir seals and may vertically compartmentalize reservoirs. Marine cementation in reefs, combined with bioerosion and internal sedimentation, can, depending on reef type and accretion rate, totally destroy the high initial porosities enjoyed by most reef sequences. The key to recognition of marine cementation in the ancient record rests with distinctive cement textures, fabrics, geochemistry, and geologic setting. Reef-related marine cementation is a major factor in the porosity evolution of a number of economically important ancient reef trends.

The deep marine environment can be especially important relative to porosity evolution along the steep margins of carbonate platforms fronting oceanic basins, where the thermocline and the carbonate compensation depth impinge on previously deposited carbonate sequences. Dissolution of aragonite, precipitation of radiaxial calcite cements, dissolution of calcite, and, finally, dolomitization can occur and modify porosity as the shelf margin encounters progressively deeper and colder waters toward the basin.

Perhaps one of the most significant recent concepts relative to marine diagenesis is the development and initial testing of the thermal convection model of dolomitization using normal marine waters to dolomitize the margins as well as the interiors of carbonate platforms and atoll reef complexes. Radiaxial calcite cements, geologic setting, and distinctive geochemical signatures and gradients help define sequences affected by deep marine diagenesis. Ancient shelf margins, such as the Lower Cretaceous of the Gulf Coast and the Devonian of western Australia and Canada, seem to bear the marks of deep marine diagenesis.

Chapter 5

EVAPORATIVE MARINE DIAGENETIC ENVIRONMENTS

INTRODUCTION

Marine and marine-related evaporites are precipitated, and/or deposited in a number of marine and marginal marine environments (Schreiber and others, 1982; Kendall, 1984). The following discussion, however, does not exhaustively survey the complex world of evaporites, but focuses on those general situations where the marine evaporative regime significantly impacts the porosity evolution of carbonate marine sequences, either by porosity enhancement related to dolomitization, or by porosity occlusion and reservoir sealing due to precipitation of evaporites.

There are two general settings that will be considered: the marginal marine sabkha, where evaporites are generally formed within fine-grained carbonate tidal flat sequences; and subaqueous hypersaline marginal marine environments, such as lagoons, where evaporites are generally formed in the water column and large volumes of evaporated marine waters are available to interact diagenetically with associated marine carbonates to form dolomites.

The following pages will be a review of the special diagenetic conditions found in conjunction with waters that have been modified by evaporation, followed by a discussion of each of the two major diagenetic settings. Those processes and products having the potential for significant impact on porosity development and evolution will be emphasized.

Introduction to diagenesis in evaporative marine environments

When marine waters are evaporated in a closed laboratory system, such as in the seminal experiments of Usiglio (1849), a predictable sequence of mineral species are incrementally precipitated, starting with $CaCO_3$ and advancing through the evaporative mineral suite to include $CaSO_4$ and $NaCl$, and ending ultimately with the bittern salts (KCl and others). During the evaporative process, the remaining waters become progressively denser (Fig. 5.1). In the more open, natural carbonate systems bordering modern oceans today, the processes become more complex with the evolving evaporative brines often interacting with adjacent sedimentary carbonate phases and mixing with other waters.

The single most important factor concerning carbonate porosity modification in

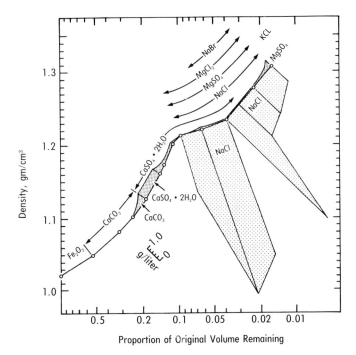

Fig. 5.1. Graph of density versus proportion of original volume remaining in closed-system evaporation of Mediterranean seawater by Usiglio in the 1840's showing physical conditions under which each mineral is precipitated. Amounts of each mineral (in grams per liter) shown by the lengths of lines drawn perpendicular to curve. Reprinted with permission from Principals of Sedimentology. Copyright (C), 1978, John Wiley and Sons.

the evaporative marine environment is the common association of dolomite with evaporite minerals. While most modern marine surface waters are supersaturated with respect to dolomite — and as seen in the last chapter, dolomite can be produced from unmodified marine waters — dolomite formation appears to be particularly favored in evaporated marine waters. Fig. 5.2 illustrates the interaction of two of the main parameters that may influence the kinetics of the precipitation of dolomite in natural surface waters: the Mg/Ca ratio and the salinity of the precipitating fluid (Folk and Land, 1975). Under conditions of elevated salinities, such as may be found in marine environments, Mg/Ca ratios must be significantly higher to overcome the propensity of Mg^{2+} to form hydrates, before dolomite can precipitate (Morrow, 1982a).

The precipitation of $CaCO_3$ as aragonite and $CaSO_4$ as gypsum during the early stages of evaporation (Fig. 5.1) is, from the standpoint of porosity modification, the most important segment of the evaporative series. While the gypsum precipitate will

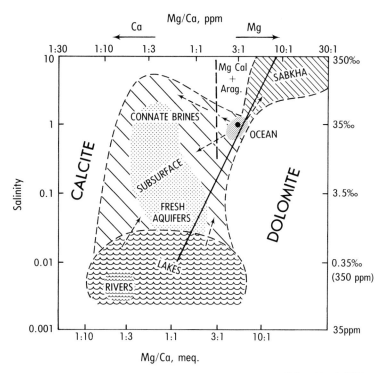

Fig. 5.2. Fields of occurrence of common natural waters plotted on a graph of salinity vs Mg/Ca ratio. Fields of preferred occurrence of dolomite, aragonite, Mg-calcite, and calcite are also shown. At low salinities with few competing ions and slow crystallization rates, dolomite can form at Mg/Ca ratios near 1:1. As salinity rises, it becomes more difficult for ordered dolomite structure to form, thus requiring progressively higher Mg/Ca ratios; in the sabkha, Mg/Ca ratios of 5 to 10:1 are necessary before dolomite can crystallize owing to abundance of competing ions and rapid crystallization. Reprinted with permission of the American Association of Petroleum Geologists.

ultimately form an aquaclude, and often acts as a reservoir seal, the progressive extraction of Ca^{2+} from the evaporating seawater to form aragonite and gypsum drives its Mg/Ca ratio ever higher, favoring dolomite precipitation. Evaporative marine waters associated with the precipitation of aragonite and gypsum, then, are not only chemically suited for the formation of dolomite, but their higher density makes them a particularly effective Mg^{2+} delivery system for the dolomitization of adjacent carbonates.

The dolomites forming today under surface evaporative marine conditions are generally poorly ordered, contain excess calcium, and have been termed *protodolomites*, or *calcian dolomites* (Lippmann, 1973; Gaines, 1977; Land, 1980; Reeder, 1983; Land,

1985). Land (1980, 1985) and Hardie (1987) point out that ancient dolomites are generally more ordered and less soluble than their modern, surface-formed, disordered counterparts. They both strongly emphasize that surface-formed dolomites tend to reorganize structurally and compositionally with time, "giving surface-formed dolomites a recrystallization potential that is increased with increasing temperature and burial" (Hardie, 1987).

The utilization of stable isotopes and trace elements to determine the characteristics of ancient dolomitizing fluids has become increasingly more popular (Land and Hoops, 1973; Land and others, 1975; Choquette and Steinen, 1980; Ward and Halley, 1985). Evaporative, surface, marine waters seem to leave a distinct geochemical fingerprint on dolomites precipitated under their influence. The stable isotopic composition of several suites of samples from a number of Holocene evaporative environments (and 1 ancient example) is presented in Fig. 5.3. Modern surface-formed dolomites are generally characterized by high ^{18}O and ^{13}C (PDB) values (Land and others, 1975; Veizer, 1983; Land, 1985). In addition, they tend to be enriched with Na and Sr (Land and Hoops, 1973; Veizer, 1983; Land, 1985).

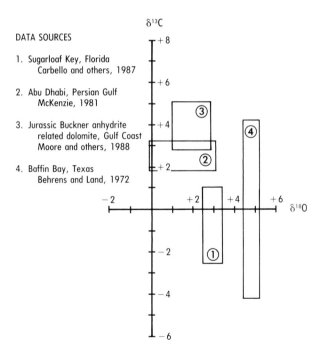

Fig. 5.3. Oxygen and carbon isotopic composition of evaporative-related marine dolomites. Data sources as indicated.

The geochemical fingerprint approach described above, and taken by many carbonate geologists, must be used with extreme caution. There is a great deal of uncertainty about trace element distribution coefficients, and fractionation factors in dolomites (Chapter 3), and the prospect of extensive recrystallization of surface-formed dolomites compounds the problem. The isotopic and trace element composition of ancient evaporite-related dolomite sequences, then, may actually only reflect a gross representation of the chemistry of recrystallizing fluids, rather than the composition of original precipitational fluids.

Therefore, the following paragraphs discuss modern as well as ancient evaporative marine diagenetic environments, and attempt to develop integrated stratigraphic-petrographic-geochemical criteria that can be used to recognize the major evaporative marine diagenetic environments and their associated dolomites.

THE MARGINAL MARINE SABKHA DIAGENETIC ENVIRONMENT

Modern marginal marine sabkhas

The marginal marine sabkha is perhaps one of the most important and best characterized depositional environments (James, 1984). Because of its transitional position between marine and continental conditions, sabkha pore fluids, while commonly highly evaporative, often have a complex origin. They may be derived from the sea by periodic surface flood recharge; from reflux of marine waters upward through sabkha sediments driven by evaporative pumping; landward, from gravity driven continental aquifer systems; and finally, vertically from coastal rainfall (Patterson and Kinsman, 1977; McKenzie and others, 1980) (Fig. 5.4).

Progressive evaporation of marine and continental waters drives sabkha pore fluids to gypsum saturation and leads to the precipitation of aragonite and gypsum within lagoon-derived aragonitic sabkha muds (Butler, 1969; Hsu and Siegenthaler, 1969; McKenzie and others, 1980) (Fig. 5.5). This precipitation causes a dramatic increase in the molar Mg/Ca ratio of the pore fluids from marine values of 5-1 to over 35-1, as reported from modern sabkhas of the Trucial Coast (Fig. 5.5). These elevated Mg/Ca ratios are favorable for the formation of dolomite, either by the replacement of aragonitic sabkha muds (Butler, 1969; Patterson and Kinsman, 1977; McKenzie, 1981) or as a direct interstitial precipitate (Hardie, 1987).

At levels of higher chlorinity, such as are found in the interior of the sabkha, waters are in equilibrium with anhydrite and may reach halite saturation. At the continental

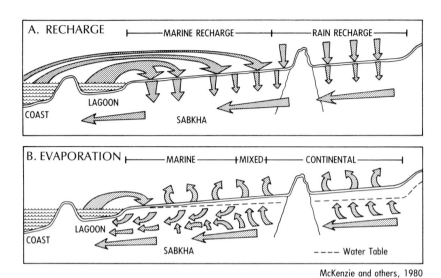

Fig. 5.4. A. Diagram of the sabkha illustrating water movement during recharge, mainly in winter or spring with shamal storms. B. Origin of water and circulation pattern during times of evaporation, which is the situation most of the year. Used with permission of SEPM.

margins of the sabkha, retrograde diagenesis, driven by continental brines, may dominate, with sulfate equilibrium swinging from anhydrite and halite saturation back into the gypsum field (Butler, 1969; McKenzie and others, 1980).

As seen in modern sabkhas along the Persian Gulf, such as at Abu Dhabi, the sabkha sequence is relatively thin, averaging little more than 3 m thick, with a lateral extent of some 12 km. Syngenetic dolomites formed at Abu Dhabi are concentrated across a 3-5 km swath of the interior of the sabkha starting about 1 km landward from the lagoon, coincident with the highest reported pore fluid salinity values. Dolomitization in this area affects a 1-2 m sequence of muddy lagoonal and intertidal sediments, starting about .5 m below the sabkha surface (Fig. 5.6). While dolomite percentages of up to 60% have been reported, 20-40% dolomitization is more common (Bush, 1973). Gypsum is generally most concentrated in the upper meter of the sabkha sequence, but can be found in quantity throughout the sequence in the interior of the sabkha in areas of highest chlorinity (Fig. 5.6). Anhydrite can occupy the entire upper .5 m of the sequence in the interior reaches of the sabkha (Fig. 5.6). Halite deposits, as crusts, are ephemeral and are generally destroyed by infrequent rains (Butler, 1969; Bush, 1973). The interior of marginal marine sabkhas are often marked by rapid intercalation with, and transition into, continental siliciclastics.

More humid conditions with greater meteoric water input will, of course, limit the production and preservation of evaporites in a supratidal environmental setting.

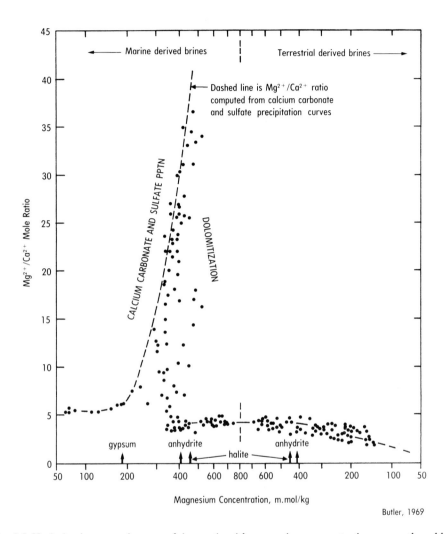

Fig. 5.5. Variation in magnesium-to-calcium ratio with magnesium concentration across the sabkha. Used with permission of SEPM.

Therefore, reduced evaporite production may diminish the dolomitzation potential within a supratidal environment. Modern humid supratidal sequences, such as those of Andros Island in the Bahamas, exhibit sparse dolomite that is generally concentrated in relatively thin surface crusts associated with exposure along the natural levees of tidal creeks, and adjacent to palm hammocks (Fig. 5.7) (Shinn and others, 1965; Shinn, 1983). There is some recent work suggesting that these limited dolomite crusts may be forming from normal marine waters aided by tidal pumping, rather than from evaporative marine

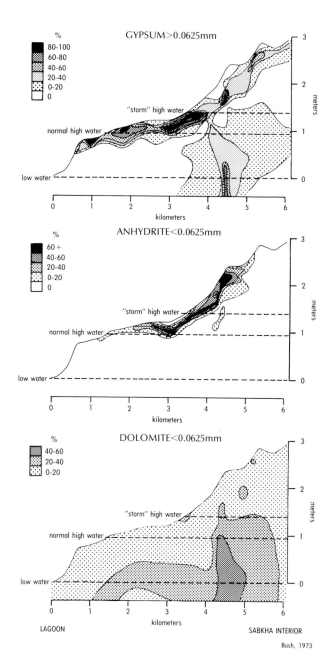

Fig.5.6. Distribution of gypsum, anhydrite, and dolomite across a sabkha traverse taken southeast of Abu Dhabi from point of low water (sea level) to 3 meters above low water, some 6 km from the lagoon shore. Reprinted with permission from the Persian Gulf. Copyright (C) 1973, Springer-Verlag, Berlin.

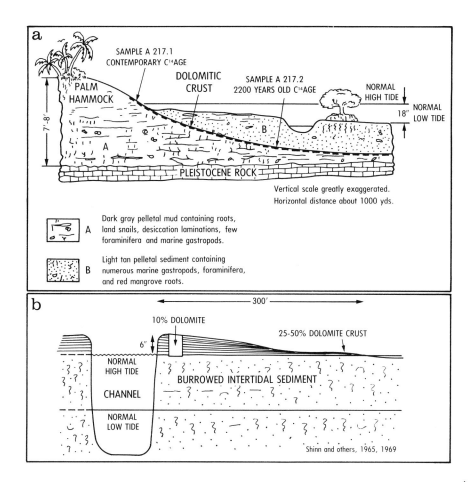

Fig. 5.7. a. Stratigraphic cross section across the flank of a palm hammock on Andros Island, the Bahamas. Note variation in age along the dolomitic crust. b. Schematic cross section across a tidal channel natural levee on the Andros Island tidal flats. Note that dolomite % is greater in the thin flanks of the levee than in the thicker channel margin part of the levee due to a dilution effect caused by the greater rate of sedimentation along the channel margins. Used with permission of SEPM.

waters (See Chapter 4, this book; Carbello and others, 1987; and similar dolomite crusts located in Belize, as described by Mazzulo and Reid, 1987).

Diagenetic patterns associated with ancient marginal marine sabkhas

Ancient sabkha sequences are important oil and gas reservoirs, with dolomitized supratidal and subtidal sediments forming the reservoir, and associated evaporites often forming the seal. When a variety of these economically important ancient sabkhas are compared to their modern analogues from the Persian Gulf, the volume of dolomite present in the ancient sequences is surprisingly greater than that of their modern counterparts. In the typical ancient shoaling upward sequence capped with sabkha deposits, the muddy subtidal and more open marine materials, as well as the overlying sabkha, are often totally dolomitized (Figs. 5.12, 5.15, 5.18) (Loucks and Anderson, 1985; Ruzyla and Friedman, 1985; Chuber and Pusey, 1985). As was mentioned above, dolomitization on modern Persian Gulf sabkhas is restricted laterally, as well as vertically, and seldom exceeds 60% of the sediment by volume. The lateral distribution of dolomitization may be extended by the sedimentologic progradation of the sabkha environment in a seaward direction. However, the vertical extension of dolomitization to previously deposited subtidal marine sediments below the sabkha seems to necessitate the downward and lateral reflux of evaporite-related brines with elevated Mg/Ca ratio through the porous, but muddy subtidal sediments. While the modern sabkha hydrological system seems compatible with such a scenario (McKenzie and others, 1980; McKenzie, personal communication; Land, 1985), and modern muddy subtidal sediments display extremely high porosities that could support the displacement of marine pore fluid by heavy evaporative brines (Chapter 3), the system's operation has not been adequately documented.

Indeed, the reported skewed distribution of dolomites through geologic time, with their apparent increasing dominance into the Lower Paleozoic and Precambrian (Fig. 5.8a), indicates that the carbonate chemistry of ancient oceans was possibly more favorable to the dolomitization process than that of the marine waters of the present (Ingerson, 1962; Zenger, 1972a; Zenger and Dunham, 1980). However, Given and Wilkinson (1987) recently reevaluated much of the data on dolomite distribution versus age and have developed a compelling case for a cyclic distribution of dolomite through time, rather than the exponential increase with age reported by previous authors (Fig. 5.8b). Distinct modes of high dolomite percentage in the Mesozoic and Lower Paleozoic correlate well with extensive continental marine flooding expected during times of relatively high sea level. Given and Wilkinson (1987) assumed (see Chapter 4, and discussion above) that most dolomitizaton is accomplished by marine waters, and related these patterns of dolomitization to periods of higher P_{CO_2}, lower CO_3^{-2} concentrations, and hence lower levels of calcite saturation in marine waters that would act to promote dolomitization (see also Machel and Mountjoy, 1986).

Porosity development and modification in sabkha-related reservoirs centers around the dolomite-rich facies of the original mud-dominated sediments characteristic of the supratidal sabkha and adjacent subtidal marine-to-lagoonal environments. The common occurrence of moldic porosity developed by dissolution of bioclasts, pellets, and ooids, in association with fine crystalline dolomite in subtidal and sabkha sequences

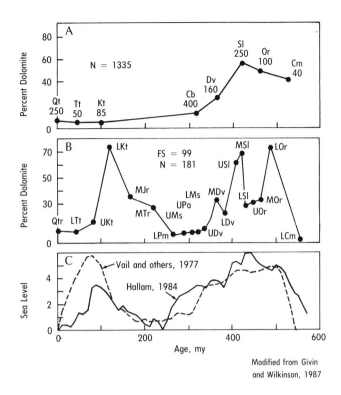

Fig. 5.8. A. Percentage dolomite as a function of age as reported by Chilingar (1956). B. Phanerozoic dolomite abundances recalculated in part by Given and Wilkinson (1987) from data in Chilingar (1956); C. Estimates of global sea level from Vail and others (1977) and Hallam (1984). Used with permission of SEPM.

(Moore and others, 1988), suggests preferential dolomitization of muddy matrix, as well as the introduction of fresh water into the environment, perhaps shortly after dolomitization. Indeed, fresh water flushing of a partially dolomitized muddy sediment would tend to dissolve the remaining undolomitized aragonite, magnesian calcite, or calcite, concentrating the initially floating dolomite rhombs into a crystal-supported fabric, and resulting in the porous sucrosic dolomite texture so common to many ancient sabkha sequences (Fig. 5.9) (Ruzyla and Friedman, 1985). In addition, fresh water would tend to dissolve evaporites associated with the dolomites, leading to a dramatic increase in porosity by the development of vugs after gypsum and anhydrite. In the upper parts of sabkha sequences, where gypsum and/or anhydrite are more concentrated, sulfate removal can lead to solution collapse breccias and the development of significant additional porosity (Loucks and Anderson, 1985; Fig. 5.19, this book). Finally, the common introduction of fresh meteoric water shortly after or during sabkha progradation is the ideal vehicle to stabilize, through recrystallization, the metastable dolomites commonly associated with marginal marine sabkhas (Land, 1985).

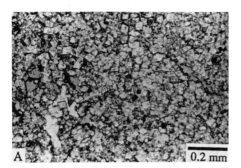 

Fig. 5.9. (A) Sucrosic dolomite from a surface exposure of a Texas Lower Cretaceous supratidal sequence occurring above the beach-shoreline complex described in Chapter 4. Dolomite is thought to have been flushed by meteoric water during its early diagenetic history. Porosity is 34%, permeability 18 md (see Fig. 4.9 for a mercury capillary curve of this sample). (B) SEM photomicrograph of same sample showing loose arrangement of individual dolomite rhombs.

Fresh water influence in a sabkha sequence is the usual result of the transitional nature of the sabkha, a dominantly subaerial depositional environment, standing midway between continental and marine conditions. Roehl (1967) coined the term *diagenetic terrain* for the subaerially exposed landward margins of the sabkha perpetually under the influence of continental and meteoric waters. In a regressive situation, the diagenetic terrain would be expected to follow, and to diagenetically modify the marginal marine sabkha as it progrades over previously deposited subtidal marine sediments. Perhaps the single most important factor controlling the development of economic porosity within sabkha-related sequences is the common development of diagenetic terrains at the end of each sabkha cycle.

Ordovician Red River marginal marine sabkha reservoirs, Williston Basin, U.S.A.

Probably the best known, economically important ancient sabkha sequence is the Ordovician Red River Formation producing from a number of fields in the U.S. portion of the Williston Basin. The Red River is preferentially dolomitized around the margins of the Williston Basin, and across paleohighs such as the Cedar Creek Anticline, coincident with thick cyclical evaporite-bearing sabkha sequences (Roehl, 1967; Wilson, 1975; Ruzyla and Friedman, 1985). The Cabin Creek field, developed along the crest of the Cedar Creek Anticline in southeast Montana (Fig. 5.10.), is a typical Red River sabkha-related reservoir.

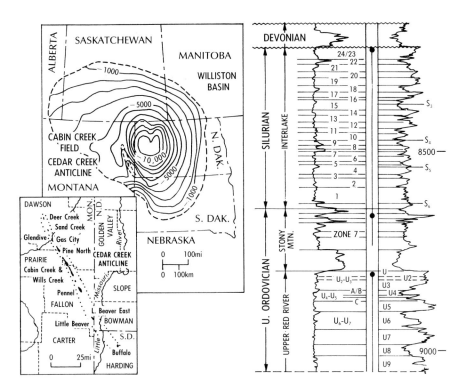

Fig. 5.10. (Left) Location map of the Cabin Creek field on the Cedar Creek Anticline in the southwestern margins of the Williston basin (from Roehl, 1985).(Right) Type radioactivity log of Lower Paleozoic formations associated with the Cedar Creek Anticline. Major zones and markers are shown. Solid circles and lines in the center of the log indicate productive zones (from Roehl, 1985). Reprinted with permission from Carbonate Petroleum Reservoirs. Copyright (C) 1985, Springer-Verlag, New York.

The Cabin Creek field has produced over 75 million barrels of oil from dolomitized Lower Ordovician Red River and Silurian Interlake sequences with an average porosity of some 13%. The field is a combined structural-stratigraphic trap occurring at an average depth of 9000 ft (2750 m). The upper 50 m of the Red River at Cabin Creek consist of three shoaling-upward sequences capped by sabkha complexes with diagenetic terrains developed at the intercyle unconformities (Fig. 5.11) (Ruzyla and Friedman, 1985). While all the supratidal sequences are dolomitized, subtidal carbonates exhibit both limestone and dolomite (Fig. 5.11). Although most of the effective porosity is concentrated within the dolomitized supratidal cap, good porosity occasionally occurs in dolomitized subtidal sequences toward the base of a cycle (Fig. 5.11). A schematic diagram of a typical Red River sabkha-capped cycle outlines the general distribution of diagenetic features, porosity, and pore types that might be expected in similar sequences

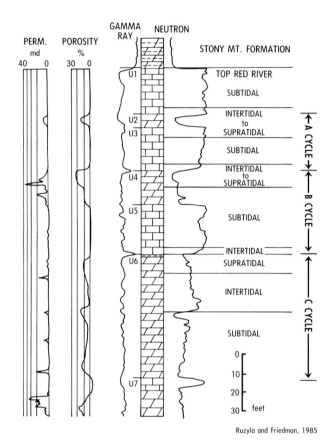

Ruzyla and Friedman, 1985

Fig. 5.11. Typical columnar section of the upper Red River showing depositional environments, porosity, and permeability determined by core analysis and gamma ray-neutron curves. Reprinted with permission from Carbonate Petroleum Reservoirs. Copyright (C) 1985, Springer-Verlag, New York.

Zone & Environment	Porosity %	Perm (md)	N_c	Oil Sat. %	Water Sat. %	Total Sat. %	Pore Size (mm)	Relative Amounts of Each Pore Type[1]				
								V	M	Ip	Ic	N_{ts}
U_4-U_5 supratidal	16.8	11.7	59	29.3	21.4	50.7	0.13	48%	13%	13%	26%	23
U_6-U_7 supratidal	7.4	0.6	110	16.3	47.4	63.7	0.04	95%	5%	—	—	38
U_4-U_5 intertidal	15.3	7.1	49	30.2	23.5	53.7	0.19	19%	81%	—	—	26
U_6-U_7 intertidal	7.1	3.6	156	17.2	39.4	56.6	0.13	31%	31%	26%	11%	57
U_4-U_5 subtidal	10.7	1.4	110	15.3	41.3	56.6	0.19	32%	64%	—	4%	33
U_6-U_7 subtidal	9.4	2.3	156	16.2	44.3	60.5	0.17	25%	44%	10%	21%	64

[1] V = vug, M = moldic, Ip = interparticle, Ic = intercrystal
N_c = number of core analyses
N_{ts} = number of thin-section analyses

Ruzyla and Friedman, 1985

Table 5.1. Porosity, permeability, saturation, and thin section data for dolomite reservoirs in the upper Red River Formation, Cabin Creek field. Reprinted with permission from Carbonate Petroleum Reservoirs. Copyright (C) 1985, Springer-Verlag, New York.

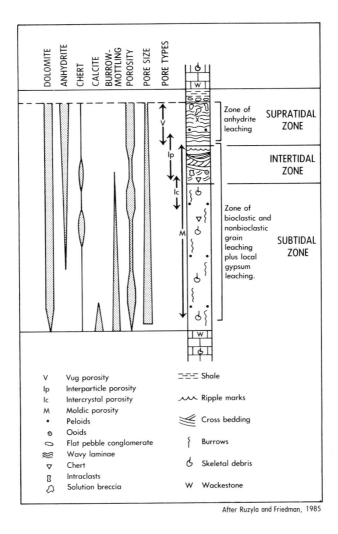

Fig. 5.12. Depositional and diagenetic features of the upper Red River dolomites in the Cabin Creek field. Reprinted with permission from Carbonate Petroleum Reservoirs. Copyright (C) 1985, Springer-Verlag, New York.

(Fig. 5.12). Porosity quality and porosity type are variable and may be facies-selective in response to fresh water leaching of evaporites, aragonitic grains, or undolomitized mud matrix. Table 5.1 summarizes the porosity, permeability, and pore type for each environment at Cabin Creek, while Fig. 5.13 outlines the diagenetic processes responsible for these pore systems. The dominance of secondary-solution-related porosity, such as vugs and molds, clearly illustrates the influence of freshwater flushing in the development of porosity at Cabin Creek.

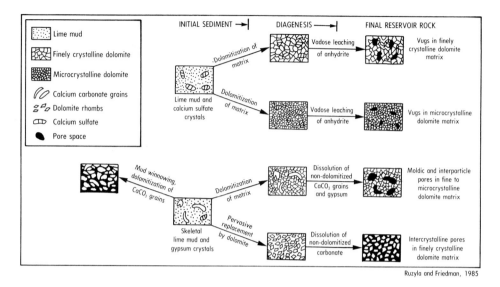

Fig. 5.13. Diagenetic processes involved in the development of upper Red River pore systems. Reprinted with permission from Carbonate Petroleum Reservoirs. Copyright (C) 1985, Springer-Verlag, New York.

Mississippian Mission Canyon marginal marine sabkha reservoirs, Williston Basin U.S.A.

The Mississippian Mission Canyon Formation is also a prolific oil producer in the Williston Basin. The Mission Canyon is a shallowing-upward regressive sequence ranging from basinal deep-water carbonates to evaporite-dominated coastal sabkhas and evaporative lagoons (Fig. 5.14) (Lindsay and Roth, 1982; Lindsay and Kendall, 1985). A stratigraphic cross section across the North Dakota portion of the Williston Basin illustrates the lateral facies change from evaporites on the northeast margin of the basin to carbonates toward the basin center (Fig. 5.14) (Malek-Aslani, 1977). The presence of thick continuous anhydrites with halite indicates that at least part of the northeast margin could have been a subaqueous, marginal marine, evaporative lagoon, separated from the marine shelf by barrier shoreline complexes. The Little Knife field, located in the center of the Williston Basin of western North Dakota (Fig. 5.14), is a typical Mission Canyon evaporite-related reservoir.

The Little Knife field has produced over 31 million barrels of oil from dolomitized Mission Canyon reservoirs, with porosity averaging 14% at an average depth of 9800 ft (3000 m). The field is a combined structural-stratigraphic trap developed on a northward-plunging, anticlinal nose, with closure to the south by lateral facies change into

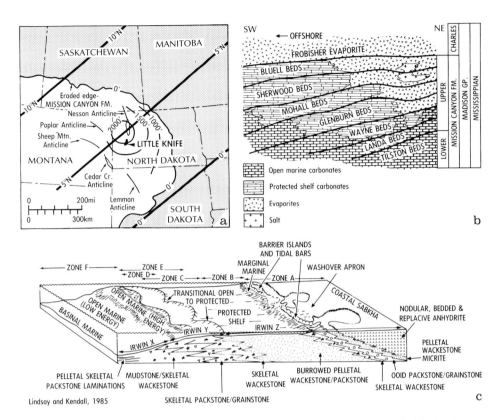

Fig. 5.14. (Upper left) Index map of the Williston Basin showing: eroded edge of the Mission Canyon Formation; major surface and subsurface structural features; isopach thickness of the Madison Group; generalized Carboniferous paleolatitude lines; and the location of the Little Knife field. (Upper right) Regional cross section of Mission Canyon Formation across the North Dakota portion of the Williston Basin. (Lower) Idealized depositional setting of the Mission Canyon Formation at Little Knife field. Informal log zonations A-F, and Irwin's (1965) epeiric sea energy zones (XYZ) illustrate respective positions occupied in the depositional system. (Lindsay and Kendall, 1985). Reprinted with permission from Carbonate Petroleum Reservoirs. Copyright (C) 1985, Springer-Verlag, New York.

nonporous lithologies (Lindsay and Kendall, 1985). Fig. 5.15 depicts a typical log profile of the Mission Canyon in the center of the Little Knife field, and correlates major porosity-modifying, diagenetic events with depositional environments. Pervasive dolomitization is present at the top of the cycle, closely associated with evaporites of the capping sabkha-lagoonal sequence, and decreases dramatically toward the base, or toward the more normal marine portions of the sequence. In addition, anhydrite replacement of allochems is common in the upper three zones of the cycle. Dolomitization and anhydritization of the sequence are clearly the result of reflux of evaporative brines generated on adjacent marginal marine sabkhas, or within marginal marine,

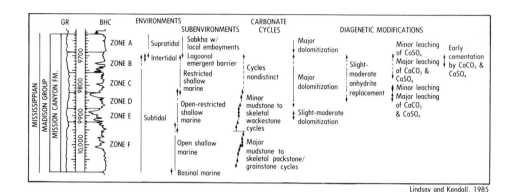

Fig. 5.15. Type gamma-ray/sonic log of Mission Canyon Formation in the central portion of the Little Knife field. Depositional environments, cycles, and diagenetic modifications are shown to the right of the log. Reprinted with permission from Carbonate Petroleum Reservoirs. Copyright (C) 1985, Springer-Verlag, New York.

evaporative lagoons. The lagoonal model would certainly provide a larger volume of dolomitizing brines than a sabkha for reflux through adjacent mud-dominated marine sequences. The distribution of dolomite seems to have been controlled by porosity-permeability trends developed in the original sedimentary sequence by burrowing and winnowing (Lindsay and Kendall, 1985). Again, early, fresh water influx, probably associated with subaerial exposure at the end of the main cycle, had a major impact on porosity development, dissolving remnant undolomitized matrix, as well as metastable and anhydritized allochems, and resulting in significantly enhanced porosity of the reservoir.

Ordovician Ellenburger marginal marine, sabkha-related dolomite reservoirs, west Texas, U.S.A.

The Ordovician Ellenburger Dolomite is a major exploration target in southwest and west-central Texas. The Ellenburger was deposited on a broad, shallow marine shelf of some 800 km width during a major Cambrian-Ordovician transgression over the Precambrian basement (Barnes and others, 1959) (Fig. 5.16). The lower Ellenburger passes conformably into the Cambrian Bliss Sandstone below. Fig. 5.17 portrays the general depositional model for the Bliss-Ellenburger, as conceived by Loucks and Anderson (1985). The Bliss and lower Ellenburger represent a series of shallow marine, shallowing-upward sequences capped with coastal sabkha complexes that were associated with alluvial fans and barrier-protected restricted lagoons (Fig. 5.17). The middle and upper Ellenburger record a series of cycles consisting of shallow marine shelf sequences building upward into a marginal marine, supratidal-tidal channel complex

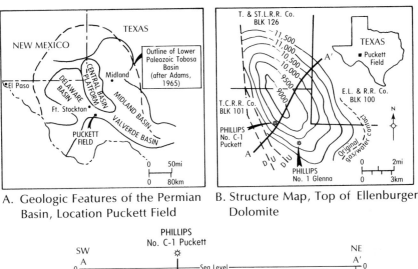

Fig. 5.16. Geologic setting of the Puckett field, Permian Basin, U.S.A. Reprinted with permission from Carbonate Petroleum Reservoirs. Copyright (C) 1985, Springer-Verlag, New York.

generally capped with diagenetic terrains developed during terminal exposure of the cycle prior to the development of the subsequent cycle (Fig. 5.17). The Puckett field, located in the transition between the Delaware and Val Verde basins in far west Texas, is a representative Ellenburger field (Fig. 5.16).

The Puckett field is a structural trap developed along a large, faulted, anticlinal feature with more than 760 m of closure (Fig. 5.16). The field has produced more than 2.6 trillion cu ft of gas from Ellenburger dolomites, with an average porosity of 3.5% at a depth averaging 3600 m. The original gas column was over 490 m thick (Loucks and

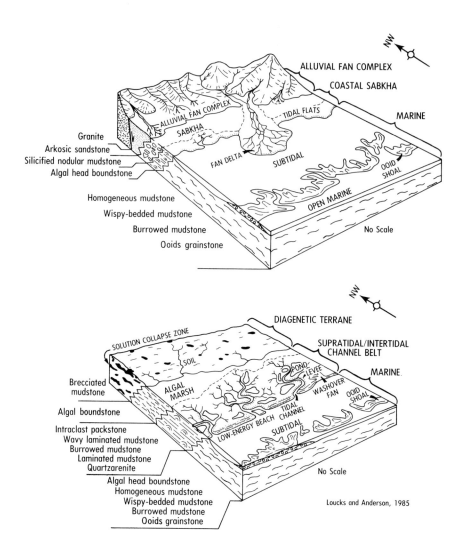

Fig. 5.17. (Top) Depositional model for the Bliss and lower Ellenburger at Puckett field. (Bottom) Depositional model for the middle and upper Ellenburger at Puckett field. Reprinted with permission from Carbonate Petroleum Reservoirs. Copyright (C) 1985, Springer-Verlag, New York.

Anderson, 1985). The entire Ellenburger at Puckett field is dolomitized with subtidal facies dolomitization at least in part related to reflux of evaporative brines from adjacent marginal marine sabkha and supratidal sequences. Economic porosity and permeability, however, occur in a cyclic fashion through the entire sequence (Fig. 5.18). A detailed look at typical cored sequences at Puckett reveals that major porosity zones in the lower

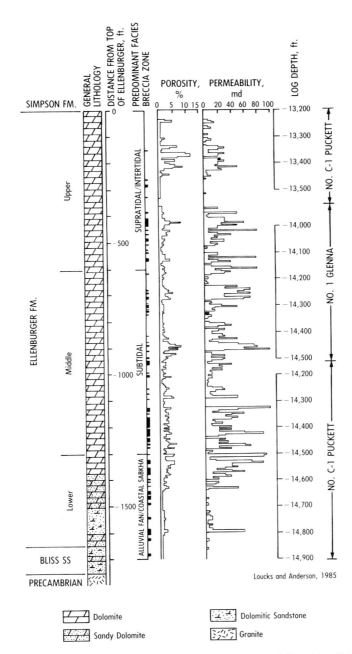

Fig. 5.18. Profiles of porosity and permeability vs depth in the Phillips No. C-1 Puckett well. Predominant facies refers to the facies comprising the largest proportions of individual cycles. Reprinted with permission from Carbonate Petroleum Reservoirs. Copyright (C) 1985, Springer-Verlag, New York.

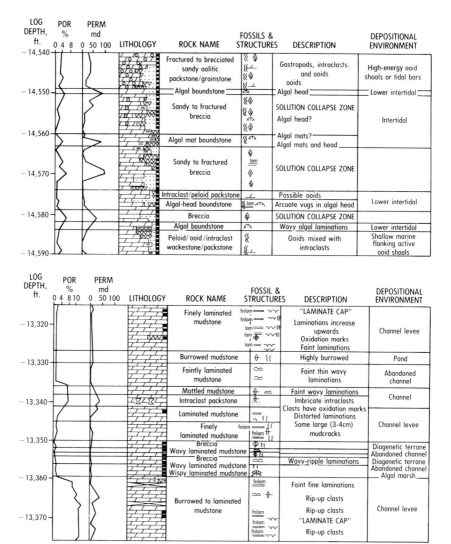

Fig. 5.19. (Top) Core description, environmental interpretation, and porosity-permeability profile of a section of the alluvial fan/sabkha sequence in the lower Ellenburger and upper Bliss. Phillips No. C-1 Puckett well, 14,797 to 14,846 feet (4510-4525 m. (Bottom) The same for a section of the low-energy subtidal sequence in the middle Ellenburger. Phillips No. 1 Glenna well, 14,306 to 14,359 feet (4361-4377 m). Reprinted with permission from Carbonate Petroleum Reservoirs. Copyright (C) 1985, Springer-Verlag, New York.

Ellenburger generally coincide with solution-collapse breccias, while porosity zones in the upper Ellenburger are developed below diagenetic terrains or within channel sequences (Fig. 5.19). The solution collapse breccias in the Lower Ellenburger seem to be related to the removal of marginal-marine sabkha evaporites during intercycle exposure. Porosity development in the upper Ellenburger is enhanced by mold and vug generation associated with intercycle diagenetic terrains and may also be facies controlled, such as the better porosity development seen in grainstone-packstone channel sequences (Fig. 5.19). Fractures in all reservoir facies significantly enhance permeability.

Criteria for the recognition of ancient marginal marine sabkha dolomites

Table 5.2 outlines some of the major criteria that should be satisfied before an ancient dolomite is assigned to a marginal marine sabkha origin. The criteria are listed

Table 5.2.

Criteria for the recognition of ancient marginal marine sabkha dolomites, listed in decreasing order of reliability.

1. Dolomite must be syngenetic

2. Dolomite must be associated with a sequence exhibiting sedimentologic and mineralogic characteristics of the marginal marine sabkha depositional environment.

3. Patterns of dolomitization should reflect the complex but localized sedimentologic, diagenetic, hydrologic setting of the sabkha. Dolomitization tends to be spotty and interbedded with evaporites, limestones and siliciclastics.

4. Marginal marine sabkha dolomites crystal sizes tend to be small, but may gain significant size during later diagenesis.

5. Dolomite geochemistry may reflect original evaporative pore fluids if sequence buried in relatively closed system. Normally, however chemical and isotope composition seem to have been re-set by later recrystallization.

in decreasing order of reliability.

Perhaps the most important constraint on dolomite origin, and usually the most difficult to determine, is the timing of the dolomitization event. Fabric and textural destruction often preclude the dolomite's precise assignment in a relative paragenetic sequence. The following have all been used, with varying success, to suggest the penecontemporaniety of various dolomitization events: the close relationship of dolomite fabrics to recognizable early diagenetic features, such as demonstrable vadose cements (Ward and Halley, 1985); the absence of two-phase fluid inclusions (O'Hearne and Moore, 1985; Moore and others, 1988); dolomitization fabric selectivity (Beales, 1956; Kendall, 1977b); and relationship to early compactional phenomena (Moore and others,1988). The easiest and most widely applied criterion is the incorporation of the dolomite into a sequence that exhibits the sedimentological, and mineralogical characteristics of the marginal marine sabkha depositional environment, such as algal laminations; mudcracks; ripup clasts; thin, interbedded, nodular anhydrite; or in their absence, solution collapse breccias. As an example, Wood and Wolfe (1969) were among the first to use the sedimentological benchmarks erected by Illing and others (1965), Shearman (1966), and Kinsman (1966) in their early studies of the sabkhas at Abu Dhabi. They suggested that producing subsurface dolomites of the Mesozoic Arab D Formation in the offshore Trucial Coast of Arabia were early syngenetic, and related to the sabkha diagenetic environment.

The patterns of dolomitization associated with a marginal marine sabkha should reflect the complex, but localized sedimentologic and diagenetic environment normally found in this setting. For an example, ancient sabkha-related dolomitization should normally exhibit local, rapid shifts into undolomitized limestones, anhydrites, and siliciclastics, both laterally and vertically. Thick, regional, pervasive, dolomitized sequences should not be expected in the sabkha environment.

Modern sabkha dolomites often exhibit small crystal sizes (less than 5 microns, with occasional aggregates ranging up to 20 microns) believed to reflect the numerous nucleation sites available in mud-dominated, sabkha-related sequences (Land, 1980, 1985; McKenzie, 1981). Ancient sabkha-related dolomites commonly display much coarser crystal sizes, perhaps because of later recrystallization of the initial metastable calcian dolomite. For example, most of the Mesozoic Arab-D sabkha dolomite studied by Wood and Wolfe (1969) ranged in crystal size from 20-80 microns. As has been discussed earlier, the trace element and isotopic composition of ancient sabkha-related dolomites may not adequately reflect the evaporative nature of the dolomitizing fluids because of the probability of recrystallization of the original metastable dolomite.

Land (1980), further suggests that most carbonate platforms have been flushed by meteoric waters early in their history as a result of carbonate platform sedimentation generally being punctuated by shoaling-upward cycles separated by subaerial exposure surfaces. This propensity for frequent platform subaerial exposure leaves sabkha terminal sequences susceptible to recrystallization in dilute meteoric waters. This recrystallization is generally reflected in the chemistry of the dolomites by low trace-element compositions and light, stable, isotopic ratios (Mueller, 1975; Land, 1980, 1985). These relationships have no doubt contributed to the tremendous rush toward the utilization of the Dorag, or schizohaline model of dolomitization (Hardie, 1987). After all, most dolomites found on carbonate platforms might well exhibit a meteoric water, geochemical fingerprint, regardless of origin.

MARGINAL MARINE, EVAPORATIVE LAGOONS AND BASINS (REFLUX DOLOMITIZATION)

The marginal marine, evaporative lagoon as a diagenetic environment

Structural or sedimentological barrier or sill development in a marginal marine setting often modifies free hydrologic exchange between adjacent marine waters and back-barrier, shelf-lagoonal environments. It also sets the stage for the development of hypersalinity, evaporite precipitation, and potentially strong density gradients. Fig. 5.20 illustrates the basic characteristics of a barred basin or lagoon. Given the proper climatic conditions, inflowing seawater is progressively evaporated as it moves from the sill toward the shoreline, setting up strong horizontal concentration (density) gradients. The heavier brines ultimately sink as they approach the landward margins of the lagoon, leading to a return flow of heavy brines toward the sill, and establishing a seaward-sloping pycnocline separating the inflowing marine waters from seaward-flowing dense brines (Scruton, 1953; Dellwig, 1955; Adams and Rhodes, 1960; Logan, 1987).

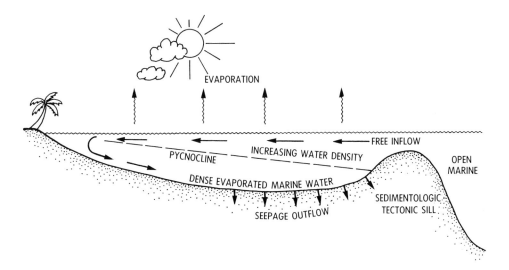

Fig. 5.20. Schematic diagram of an evaporitic silled basin, or silled shelf lagoon model. As water is evaporated, increasing water density, a pycnocline is established, separating dense evaporative fluid below, from less dense marine waters above. The dense evaporative brines may reflux through the basin, or lagoon floor, or through the sill if porosity and permeability allow. The sill may be either sedimentologic such as a reef build-up, or high - energy sand complex, or tectonic, such as a horst block.

Depending on the position of the sill relative to the pycnocline, the return brine will either flow over the sill, or become trapped behind the sill to be refluxed laterally through the sill, or downward through subjacent units forming the lagoon, or basin floor.

Adams and Rhodes (1960) used the fundamental elements of the barred basin model to develop their concept of dolomitization by seepage refluxion based on the Permian of west Texas. In their model (Fig. 5.21), sea water is progressively evaporated as it moves from the sill toward the shoreline across a relatively shallow, but wide, barred-shelf lagoon. Gypsum saturation is reached in the lagoon interior, and evaporites are deposited along the shoreward margins of the lagoon. At this point, brine density is high enough to set up density-driven, brine return flow toward the sill, and horizontal density stratification is established with a seaward-dipping pycnocline. While the landward margin of the lagoon floor is sealed by the precipitation of evaporites, the dense brines are able to reflux down through the lagoon floor seaward of the zone of evaporite precipitation. The refluxing dense brines exhibit elevated Mg/Ca ratios, and proceed to dolomitize the lagoon floor and subjacent porous limestone units as they displace lighter marine pore fluids.

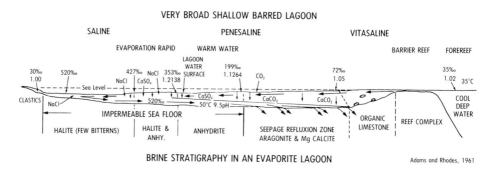

Figure 5.21. Idealized section of brine stratigraphy in evaporite lagoon based on the Permian reef complex of New Mexico and Texas, U.S.A. Reprinted with permission of the American Association of Petroleum Geologists.

If the sill is sedimentological in nature, such as a reef tract or ooid shoal complex, and an equilibrium is established among sedimentary accretion, relative sea level rise, and evaporite precipitation, thick sequences of evaporites can be precipitated along the margins of shelf lagoons. In addition, enormous volumes of dolomitizing fluids can be available to reflux through and pervasively dolomitize adjacent porous units (Moore and others, 1988).

One of the major difficulties with the silled lagoon-reflux model of dolomitization has been the general lack of an adequate modern analogue. Initially, the Pekelmeer, a salt pond on the island of Bonaire barred from the sea by a beach ridge, was believed to be a reflux-driven system resulting in dolomitization of adjacent Plio-Pleistocene sequences (Defeyes and others, 1965; Murray, 1969; Bandoian and Murray, 1974). Subsequent work failed to provide direct evidence for the involvement of refluxing evaporative brines. Sibley (1980) called on a mixed, fresh-marine water as the dolomitizing fluid. In 1977, Aharon and others described a small-scale seepage reflux dolomitization system associated with a solar pond near Elat on the Gulf of Aqaba. Potentially, the most compelling modern analogue for the evaporative silled lagoon and reflux dolomitization model is the MacLeod salt basin near Carnarvon in western Australia, as recently described in detail by Logan (1987).

The MacLeod salt basin

The MacLeod is a graben basin some 30 km wide and 60 km long oriented approximately north-south and separated from the Indian Ocean on the west by a relatively porous Pleistocene carbonate ridge attaining elevations of up to 60 m, and on the south by a low Holocene dune field (Fig. 5.22). During the Holocene sea level rise, from approximately 8000 to 5000 years BP, the MacLeod was open to the sea across a narrow southern sill. It acted as a silled basin or lagoon with the precipitation of both aragonite and gypsum and with the seepage outflow of dense brines through the lagoon floor (Fig. 5.23A) (Logan, 1987). At approximately 5000 years BP, sea level began to drop, and the southern sill was sealed by a parabolic dune field. At this point, the brine surface was depressed some 10 m to the lagoon floor (Fig. 5.23B). Fresh marine waters continued to be furnished to the MacLeod in the form of seepage along the margins of the lagoon through the porous barrier, ultimately filling the MacLeod with up to 10 m of gypsum and halite (Logan, 1987) (Fig. 5.23C). Free reflux of evaporative brines was restricted by the development of a seal across most of the lagoon floor caused by gypsum precipitation associated with the initial evaporative drawdown when the southern sill was closed (Fig. 5.23B). However, dense brines continue to reflux down into the underlying sedimentary units through well-defined brine sinks subsequently developed along the margins of the lagoon (Logan, 1987) (Figs. 5.23C and D). A recent series of exploratory core holes have been drilled through the MacLeod evaporite sequence to determine whether seepage refluxion dolomitization is operative at MacLeod. While no details are presently available, Logan (1987; personal communication) does report that extensive dolomitization is occurring in several environments within the MacLeod system.

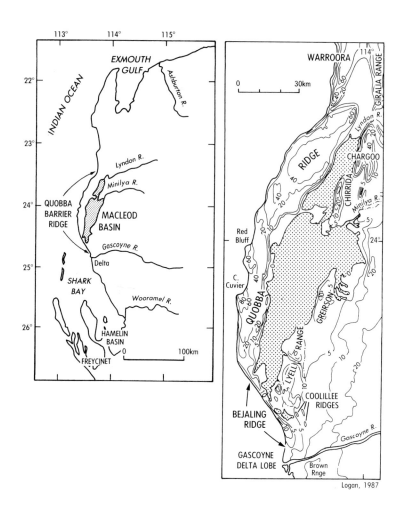

Fig. 5.22. (Left) Map showing the location of the MacLeod basin. During Quaternary marine phases the basin was a marine gulf connected southward to Shark Bay. (Right) Contour map showing the topography of the MacLeod region. The stippled area lies between sea level and -4.3 m. Reprinted with permission of the American Association of Petroleum Geologists.

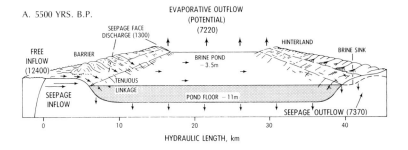
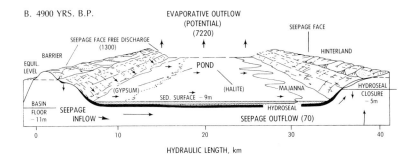
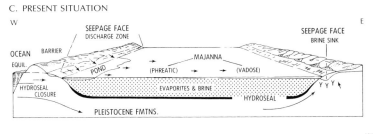
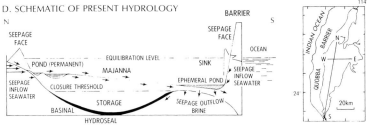

Fig. 5.23. Schematic of the MacLeod basin system from 5500 years B.P. (A) until the present (C), illustrating the effect of sea level variation on the effectiveness of the barrier, the development of the hydroseal at 4900 years B.P. (B), and the ultimate filling of the basin with evaporites (C). The present hydrology of the MacLeod basin is shown in D. Location of cross sections shown at lower right. Reprinted with permission of the American Association of Petroleum Geologists.

The Upper Permian Guadalupian of west Texas, U.S.A.: an ancient marginal marine, evaporative lagoon complex

Perhaps the world's best exposed carbonate shelf margin-to-evaporite transition occurs in the Upper Permian outcrops along the Guadalupe Mountains of New Mexico and Texas. Here, the massive biogenic limestones of the Capitan Formation mark the margin of the Delaware Basin, which extends across the southeastern corner of New Mexico into far west Texas (Fig. 5. 24). Landward from the steep, abrupt Capitan shelf margin, extensive Upper Permian shelfal carbonates and evaporites, called the *Artesia Group*, composed of the Seven Rivers, Yates, and Tansill formations, were deposited.

Fig. 5.24 illustrates the facies distribution across the Capitan shelf margin during lower Seven Rivers time, extending across a portion of the northwestern shelf at Walnut Canyon near Carlsbad, New Mexico. The carbonates of the shelf margin give way to evaporites (gypsum) within 15 km in a shelfward direction. Most of the carbonates shelfward of the Capitan margin consist of replacive, fine crystalline dolomite that preserves the fabrics and textures of the original packstones, grainstones, and wackestones that were deposited along the margins of an extensive evaporative shelf lagoon (Sarg, 1981).

In his studies of the carbonate-evaporite transition within the Seven Rivers along the Guadalupe Mountains, Sarg (1976, 1977, and 1981) concluded that the evaporites were dominantly precipitated under subaqueous conditions and that the evaporite facies of the Guadalupian shelf represented an enormous shelf lagoon. This interpretation supports the earlier interpretations of Lloyd (1929) and Bates (1942), but conflicts directly with the interpretations of Kendall (1969), Silver and Todd (1969), Hills (1972), and Meisner (1972), who proposed that the Seven Rivers evaporites and related dolomites were of sabkha origin.

Sarg (1981) based his interpretation on the considerable thickness of the gypsum sequences relative to the thin evaporite sequences seen in modern sabkhas; lateral facies relationships with mesohaline facies that show no evidence of subaerial exposure; lack of any evidence of subaerial exposure either in the evaporite or related carbonate facies; the maintenance of the gypsum carbonate transition at the same geographic location through most of Seven Rivers time, a relationship not compatible with a prograding sabkha sequence; and finally, the lack of any characteristic sabkha vertical sequences.

Consequently, Sarg (1981) calls upon evaporative reflux from a regional evaporative shelf lagoon, as originally envisioned by Adams and Rhodes (1960), to dolomitize the associated carbonate facies of the Seven Rivers as well as the Queen Formation below and the Yates and Tansill formations above. In the nearby subsurface these units form some of the most prolific oil-producing reservoirs in west Texas and are of enormous

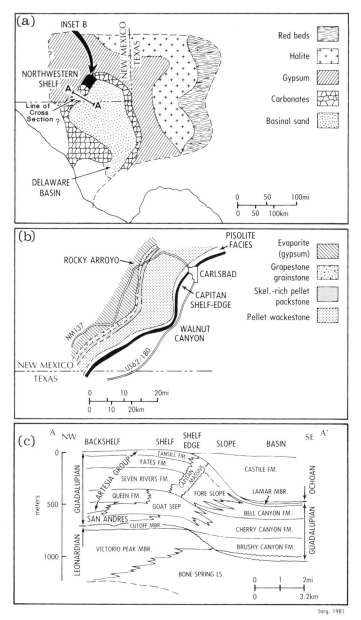

Fig. 5.24. Geologic setting of the Upper Permian Guadalupian of west Texas and New Mexico. (a.) Major facies of the Guadalupian shelf surrounding the Delaware basin. Note that the carbonate shelf is rather narrow, and gives way to evaporites toward the land, and siliciclastics into the basin. (b) Lithofacies map of the Guadalupian shelf margin in the vicinity of Carlsbad, New Mexico showing the transition from shelf limestones to evaporites. (c) The stratigraphy of the Upper Permian, shelf-to-basin along line of section A-A' (see a above for location). Used with permission of SEPM.

economic importance (Silver and Todd, 1969). The Lower Cretaceous Ferry Lake Anhydrite in east Texas was deposited in a similar setting.

Ferry Lake Anhydrite, central Gulf of Mexico Basin, U.S.A.

The Ferry Lake was formed in a shallow, subtidal, hypersaline lagoon barred from the sea by shelf-margin, rudist reefs and perhaps a tectonic sill (Fig. 5.25). The lagoon was up to 260 km wide, and extended across much of east Texas and into adjacent Louisiana (Loucks and Longman, 1982). Subjacent limestone units, however, seemingly suffered little dolomitization. Perhaps the lagoon was too shallow and too agitated because of the large fetch to allow the formation of a pycnocline and to establish an underflow of heavy brines toward the sill. Nevertheless, the Ferry Lake provides an excellent seal to important Lower Cretaceous reservoirs throughout its extent.

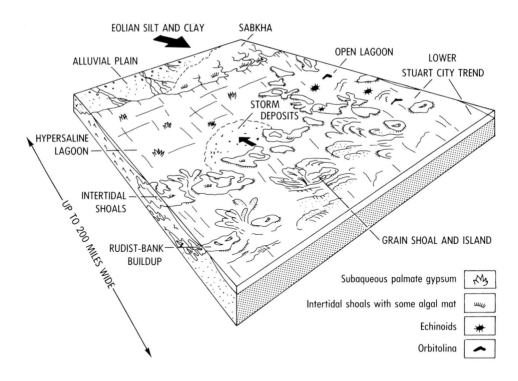

Fig. 5.25. *Depositional model for the Ferry Lake Anydrite in east Texas (from Loucks and Longman, 1982). Used with permission of SEPM.*

Upper Jurassic Smackover platform dolomitization, east Texas, U.S.A.: a reflux dolomitization event

The Upper Jurassic Smackover Formation is an important oil and gas trend across the northern Gulf of Mexico, from Mexico to Florida. Reservoir rocks are developed in ooid grainstones accumulated on high-energy platforms developed along the landward margins of a series of interior salt basins (Fig. 5.26) (Moore, 1984). Upper Smackover ooid grainstones have suffered pervasive platform dolomitization over a large area of east Texas, while the upper Smackover in adjacent areas of Louisiana and Mississippi is still limestone (Fig. 5.26).

The stratigraphic setting of the Smackover is shown in Fig. 5.27. The upper

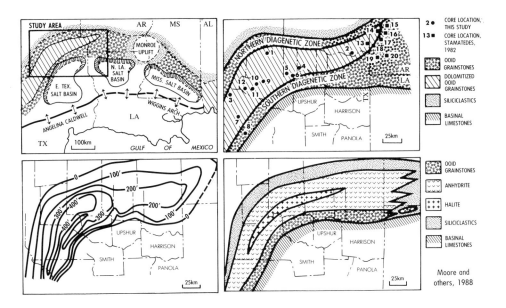

Fig. 5.26. Geologic setting of the Upper Jurassic, Gulf Coast, U.S.A. (Upper left) Structural setting at the time of deposition of the Upper Jurassic. High-energy, ooid grainstone shelves developed landward of interior salt basins. These grainstones dolomitized to the east in Alabama, and to the west in Texas. Location of study area in east Texas indicated. (Upper right) Upper Smackover shelf lithofacies, emphasizing distribution of dolomitized grainstones in east Texas. Sample control indicated. (Lower left) Isopach map of the Buckner evaporite (gypsum plus halite). (Lower right) Lithofacies developed in the Buckner Formation. Used with permission of SEPM.

Smackover ooid grainstones represent a sea level highstand and the development of a high-energy platform. With continued sea level rise in the Kimmeridgian, a rimmed evaporative shelf lagoon was developed. In east Texas, these evaporites are referred to as the *Buckner Formation*. Fig. 5.26 shows an isopach of the Buckner and its lateral facies relationships, from siliciclastics in the shelf interior to an ooid grainstone barrier facies along the basin margin. Buckner evaporites consist dominantly of anhydrite, with 30 m of halite present in the Buckner depocenter (Moore and others, 1988).

The close correspondence of upper Smackover platform dolomitization and the

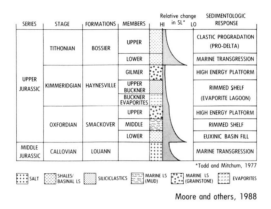

Fig. 5.27. Stratigraphy of the Jurassic, east Texas, U.S.A. Used with permission of SEPM.

distribution of evaporites in the Buckner above suggest that reflux dolomitization is the most appropriate model that can be applied to the Smackover in east Texas. Petrographic relationships and the lack of two-phase fluid inclusions in the dolomite indicate a relatively early timing for the dolomitization event. Finally, dolomite distribution patterns in the upper Smackover, showing a dolomitization gradient increasing upward toward overlying Buckner evaporites, are compatible with a reflux model (Fig. 5.28).

The reflux model proposed by Moore and others (1988) for the upper Smackover platform dolomites is shown in Fig. 5.29. A high-energy Buckner barrier developed during the Kimmeridgian sea level rise, generating an evaporative shelf lagoon. Anhydrites precipitated in the shelf interior, density stratification developed in the lagoon, and magnesium-rich brines refluxed down through porous upper Smackover ooid grainstones below, ultimately dolomitizing the central portion of the upper Smackover platform across most of east Texas. As sea level reached a high stand in the Kimmeridgian (Fig. 5.27), sedimentation along the lagoon barrier shut off the free flow of water into the lagoon, evaporative drawdown took place, the halite facies was deposited, the floor of the lagoon was sealed, and reflux ceased.

The geochemistry of these dolomites, however, does not reflect an evaporative origin, but may be more characteristic of meteoric waters (Moore and others, 1988). Smackover dolomites display low Sr and Na compositions and low $\partial^{18}O$ values compared to dolomites encased in Buckner anhydrites above (Figs. 5.30A and B). Geochemical

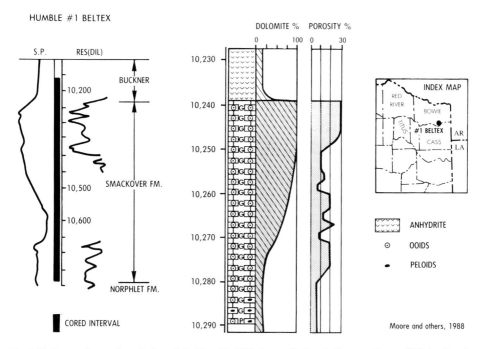

Fig.5.28. Log and core description of the Humble #1 Beltex well, Bowie County, Texas, U.S.A., showing distribution of dolomite relative to the overlying evaporites of the Buckner Formation. Used with permission of SEPM.

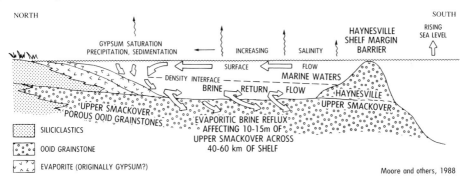

Fig. 5.29. Conceptual model of the shelf lagoon setting for the evaporative reflux dolomitization of the upper Smackover in east Texas. Used with permission of SEPM.

gradients, particularly in Fe, Mn, and $^{87}Sr/^{86}Sr$, established by Moore and others (1988), in the Smackover platform dolomites indicate the influence of a regional meteoric water system acting from platform interior to platform margin (Fig. 5.30C).

These relationships strongly suggest a major recrystallization event of an original

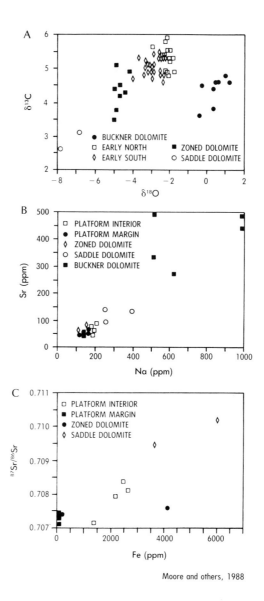

Fig. 5.30. Isotopic and trace element composition of Smackover and related dolomites, east Texas. Used with permission of SEPM.

evaporative reflux platform dolomite by a major meteoric water system. The meteoric aquifer probably developed during exposure of the landward, siliciclastic facies of the upper Smackover carbonate sequence associated with a major unconformity developed during the Cretaceous. The Smackover platform dolomites of east Texas dramatically illustrate the problems facing the modern geologist dealing with regional platform dolomitization events.

The Elk Point Basin of Canada

The Elk Point is an enormous middle Devonian (Givetian) basin extending some 1700 km (1100 mi) from northern Alberta, Canada, to North Dakota and Montana in the northern U.S. (Fig. 5.31). It is 700 km (450 mi) wide, extending into Saskatchewan and Manitoba on the east. The basin is rimmed with structural highs on the west (Peace River

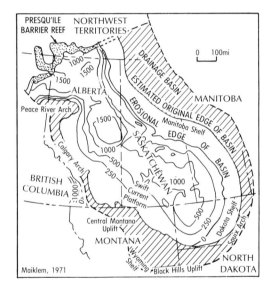
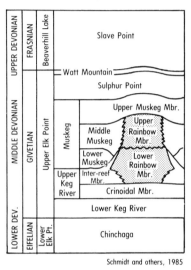

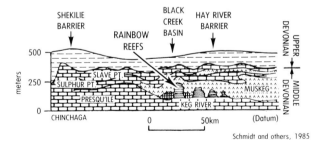

Fig. 5.31. (Upper left) Map of the Elk Point basin. Contours on the Elk Point Group in meters. (Upper right) Stratigraphic chart of Devonian formations in the Rainbow region, north-western Elk Point basin. (Lower) Idealized regional stratigraphic cross section of Middle Devonian formations across the Presqu'ile barrier reef in the northwestern Elk Point basin. Reprinted with permission of the Society of Canadian Petroleum Geologists.

Arch, Central Montana Uplift) and the Precambrian shield and carbonate shelves on the east and south (Manitoba and Dakota shelves) (Fig. 5.31).

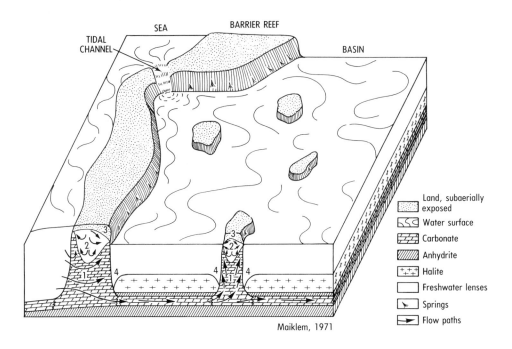

Fig. 5.32. *Conceptual model of the Elk Point Basin. The sill is a barrier reef, shelfal pinnacle reefs developed behind the barrier. Diagram drawn at time of evaporative draw-down causing hydraulic head between the sea and the shelf lagoon. The numbers represent diagenetic processes and environments. 1. Normal marine waters driven through the barrier along the porous shelf sequences and up the porous pinnacle reefs, dolomitizing as it goes. 2. Floating meteoric lenses beneath barrier and pinnacles, solution and precipitation of calcite. 3. Vadose diagenesis in the upper reaches of the barrier and pinnacles, solution and precipitation. 4. Solution of evaporites by normal marine water along margins of the barrier and the pinnacles. Reprinted with permission of the Society of Canadian Petroleum Geologists.*

The Elk Point Basin is barred from the open sea to the north by the "Presqu'ile" barrier reef complex, and is filled with a thick (up to 450 m or 1500), evaporative sequence consisting of both anhydrite and halite, referred to as the *Muskeg evaporites*. The "Presqu'ile" barrier reef is part of the Keg River carbonates (Fig. 5.31). A series of economically important pinnacle reefs (Rainbow) encased in Muskeg evaporites are developed just southeast of the "Presqu'ile" barrier (Fig. 5.31). The carbonates closely associated with the "Presqu'ile" are also hosts to significant lead-zinc mineralization (the Pine Point complex) (Skall, 1975).

Shearman and Fuller (1969) interpreted the bulk of the massive evaporites of the Elk Point Basin as supratidal sabkha. But Maiklem (1971) and Maiklem and Bebout (1973) reinterpreted the Muskeg evaporites to be dominantly subaqueous, with most of the evaporite deposition taking place during periodic drawdown episodes that resulted in the recognition of a number of sedimentologic cycles within the Muskeg and associated carbonates (Maiklem and Bebout, 1973).

Maiklem (1971), using the Elk Point Basin, presented an interesting diagenetic model based on an evaporative drawdown episode within a barred basin (Fig. 5.32). As the water in the basin evaporates and its water level drops, a hydrologic head between the sea and the basin is established, forcing marine waters to reflux through the barrier into the basin as well as through platform carbonates, perhaps involving pinnacle reefs such as the Rainbow Group (Fig. 5.32). The constant refluxion of large volumes of normal marine waters through the barrier, platform, and adjacent pinnacles during evaporative drawdown could lead to pervasive dolomitization of these associated limestones. Since seawater is undersaturated with respect to gypsum and halite, nearby evaporites would be dissolved as well. Schmidt and others (1985) reached a similar conclusion for early dolomitization of the Rainbow reefs southeast of the barrier (Fig. 5.31).

Maiklem's model of evaporative drawdown is very similar to the situation found by Logan (1987) in the MacLeod salt basin (as described above), including the dissolution of halite by refluxing marine waters. While Logan has not yet found significant dolomitization, the results of the recent drilling program should test the viability of this model.

Michigan Basin, U.S.A.

A number of large barred basins, such as the Michigan Basin in the northern U.S., and the Permian Zechstein Basin of western Europe, exhibit successions of thick, higher-order evaporites such as halite and, in some cases, thick sequences of potash salts and bitterns, such as in the west Texas, Permian Castile Formation. Carbonate sequences such as the pinnacle reefs of the Michigan Basin and the Zama reefs of Alberta are often encased in these evaporites (Sears and Lucia, 1980; Schmidt and others, 1985).

In the Michigan Basin, the Silurian Niagaran pinnacle reefs are partially dolomitized by reflux of brines from anhydrites forming in shallow-water-to-supratidal conditions near the crests of the reefs during evaporative drawdown. However, brines associated with the thick halites encasing the reefs tend to precipitate porosity-destructive halite in peripheral reef pore systems. These halites represent one of the major problems to be faced in the economic development of these reef reservoirs (Sears and Lucia, 1980).

Criteria for the recognition of ancient reflux dolomites

Major criteria for the recognition of ancient reflux dolomites are outlined in Table 5.3. These criteria are arranged in a general decreasing order of reliability. The most critical criterion that must be satisfied, if a dolomite is to be interpreted as originating from the reflux of evaporated marine waters, is the close spatial relationship of the dolomite to evaporites. These evaporites must have the sedimentary structures, fabrics, mineralogical facies, and geological setting appropriate to a subaqueous origin for precipitation (see, for example, Kendall, 1984). Even though reflux undoubtedly occurs in the marginal marine sabkha, the hydrologic setting and the volume of waters available for dolomitization of an entire platform is inadequate in a sabkha environment.

The stratigraphic-tectonic and sea level setting of the associated evaporites must

Table 5.3

Criteria for the recognition of ancient evaporative reflux dolomites, listed in decreasing order of reliability.

1. Close spatial relationship of dolomite to evaporite sequences that can reasonably be deduced as being subaqueous in origin.

2. The stratigraphic, tectonic, and sealevel setting of the associated evaporites must be appropriate for the development of a barred basin or lagoon.

3. The dolomitized reflux receptor units must have had high porosity-permeability, and suitable hydrologic continuity at the time of dolomitization.

4. Reflux dolomitized units should exhibit dolomitization gradients, with increasing volumes of dolomite toward the source of refluxing brines.

5. Reflux dolomite crystal size are generally coarser than those found associated with sabkha sequences, because coarse receptor beds dictate fewer nucleation sites.

6. While geochemical information relating reflux dolomite to evaporative pore fluids might be lost by recrystallization, gross vertical geochemical gradients, such as increasing Sr, Na trace elements, and stable isotopes enriched in ^{18}O toward the evaporataive brine source might be preserved.

be appropriate for the development of a restricted or barred lagoon or basin. For example, the basin or lagoon must have a suitable sedimentologic or tectonic barrier such as a reef, grainstone complex, or horst block to restrict the interchange of marine waters between the sea and basin or lagoon. It seems that a rising sea level is the most favorable setting for the development of a sedimentologic barrier that will allow the accumulation of significant evaporites in the basin or lagoon.

The dolomitized reflux receptor beds must have exhibited high porosity and permeability, and suitable hydrologic continuity at the time of dolomitization. Platform grainstone blankets, coarse high-energy shoreline sequences, and reefs are all favorable conduits for the regional dispersal of reflux brines.

Sequences believed to have been dolomitized by evaporative brine reflux should exhibit dolomitization gradients, with increasing volumes of dolomite toward the source of refluxing brines. See, for example, the dolomitization gradients established in the platform dolomites of the Gulf Coast Upper Jurassic Smackover Formation (Fig. 5.28). These gradients can be modified by porosity and permeability variations in the receptor units.

Dolomite crystal sizes may be considerably coarser than those found in modern sabkha environments because coarse receptor beds dictate fewer nucleation sites than are found in the mud-dominated sabkha. However, the recrystallization potential of these surface-formed dolomites may negate any original crystal-size differences related to the original dolomitization environment.

While geochemical information specifically relating the reflux dolomite to an evaporative pore fluid might be lost by recrystallization, gross vertical geochemical gradients, such as increasing Sr and Na and stable isotopes with increasingly higher ^{18}O values toward the evaporative brine source, might be preserved. In the case history of the Jurassic detailed above, all original geochemical gradients were apparently wiped out and overprinted by a major meteoric water system. Climate considerations relative to the availability of meteoric water might well determine the preservability of relevant original geochemical gradients.

SUMMARY

Dolomitization and evaporite precipitation dominate the evaporative marine environment. Evaporites seal reservoirs and occasionally destroy porosity by cementation. The precipitation of gypsum is directly linked to dolomitization by the dramatic increase in the Mg/Ca ratio in the associated fluids. Effective porosity development in evaporite-related dolomites seems to depend on post-dolomitization fresh water flushing as a consequence of exposure and formation of intercycle diagenetic terrains.

Marginal marine, evaporative sequences occur in two major settings: the sabkha, where evaporites and dolomites form under dominantly subaerial conditions; and the barred evaporite lagoon or basin, where evaporites are formed subaqueously. Sabkha dolomitization is limited by a restricted hydrologic system, although adjacent porous limestones often seem to be dolomitized by refluxed brines. The barred lagoon-basin seems to be a more effective dolomitization instrument for large-scale platform dolomitization because of the large quantities of brines with high Mg/Ca ratios that are produced and the favorable gravity-driven, hydrologic head provided by the heavy brines that results in the reflux dolomitization of porous adjacent limestones.

Geologic setting, timing of dolomitization, relationship of dolomite and evaporite sequences, as well as traces of leached evaporites, such as collapse breccias, are fundamental to the recognition of evaporative marine diagenesis. Although the chemical and isotopic signatures of modern evaporative marine dolomites are distinctive, great care must be exercised in using dolomite geochemistry because of the possibility of metastable dolomite recrystallization.

Some of the most important Paleozoic fields and production trends around the world are developed in dolomite reservoirs related to diagenesis in evaporative marine environments.

Chapter 6

INTRODUCTION TO DIAGENESIS IN THE METEORIC ENVIRONMENT

INTRODUCTION

Most shallow marine carbonate sequences inevitably bear the mark of meteoric diagenesis (Land, 1986). This common diagenetic environment is also one of the most important diagenetic settings in the development and evolution of carbonate porosity because of the general aggressiveness of meteoric pore fluids toward sedimentary carbonate minerals. This chemical aggressiveness ensures the relatively rapid dissolution of carbonate grains and matrix, and the creation of secondary porosity. The carbonate derived from this dissolution can ultimately be precipitated nearby, or elsewhere in the system, as porosity destructive carbonate cement, occluding primary as well as secondary pores. Therefore, meteoric diagenesis can lead to a dramatic and early restructuring of original depositional porosity if it occurs shortly after deposition and is applied to unstable mineral suites. Conversely, it can rejuvenate and enhance the porosity of older, mature carbonate sequences if it occurs during exposure associated with unconformities.

Some of the fundamental geochemical, mineralogical, and hydrological aspects of the meteoric diagenetic environment relative to the development and evolution of porosity will be reviewed in this chapter.

GEOCHEMICAL AND MINERALOGICAL CONSIDERATIONS

Geochemistry of meteoric pore fluids and precipitates

In Table 3.2, the chemical composition of meteoric waters, is compared to that of marine and subsurface pore fluids. Meteoric waters are dilute and contain small amounts of ions commonly involved in the carbonate system, such as Ca^{+2}, Mg^{+2}, Sr^{+2}, $Mn^{+2 \text{ and } +3}$ and $Fe^{+2 \text{ and } +3}$.

Meteoric waters, while exhibiting a wide range of carbonate saturation states, are often strongly aggressive (undersaturated) with respect to most carbonate mineral species. This aggressiveness is attributed to the accessibility of meteoric waters to large reservoirs of CO_2 present in the atmosphere, and particularly in the soils of the vadose zone. The Pco_2 of soils can often reach concentrations of 10^{-2} atm, two orders of magnitude higher than atmospheric Pco_2 ($10^{-3.5}$ atm) (Matthews, 1974). Because meteoric waters must pass through the atmosphere on their journey to the surface of the earth, and often will pass through the vadose zone and its attendant soils as they recharge the phreatic zone, these waters have ample opportunity to dissolve significant quantities of CO_2, become more strongly acidic and undersaturated with respect to $CaCO_3$. Any carbonate dissolved by these aggressive fluids becomes available for later precipitation as carbonate cements close by, or elsewhere, in the system as the waters evolve chemically during rock (or sediment)-water interaction.

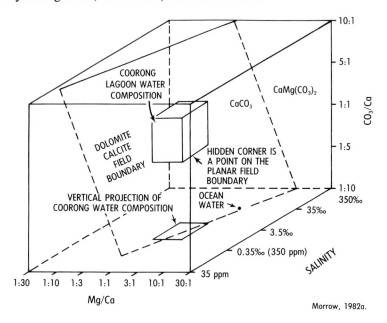

Fig. 6.1. Block diagram showing the effect of variation in the three parameters, the Mg/Ca solution ratio, the salinity, and the CO_3 / Ca ratio. The plane represents the kinetic boundary between dolomite and calcite or aragonite and it includes the hidden corner of the Coorong Lagoon waters as a point on the plane. The basal plane is after Folk and Land (1975). Note that the vertical projection of Coorong Lagoon waters falls largely on the calcite-aragonite side of the stability boundary on the basal plane. A decrease in salinity, an increase in the Mg/Ca ratio or an increase in the CO_3 / Ca ratio favours the precipitation of dolomite. Reprinted with permission of the Geological Association of Canada.

The low Mg/Ca ratios and salinities of most meteoric waters favor the precipitation of calcite (Fig. 6.1). The obvious exception is the common occurrence of aragonite and Mg-calcite speleothems forming in caves. Gonzalez and Lohmann (1988) suggest that aragonite speleothems form as a result of high Mg/Ca ratios present in some cave waters. Cave fluids seep into large vadose cavities with lower Pco_2 and quickly degas; calcite precipitates; and assumably concomitant evaporation drives Mg/Ca ratios to levels that support aragonite precipitation (Fig. 6.2A). Analyses of cave waters associated with calcite and aragonite speleothems suggest that aragonite begins to precipitate at elevated CO_3^{-2} levels with Mg/Ca ratios higher than 1.5 (Fig. 6.2B).

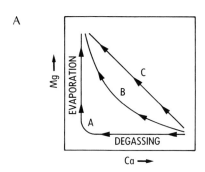

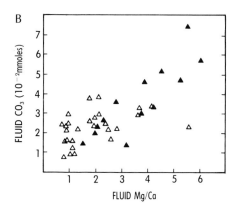

Gonzalez and Lohmann, 1988

Fig. 6.2. (A). Possible changes in Ca and Mg concentration during degassing, evaporation, and carbonate precipitation. Precipitation induced by fluid degassing results in decreasing Ca concentrations; evaporation after degassing results in an increase in Mg concentrations (line A). If degassing and evaporation occur simultaneously and at a constant rate during carbonate precipitation, trend C will result. If degassing rate decreases at a constant rate of evaporation, all trends will lie under line C (e.g., line B).
(B) Fluid Mg/Ca ratios and CO_3 concentrations for waters associated with monomineralic cave samples. Aragonites are solid diamonds; calcites are open diamonds. Aragonite precipitation seems to occur only at Mg/Ca ratios higher than 1.5, indicating the importance of elevated Mg concentration or Mg/Ca ratio for aragonite precipitation to occur. Reprinted with permission from Paleokarst. Copyright (C), 1988, Springer-Verlag, New York.

Dolomites also form in continental meteoric waters associated with cave environments (Gonzalez and Lohmann, 1988) as well as in evaporative lakes in the Coorong of southeastern Australia (von der Borch and others, 1975). While the majority of the Coorong waters fall within the calcite field of the Folk and Land's (1975) calcite/dolomite stability diagram (Fig. 6.1), Morrow (1982a) suggests that the Coorong dolomites (and perhaps the spelean dolomites of Gonzalez and Lohmann) are the result of elevated CO_3/Ca ratios combined with relatively low salinities and intermediate Mg/Ca ratios (Fig. 6.1).

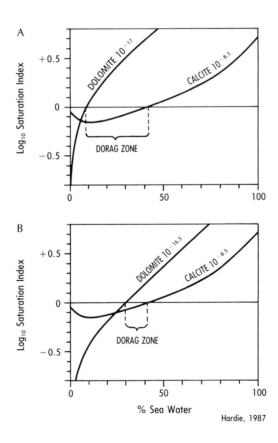

Fig. 6.3. Theoretical saturation relations of dolomite and calcite in mixtures of seawater and meteoric water at 25°C. (A) The "Dorag Zone", where mixtures are supersaturated with dolomite but undersaturated with calcite, for dolomite with $K=10^{-17}$. (B) The "Dorag Zone" for disordered dolomite with $K=10^{-16.5}$. Used with permission of SEPM.

The calcite solubility curve does not covary in a linear fashion with salinity (Runnells, 1969; Matthews, 1974; Plummer, 1975; Back and others, 1979; Bathurst, 1986). When the theoretical dolomite and calcite solubility curves are plotted on a solubility/salinity diagram, a zone in which the waters will be saturated with respect to dolomite, and undersaturated with respect to calcite can be seen (Fig. 6.3A). This relationship implies that mixed waters can be responsible for extensive carbonate dissolution and the development of significant secondary porosity (see Chapters 7 and 8), and is the foundation for the popular Dorag, or meteoric-marine water mixing zone model of dolomitization (see Chapter 8).

Isotopic composition of meteoric waters and carbonates precipitated from meteoric waters

The isotopic compositions of meteoric waters and the calcite cements and dolomites precipitated from these waters are highly variable. Fig. 3.10 illustrates the range of isotopic compositions that might be expected in natural geologic situations. The oxygen isotopic composition of meteoric waters is strongly latitude-dependent and can vary in $\partial^{18}O$ composition from -2 to -20 permil (SMOW) (compare the oxygen isotopic compositions of Enewetak and Barbados meteoric cements as seen on Fig. 3.10) (Hudson, 1977; Anderson and Arthur, 1983). At any one geographic site, however, the latitudinal effect is zero and the initial $\partial^{18}O$ composition of meteoric water should be relatively constant. But if these waters interact with limestones and sediments, the water's ultimate oxygen isotopic composition will be determined by the ratio O_L (number of moles oxygen from limestone) to O_W (number of moles oxygen from water)(Allen and Matthews, 1982). The solubility of calcium carbonate in surface meteoric waters is very low (1 liter of water at 25° C, $P_{CO_2} = 10^{-3.5}$, contains just 1×10^{-3} moles of total CO_2 in 55.5 moles of H_2O). Because of this low solubility, the O_L/O_W will be much less than 1, and the oxygen isotopic composition of the water should remain approximately the same, regardless of the amount of rock/water interaction or the distance down hydrologic gradient from recharge areas (Allen and Matthews, 1982). Subsequent work by Lohmann (1988) suggests that during the initial stages of mineral stabilization by meteoric water, the O_L/O_W may be great enough to buffer the oxygen isotopic composition toward positive values.

The range of $\partial^{13}C$ values available within meteoric diagenetic environments is also enormous, ranging from -20 permil (PDB), characteristic of soil gas CO_2 to +20 permil (PDB), associated with organic fermentation processes (Fig. 3.10). But in most meteoric water environments, soil gas and the CO_3 derived from the dissolution of marine limestones are the prime sources for the ultimate $\partial^{13}C$ composition of meteoric diagenetic waters (Allen and Matthews, 1983; Lohmann, 1988). The carbon isotopic composition of meteoric diagenetic fluids (and diagenetic carbonates precipitated from these fluids) is determined by the ratio of C_L (number of moles of carbon from limestone) to C_{SG} (number of moles of carbon from soil gas). At the surface in vadose recharge zones, the bicarbonate in meteoric waters comes equally from soil gas and limestone carbonate and the ratio is generally near 1. In this case, the meteoric waters will have low ^{13}C values

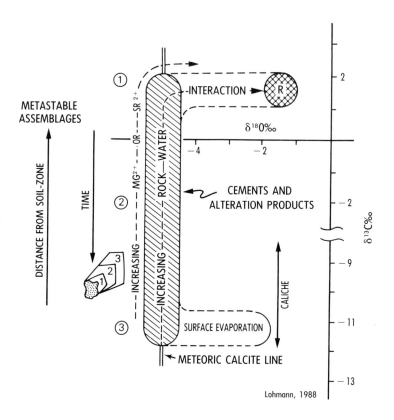

Fig. 6.4. Idealized plot of variation in $\partial^{18}O$ and $\partial^{13}C$ characteristic of meteoric vadose and phreatic carbonates. The constancy of $\partial^{18}O$ and the variable $\partial^{13}C$ define a trend termed the <u>meteoric calcite line</u> (2). Deviations from this line will generally take place where: there is increased rock-water interaction with polymineralic suites distal from meteoric recharge and water may be buffered by rock-derived carbon, oxygen, and trace elements, such as Mg and Sr (1); and at exposure surfaces where surface evaporation may drive $\partial^{18}O$ compositions toward more positive values. Reprinted with permission from Paleokarst. Copyright (C), 1988, Springer-Verlag, New York.

because of the strongly negative carbon isotopic composition of the soil gas (Fig. 3.10). As meteoric waters percolate through the vadose into the phreatic zone dissolving metastable carbonate sediments and moving away from the soil gas reservoir during their passage, the contribution of heavy carbon from the sediments (or limestones) is increased, the C_L/C_{SG} ratio becomes larger, and therefore the waters have progressively higher ^{13}C values (Allen and Matthews, 1983).

Fig. 6.4 is an idealized cross plot of the variation in $\partial^{18}O$ and $\partial^{13}C$ that might be expected in the meteoric diagenetic environment based on the concepts outlined above. The main variational pattern is one of constant $\partial^{18}O$ and rapidly changing $\partial^{13}C$. Lohmann (1988) calls this characteristic trend *the meteoric calcite line*. During increasing rock-water interaction involving metastable carbonate mineral species at sites removed from the recharge area, the Mg and Sr compositions of diagenetic calcites are expected to increase, reflecting the sequential stabilization of Mg-calcite and aragonite (Benson and Matthews, 1971; Lohmann, 1988; and Chapter 3, this book). Additionally, the isotopic composition of a diagenetic calcite crystal at a site removed from the recharge area should mirror the isotopic evolution of the water toward more depleted carbon isotopic compositions through time. This progressive depletion results from diminished rock-water interaction as the terrain matures and metastable components disappear (Lohmann, 1988). Local increases in ^{18}O composition of meteoric water (or diagenetic calcites) can take place by evaporation at exposure surfaces or within caliche soil profiles as the result of the preferential removal of the lighter ^{16}O from the water during evaporation (Fig. 6.4).

In zones of mixed meteoric-marine waters, the carbon and oxygen isotopic composition of diagenetic calcites is expected to exhibit positive covariance (Fig. 3.11) (Allen and Matthews, 1982). Lohmann (1988), on the other hand, suggests that the isotopic covariation of diagenetic cements and mineral replacements formed by mixed meteoric-marine waters will generally follow the meteoric calcite line. This is because the mixing of marine and meteoric waters results in undersaturation and dissolution, rather than precipitation, over most of the range of mixing. Therefore the calcites or dolomites precipitated in the mixing zone will represent only a small segment of the range of mixing, and the resulting isotopic composition will "...define vertical trends of variation parallel to the meteoric calcite line rather than a single hyperbolic mixed-water curve" (Lohmann, 1988, p.74).

Mineralogic drive of diagenesis within the meteoric environment

Modern carbonate sediments consist of metastable mineral suites dominated by Mg-calcite and aragonite. The solubility of aragonite is twice that of calcite, and the solubility of Mg-calcite is approximately ten times that of calcite. These differences provide a strong drive in the destination and intensity of meteoric diagenesis, because a water in contact with two solid phases of differing solubility simply cannot be in equilibrium with the rock (Matthews, 1974; James and Choquette, 1984).

If the water is near saturation with respect to aragonite, it will be supersaturated with respect to calcite, and should precipitate calcite. If the water is near saturation with respect to calcite, it should dissolve aragonite. Ultimately equilibrium in this system can only be achieved by completely destroying the most soluble phase (Matthews, 1974). This represents a drive toward stabilization by dissolution and precipitation. These processes obviously play a major role in porosity modification.

Meteoric waters interacting with a mature carbonate terrain consisting of a single stable mineral species such as calcite will rapidly move toward equilibrium with the rock, and the diagenetic processes affecting porosity modification (i.e., dissolution and precipitation) will soon cease.

Implications of the kinetics of the $CaCO_3$-H_2O-CO_2 system to grain stabilization and to porosity evolution in meteoric diagenetic environments

If a fluid is said to be oversaturated or undersaturated with respect to a particular mineral phase, it simply means that the system is not at equilibrium, and that precipitation or solution *should* occur. No particular time frame for the reaction is established. Reaction kinetics, on the other hand, indicates how the reaction *should* develop within a finite length of time (Matthews, 1974).

Chemical reactions such as the solution of aragonite or the precipitation of calcite can be broken down into a series of steps that must occur for the reaction to achieve equilibrium. For any step in the reaction, the greater the departure from equilibrium, the faster the reaction step (or the entire reaction) will be able to proceed. The overall reaction cannot proceed faster than the slowest step within the reaction chain. This slowest step is termed the *rate step* (Matthews, 1974).

Fig. 6.5 outlines the reaction steps for carbonate solution and precipitation reactions during the stabilization of a metastable mineral suite to calcite. The most significant reaction relative to porosity-permeability development is the dissolution of metastable solid phases (aragonite and Mg-calcite) and the precipitation of a stable solid phase (calcite). One of the most important rate steps in this reaction is the *spontaneous nucleation* (nucleation without benefit of previous seed crystals, or nuclei) of stable calcite relative to the dissolution of metastable aragonite. The rate of calcite growth by spontaneous nucleation from a solution in equilibrium with aragonite is at least 100 times slower than the rate at which aragonite will dissolve to replace the $CaCO_3$ lost from the solution to calcite precipitation (Matthews, 1968; Wollast, 1971; Matthews, 1974). This spontaneous nucleation step can only be circumvented by precipitation of calcite onto preexisting calcite sedimentary particles at nucleation sites.

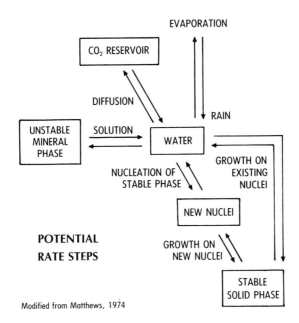

Fig. 6.5. *Potential rate steps in the stabilization of unstable carbonate mineral phases to stable carbonate mineral phases by dissolution and precipitation. Used with permission of SEPM.*

The large kinetic differences between aragonite dissolution and calcite precipitation suggest that $CaCO_3$ dissolved from aragonite is generally transported away from the site of solution. If water flux in the diagenetic system is large, and the water is strongly undersaturated with respect to aragonite, aragonite grains will undergo total dissolution, moldic voids will be formed, and all internal structure within the aragonite grains will be destroyed. The $CaCO_3$ from this grain dissolution may be carried a significant distance down hydrologic gradient before calcite is precipitated as a void-fill, porosity-occluding cement. This style of aragonite-to-calcite stabilization, termed *macroscale dissolution,* by James and Choquette (1984), leads to massive rearrangement of pore space and the development of cementation gradients in major meteoric phreatic water systems (Figs. 6.6 and 6.7C and D, and discussion in Chapter 7).

If the water flux is slow and meteoric pore fluids only slightly undersaturated with respect to aragonite, stabilization of aragonite grains may proceed by *microscale dissolution,* or *replacement* (James and Choquette, 1984; Land, 1986). In this case, stabilization of aragonite is accomplished by dissolution and precipitation across a thin reaction film within the grain (Kinsman, 1969; Sandberg and others, 1973; Pingitore, 1976). The end product is a calcite mosaic with aragonite inclusions outlining elements of the original aragonite grain ultrastructure. $CaCO_3$ seems to be conserved within the grain, and microscale dissolution does not lead to appreciable secondary porosity development or to significant calcite cementation outside the grain (Figs 6.6 and 6.7A).

The calcite precipitation rate step may be avoided if the metastable phase is Mg-calcite, because appropriate nucleation sites for calcite precipitation are available within the Mg-calcite particle. During the stabilization of Mg-calcite to calcite in a meteoric water system, metastable $MgCO_3$ components dissolve, and calcite is precipitated on readily available $CaCO_3$ nucleation sites within the grain (Matthews, 1974; James and Choquette, 1984; Land, 1986). In this case, the ultrastructure of the original Mg-calcite

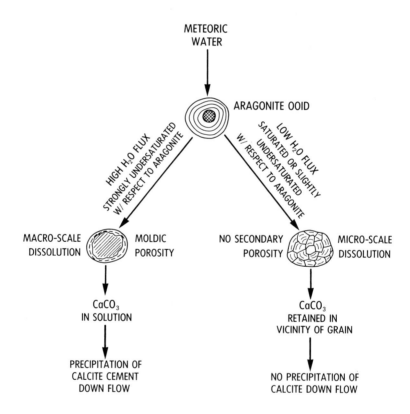

Fig. 6.6. Schematic diagram showing the two basic pathways, macroscale dissolution versus microscale dissolution, for the stabilization of aragonite grains to calcite. Major conditions controlling the path that the metastable grain will take, and the resultant porosity implications are indicated.

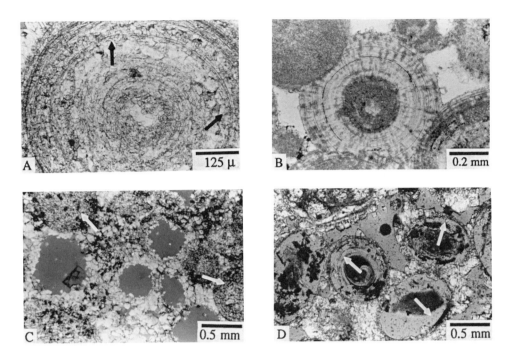

Fig. 6.7. (A) Inferred aragonite Jurassic Smackover ooid that shows evidence of stabilization by microscale dissolution. Note the ghosts of original ooid lamellar structure preserved in the fine calcite crystal mosaic.(arrows) (See Moore and others, 1986). Murphy Giffco #1, Miller Co. Texas, 7784' (2373 m). Plain light. (B) Calcite Jurassic Smackover ooid that shows no mosaic calcite and perfect preservation of internal structure (see Figs. 4.6A and B) suggesting an original calcite mineralogy (See Moore and others, 1986). Arco Bodcaw #1, Columbia Co. Arkansas, 10,875' (3316 m). Plain light. (C) Macroscale dissolution of inferred aragonite Jurassic Smackover ooids forming moldic porosity. Note that some ooids have stabilized by microscale dissolution (arrows). Same sample as (A). Plain light. (D) Macroscale dissolution of aragonite ooids (arrows) leading to oomoldic porosity, Pleistocene, Shark Bay, western Australia. Plain light.

grain is preserved (Fig. 6.7B), no moldic void is developed, and no $CaCO_3$ is released to the water to be precipitated as cement elsewhere in the system. Stabilization of Mg-calcite sediments to calcite, then, generally results in only minimal modification of the original sedimentary pore system. In certain cases, however, Mg-calcite grains may undergo macroscale dissolution, produce moldic secondary porosity, and release $CaCO_3$ into the water for potential calcite cements elsewhere. Moldic porosity development in Mg-calcites may develop under conditions of very high water flux and strongly undersaturated waters (Schroeder, 1969; James and Choquette, 1984; Saller, 1984a; Budd, 1984).

Hydrologic setting of the meteoric diagenetic environment

The meteoric diagenetic environment consists of three major hydrologic regimes, or zones: the vadose, phreatic, and mixed zones (Fig. 6.8). The water table marks the boundary between the vadose and phreatic zones. Below the water table in the phreatic zone, pore spaces are saturated with water. The phreatic aquifer is often unconfined and open to the atmosphere with a well-defined vadose zone between the water table and the surface. In the vadose zone both gas and water are present in the pores, and water is generally concentrated by capillarity at the grain contacts. In the zone of infiltration near the upper surface of the vadose zone, water is under the influence of both evaporation and plant transpiration, and is actively involved in soil and caliche formation (James and Choquette, 1984). Water moves through the vadose zone in two contrasting styles: *vadose seepage*, where water slowly trickles through a network of small pores and fractures; and *vadose flow*, where water rapidly moves through the vadose zone via solution channels, sinkholes, joints, and large fractures flowing directly to the water table (Harris and Matthews, 1968; Matthews, 1974).

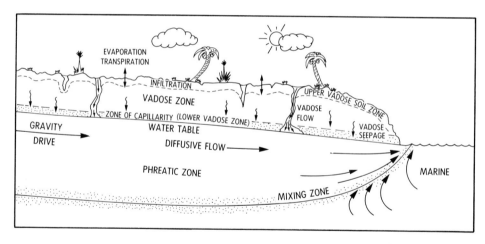

Fig. 6.8. Conceptual model of the major diagenetic environments, and hydrologic conditions present in the meteoric realm.

Water moves through the phreatic zone either by: *diffusive flow*, where there is a well-defined water table and flow conditions can be approximated by Darcy's Law; or *free flow*, where water moves through channels and integrated conduits as a free flowing subterranean river, and water tables are difficult to recognize. Diffusive flow is characteristic of young carbonate sediments newly exposed and in the early stages of stabilization. Free flow is often found in mature, stabilized carbonate terrains associated with

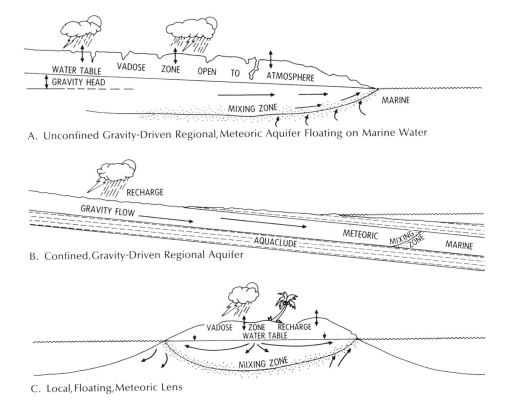

Fig. 6.9. Common hydrologic settings found in the meteoric diagenetic realm. (A) Unconfined regional meteoric aquifer floating on marine water. System is gravity driven and vertical recharge occurs throughout. (B) Confined gravity-driven regional meteoric aquifer. Recharge only occurs in the updip unconfined portions of the aquifer. (C) Local, floating meteoric lens associated with island situations. Recharge is vertical, local and limited.

karst development (James and Choquette, 1984).

Confined aquifers are sandwiched between impermeable layers and are not connected directly to the atmosphere or to a vadose zone, except in the updip recharge areas (Fig. 6.9B). As differential subsidence between recharge and discharge areas increase through time, the difference in head between discharge and recharge areas increases and the system becomes a pressure-driven artesian system. Confined aquifer systems may act as efficient conduits for the distribution of meteoric waters deep into sedimentary basins, as well as beneath the continental shel(James and Choquette, 1984; Land, 1986).

The meteoric phreatic zone either sits on an impermeable layer, or floats on more saline marine, or modified marine waters. Because of significant density differences between marine and meteoric waters, marine waters can support a column of fresh water some 40 ft (12 m) thick for each 1 ft (30 cm) of elevation of the meteoric water table (the Ghyben-Herzberg principle) (Buddemeier and Holladay, 1977). Therefore, significant meteoric lenses can be developed beneath islands with relatively small elevations above sea level (Fig. 6.9C).

The meteoric-marine water mixing zone found at the boundary of floating meteoric water lenses and at meteoric-phreatic discharge areas along sea coasts, is an important, complex, and poorly understood hydrologic domain. Meteoric and marine waters mix either by diffusion or by physical mixing, and the mixing zone can be quite thick in highly permeable sequences beneath oceanic islands such as Bermuda and the atolls of the central Pacific (Vacher, 1974; Buddemeier and Holladay, 1977).

Floating meteoric lenses occur in two basic scales and configurations as shown in Figs. 6.9A and C. Insular lenses are generally small and sometimes ephemeral with elevation on the water table measured in centimeters, and the resulting meteoric lens generally measuring less than 5 m thick (see Buddemeier and Holladay, 1977, for Enewetak Atoll, Pacific; Halley and Harris, 1979, for Joulters Cay, Bahamas; Budd, 1984 for Schooner Cays, Bahamas). In these situations, the actual flux of meteoric water through the lens may well be controlled by tidal dispersion, rather than by gravity flow associated with water table elevation (Budd, 1984). In tropical areas, meteoric lens development may be episodic, waxing and waning in concert with the strong seasonality of tropical weather systems (Moore, 1977; Budd, 1984). Based on our Holocene sample, insular meteoric water lenses generally contain relatively small volumes of water that are slowly moving through the system in response to tidal pumping and episodic recharge.

Large carbonate platforms may support significant floating meteoric lenses, particularly in areas of high rainfall, such as the Yucatan Peninsula, Mexico. This 65,500 sq km, mature karstic platform supports a 70 m meteoric lens floating on marine waters, and discharging at the coast at a rate of 8.6×10^6 m^3 per year, per km of coastline. This discharge is supported by a recharge of some 150 mm per year across a water table that attains no more than 1.5 m elevation above sea level (Hanshaw and Back, 1980).

In the next chapter, there will be a discussion of the diagenetic processes active in each of these hydrologic regimens and diagenetic subenvironments as they affect the modification and evolution of depositional, as well as secondary, porosity systems.

SUMMARY

The meteoric diagenetic environment is one of the most important diagenetic settings relative to the development and evolution of carbonate porosity. Meteoric waters are chemically dilute, though highly aggressive with respect to most carbonate mineral phases because of the ready availability of CO_2. The low Mg/Ca ratios and salinities found in most meteoric waters generally favor the precipitation of calcite. Mixed meteoric and marine waters, while often undersaturated with respect to calcite, may remain saturated with respect to dolomite. This relationship results in extensive calcite dissolution and the possibility of dolomitization.

The stable isotopic composition of meteoric waters and the carbonates precipitated from these waters are highly variable. In general, however, carbonates precipitated from meteoric waters have light, stable isotopic signatures and follow a meteoric trend marked by large variations in $\partial^{13}C$ and locally invarient $\partial^{18}O$.

The solubility differences commonly present in metastable carbonate sediments provide a strong drive for the nature and intensity of meteoric diagenesis, since mineralogical stabilization usually is accomplished by dissolution and precipitation, processes that carry profound consequences relative to porosity evolution.

During mineral stabilization by solution of aragonite and precipitation of calcite, the relatively slow calcite precipitation rate step may result in transport of $CaCO_3$ away from the site of aragonite solution before its precipitation as void fill calcite.

The meteoric diagenetic environment consists of three important hydrologic regimes: the vadose, phreatic, and mixed meteoric-marine zones.

Chapter 7

METEORIC DIAGENETIC ENVIRONMENTS

INTRODUCTION

As seen in the last chapter, there are three major hydrologic settings, or diagenetic environments, associated with meteoric waters: the vadose, phreatic, and mixed meteoric marine zones. In the following pages only porosity development and evolution in the meteoric diagenetic environments will be considered. The mixed meteoric marine environment will be discussed in a subsequent section.

The meteoric vadose and phreatic zones will first be considered in terms of the interaction of their unique diagenetic processes with mineralogically metastable carbonate sediments. Later, at the end of the chapter, the effect of meteoric diagenetic processes on porosity development and evolution in mineralogically stable terrains will be discussed (telogenetic zone of Choquette and Pray, 1970).

Finally, the phreatic zone will be evaluated on two scales: the local, island-floating, fresh water lens; and the regional aquifer system.

THE VADOSE DIAGENETIC ENVIRONMENT AS DEVELOPED IN METASTABLE CARBONATE SEQUENCES

Introduction

The vadose zone is important because most of the meteoric water that ultimately finds its way back to the sea must pass through the vadose environment before it is entrained into a regional meteoric aquifer system or a floating fresh water lens. The vadose environment may be divided into two zones: an *upper vadose soil*, or *caliche, zone* at the air-sediment or rock interface; and a *lower vadose* zone, which includes the capillary fringe just above the water table (Fig. 6.8).

Upper vadose soil or caliche zone

The air-sediment interface is generally an environment of intense diagenesis in metastable carbonate terrains because infiltrating meteoric waters are generally under- saturated with respect to $CaCO_3$ as a result of the high P_{CO_2} usually associated with soils. In regions of high rainfall and porous sediment, water tends to move through the zone quickly, and dissolution is probably the dominant process. In this situation, the air-sediment interface may show little evidence of subaerial exposure (Saller, 1984a). The $CaCO_3$ removed by dissolution may be transported downward into the zone of capillarity, or the lower vadose zone, where calcite cements are precipitated.

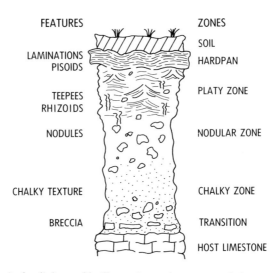

Fig. 7.1. Sketch of a typical caliche profile illustrating major zones and characteristic features.

In arid to semi-arid climates, however, the potential for carbonate dissolution, evaporation, and subsequent calcite precipitation under near-surface conditions results in the formation of distinctive surface soil crusts variously called *caliche, calcrete,* or *duricrust* (James and Choquette, 1984). Caliche soils have been the subject of intensive research over the past 10 years because of their potential usefulness in environmental reconstruction (Esteban and Klappa, 1983). Fig. 7.1 is a sketch of a typical caliche profile showing its major characteristics.

The central features of the crust include, a laminated hardpan usually developed on the top of the sequence, followed successively by a zone of plates and crusts; a nodular, or pisolitic chalky sequence; and finally a chalky transition into untouched sediments or country rock (Esteban and Klappa, 1983; James and Choquette, 1984). The caliche profile is the result of intensive dissolution of original sediments, or limestones, and the rapid reprecipitation of calcite, often driven by organic activity. Original textures and fabrics are generally destroyed and are replaced by a melange of distinctive fabrics, textures, and structures, such as: nodules, pisoids (Fig. 7.2A), rhizoids (concretions around roots), teepees (pseudo-anticlines) (Fig. 7.2B), crystal silt (microspar), and microcodium (calcified cells of soil fungi and higher plants) (Esteban and Klappa, 1983; James and Choquette, 1984).

Fig. 7.2. *(A) Pisoids developed in caliche soil profiles, Permian Yates Formation on New Mexico Hwy. 7A to Carlsbad Caverns at Hairpin Curve. (B) Teepee antiform structure in Tansill Formation at the south end of the south parking lot at the entrance to Carlsbad Caverns.*

Some modification of porosity occurs in the upper vadose zone, with potential porosity occlusion associated with extensive hardpan development and some porosity enhancement related to crystal silt formation. Successive caliche profiles in a cyclical carbonate sequence could lead to vertical compartmentalization similar to that seen in marine hardgrounds.

Lower vadose zone

The lower vadose zone is a domain of gravity percolation where water is in transit on its way to the water table. Water saturation in vadose pores increases significantly just above the water table in a region called the *capillary fringe* (Fig. 6.8). The lower vadose zone is downflow from the high organic activity and high P_{CO_2} concentrations of the upper vadose soil zone, and hence receives waters that are often saturated with respect to $CaCO_3$, as a result of carbonate dissolution in the soil zone above. This situation may result in the concentration of calcite cements immediately above, and adjacent to, the water table (Longman, 1980; Saller, 1984a).

Petrography of vadose cements

Vadose cements, away from the vadose soil zone, show an irregular pattern of distribution and are generally concentrated at grain contacts, as a result of water being held by capillarity between the grains (Chapter 3, Fig. 7.3). While most vadose cements are equant in shape and exhibit relatively small crystal sizes, they also often occur as meniscus and microstalactitic fabrics with resultant pores showing a distinctive rounded morphology (Fig. 7.3) (Halley and Harris, 1979; Longman, 1980; James and Choquette, 1984).

 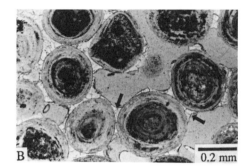

Fig. 7.3. (A) Meniscus cementation pattern developed in ooid grainstones taken from the vadose zone at Joulters Cay, Bahamas. Note the rounded shape of the pores. Crossed polars. (B) Close view of (A) showing the rounded pores as a result of meniscus cementation style (arrows). Plain light. Samples furnished by Mitch Harris.

Trace element composition of vadose cements

The trace element composition of meteoric vadose calcite cements generally reflects the dilute nature of the meteoric water from which it is derived. Vadose cements, therefore, tend to exhibit low levels of Mg, Sr, and Na (see Chapters 3 and 6, this book). If however, the rock-to-water ratio is high, as in the setting where local solution-precipitation is active, elevated levels of these trace elements may be incorporated in vadose cements (Veizer, 1983; Lohmann, 1988). Fe and Mn are generally not found in vadose cements because the vadose zone is usually an oxidizing environment and the oxidized state of these elements cannot be incorporated in the calcite lattice. As a result, vadose cements generally show no cathodoluminescence (Meyers, 1974; Niemann and Read, 1988).

Isotopic composition of vadose cements

In Chapter 6, the general stable isotopic trends found in meteoric diagenetic environments were discussed. Specifically, vadose cements tend to follow Lohmann's (1988) calcite line. They display a wide range of $\partial^{13}C$ values from positive to highly negative values and little variation in $\partial^{18}O$ (Figs. 6.4, 7.5). The $\partial^{13}C$ values of cements in the vadose zone, will tend to be more negative than those in the phreatic zone because of more direct access of precipitational waters to light soil gas CO_2 (Allen and Matthews, 1982). In the upper vadose zone, $\partial^{18}O$ values may show a distinct trend toward higher values because of the potential of evaporation near the air-sediment interface, as was discussed in Chapter 6 (Figs. 6.4, 7.5) (Lohmann, 1988).

The strontium isotopic composition of meteoric vadose cements will be dominated by the strontium isotopic composition of the sediments undergoing dissolution because there is little chance for direct input of significant volumes of radiogenic strontium from siliciclastic units in an upflow direction. Hence, Jurassic vadose cements should have the strontium isotopic signature of Jurassic seawater, because Jurassic sediments will necessarily have a Jurassic seawater strontium isotopic composition.

Porosity development in the vadose diagenetic environment

There is a consensus among carbonate geologists that dissolution in the vadose zone is balanced by precipitation. This diagenetic equilibrium results in gross porosity values remaining the same during stabilization of an unstable mineral suite under meteoric vadose conditions (Pittman, 1975; Harrison, 1975; Halley and Harris, 1979; Saller, 1984a; Budd, 1984). The small volumes of diagenetic fluids responsible for solution and precipitation in the vadose zone generally preclude the development of extensive moldic porosity in aragonitic sediments in the vadose environment. Permeability in the vadose zone, however, seems to be significantly modified because most cements are concentrated at pore throats, rather than being distributed evenly around the individual grains (Fig. 7.5) (Halley and Harris, 1979).

THE METEORIC PHREATIC DIAGENETIC ENVIRONMENT AS DEVELOPED IN METASTABLE CARBONATE SEQUENCES

Introduction

Most carbonate geologists agree that diagenetic processes, including those responsible for porosity modification, are more intense and efficient in the meteoric phreatic environment than in the vadose zone above. This higher diagenetic efficiency is a result of the much larger volume of water present in the phreatic zone (Steinen and Matthews, 1973; Matthews, 1974; Land, 1986). In the following discussions of the meteoric phreatic environment, the major differences between vadose and phreatic environments are emphasized, where appropriate.

The two major phreatic settings that will be considered are, the local floating fresh water lens, and the regional meteoric aquifer system. The discussion on the general aspects of the petrography, trace element geochemistry, isotope geochemistry, and porosity development of meteoric phreatic systems will be equally applicable to both systems.

Local floating meteoric water lens

A meteoric lens can develop below islands without the complex geologic and hydrologic setting necessary to support a regional gravity-driven meteoric water flow system (Fig. 6.9). Insular, floating lenses can range in size from less than 5 m to over 15 m (Saller, 1984a; Budd, 1984). The phreatic lens is tied directly to the vadose zone above, and all the water present in the lens must have percolated through the immediately supra-adjacent vadose zone. In this setting, the meteoric water present in these floating lenses is moved through the phreatic system by a combination of diffusion, physical mixing, and advective flow into the marine waters below, and laterally adjacent to, the lens (Budd, 1984). In small tropical islands where rainfall is strongly seasonal, the meteoric lens tends to be cyclical. During the rainy season the lens is fully developed, and water in excess of that which can be supported by the specific gravity contrast between fresh and salt water flows out of the margins of the lens. During the dry season, mixing of marine water with the meteoric water of the lens by diffusion, and particularly by tidal pumping, ultimately destroys most of the lens (Moore, 1977; Budd, 1984). Larger islands such as Bermuda, however, can support a permanent meteoric lens (Plummer and others, 1976).

Because of high CO_2 flux between the vadose and phreatic zones, most diagenetic processes in this setting are concentrated along the water table. Input from high CO_2 soil gases moves water toward undersaturation and dissolution of metastable carbonate components in the vadose zone. CO_2 degassing along the water table, aided by tidal pumping, may lead to the precipitation of calcite cements (Hanor, 1978). The zone of cementation and dissolution associated with the water table is generally no more than 1.5 m thick (Halley and Harris, 1979; Budd, 1984; Saller, 1984a). In his study of the Schooner Cays in the Bahamas, Budd (1984) observed from 10-40 wt. % cement precipitated in intergranular pores concentrated in a 1m zone along the water table. The volume of dissolution of metastable grains in the zone closely matched the volume of cement observed (>87% efficiency), so that total porosity volume did not change. While grain dissolution in the floating, meteoric phreatic lens at Schooner Cays and other sites, such as Joulters Cay in the Bahamas (Halley and Harris, 1979), generally consists of the delicate removal of outer ooid layers, Budd (1984) observed the occasional development of moldic porosity at Schooner Cays, and suggested that, given time, significant moldic porosity might evolve.

The intensity of cementation and grain dissolution sharply declines between the floating meteoric lens, and the mixing zone with marine water below. Within the mixing zone of modern, insular, floating meteoric lenses, neither cementation nor significant dissolution are observed (Halley and Harris, 1979; Budd, 1984; Saller, 1984a). While the meteoric marine mixing zone is supposed to be a site of dolomitization (see Chapter 6), no mixing zone dolomites have been found associated with modern, floating meteoric lenses (see discussion, Chapter 8).

The stabilization of metastable minerals through dissolution and precipitation in the local meteoric phreatic environment (at the water table) is relatively rapid. Halley and Harris (1979) estimated that the aragonitic ooid grainstones at Joulters Cay would be converted totally to calcite in 10,000 to 20,000 yrs, a figure similar to that calculated by Budd (1984) for the Schooner Cay area. Many authors, however, have noted that stabilization of metastable sequences is considerably slower under vadose conditions than it is in the phreatic environment (Steinen and Matthews, 1973; Saller, 1984a; Budd, 1984). Budd (1984) calculated that at Schooner Cay, the rate of mineralogical stabilization in the phreatic zone was faster than that observed in the adjacent vadose zone.

Petrography of meteoric phreatic cements

Meteoric phreatic pore-fill cements generally occur in a more homogeneous distribution pattern than their vadose counterparts. Isopachous rims consisting of equant-rhombohedral calcite crystals are the normal fabric (Figs. 3.4, 3.7, and 7.4). Crystal size

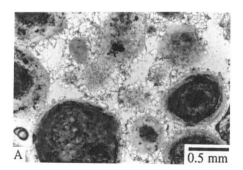 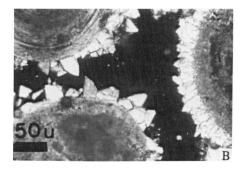

Fig. 7.4. *(A) Circumgranular crust calcite cement developed in ooid grainstone taken from phreatic zone of the floating meteoric lens at Joulters Cay, Bahamas. Plain light. Sample furnished by Mitch Harris. (B) Close view of phreatic pore from similar sample at Joulters showing rhombic-to-bladed nature of the cement. Photo from an original slide furnished by Bob Halley. Crossed polars.*

is varied, but phreatic cements can be a bit larger than vadose cements (Halley and Harris, 1979; Longman, 1980; James and Choquette, 1984; Saller, 1984a).

Trace element composition of meteoric phreatic cements from a local meteoric lens

As in the vadose environment, trace elements generally will be present in cements precipitated in meteoric phreatic lenses at relatively low levels. The trace element composition of phreatic cements in more active gravity-driven phreatic systems, such as Barbados, usually exhibits systematic intracrystalline variational patterns related to multicomponent mineral stabilization (Benson, 1974). For example, a cement being precipitated in an active meteoric phreatic system might exhibit separate zones of high Sr and Mg concentrations that would reflect the sequential dissolution of aragonite and magnesian calcite in an up-flow direction. The intracrystalline composition of cements precipitated in the more stagnant local meteoric lenses such as Schooner Cay, however, does exhibit this type of systematic variation. Budd (1984) suggested that Sr and Mg concentrations in the floating meteoric water lens came from both fresh-marine water mixing as well as mineral stabilization, which effectively destroyed any tendency toward compositional zoning.

The zone of active diagenesis in a local meteoric lens is closely associated with the water table, assuring an adequate supply of oxygen to the diagenetic waters of the lens. The resulting oxidizing conditions in the meteoric lens generally preclude the incorporation of Fe and Mn and the development of cathodoluminescence in the carbonate phases precipitated in this environment.

Stable isotopic composition of meteoric phreatic cements from a local meteoric lens

The oxygen isotopic signature of calcite cements precipitated in a local meteoric lens will generally be the same as that for vadose cements, except that it will generally not show the positive evaporative incursion sometimes associated with vadose soil exposure (Fig. 7.5). Usually, the carbon isotopic composition of phreatic cements is several permil higher than cements found in the vadose, associated with light soil gas CO_2 (Lohmann, 1988). This trend of lighter carbon for the vadose environment is useful in determining the position of paleo-water tables and the existence of a local meteoric lens in the geologic record (Allen and Matthews, 1982; James and Choquette, 1984) (Fig. 7.5). These trends will tend to reverse as one moves down into and through the meteoric-marine mixing zone (Fig. 7.5).

Porosity development in a local meteoric lens

As in the vadose, total porosity seems to be little affected by meteoric diagenesis in a local meteoric lens. While secondary porosity is generated, and cement is precipitated into intergranular porespace, the transport of $CaCO_3$ away from the site of dissolution is minimal. Because of the common circumgranular pattern of cementation in the phreatic zone, however, permeability does not seem to be reduced as it is in the vadose zone above (Halley and Harris, 1979) (Fig. 7.5, this book).

Local island model of diagenesis

Fig. 7.5 summarizes the patterns of diagenesis that can be expected to develop beneath a tropical carbonate island large enough to support at least an ephemeral floating meteoric water lens. Several important points concerning this model must be made:

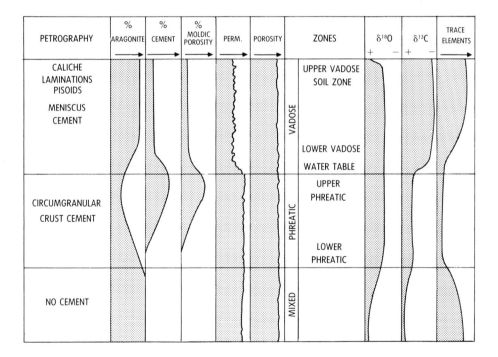

Fig. 7.5. Schematic diagram illustrating the patterns of diagenesis, porosity, and geochemical trends one might expect to find associated with a local island, floating meteoric water lens

1. Diagenetic overprint, and therefore porosity modification, is limited to two relatively thin interfaces: a) the exposed vadose soil zone, and b) the water table. The relatively untouched lower vadose and the lower phreatic-mixing zone may be buried as metastable sequences.

2. Because the island diagenetic system is basically closed, no significant transport of $CaCO_3$ takes place during stabilization; therefore, no significant gain or loss of porosity volume is accomplished, except perhaps during caliche formation in the upper vadose soil zone where calcite cements have a tendency to accumulate. Permeability, however, may be affected in the vadose zone by cementation at grain contacts that tends to affect pore throats, while not having a significant impact on total porosity.

3. The mixing zone at the base of the meteoric lens in the local island setting seems to be a zone of no significant diagenesis.

4. Stabilized water table sequences may be more resistant to pressure solution during subsequent burial than adjacent vadose and mixing zone successions that have not been stabilized to calcite, leading to the preferential preservation of porosity at the top of local meteoric lenses.

The local island model through time

What happens to the island model in response to sea level fluctuations? Is it reasonable to expect thick, sedimentary sequences to be diagenetically altered by moving an island meteoric lens up or down in response to a rising or falling sea level? Based on present knowledge of the Pleistocene, where carbonate sequences have been influenced by multiple eustatic sea level fluctuations, significant meteoric diagenetic overprints, as a result of floating meteoric lenses, are only established during sea level standstills. The resulting diagenetic pattern is one of thin, mineralogically stabilized zones (calcite-rich) associated with meteoric water tables established during sea level standstills. These thin calcite-rich zones are separated by thick sequences of relatively untouched metastable sediments (Steinen and Matthews, 1973; Saller, 1984a). Observations in ancient carbonate rock sequences indicate that the static local island model developed during sea level standstills may represent the rule, rather than the exception (Wagner and Matthews, 1982; Humphrey and others, 1986; Matthews, 1987).

The Walker Creek field : A Jurassic example of the local island model?

The Jurassic Smackover Formation is an important hydrocarbon reservoir around the U.S. Gulf of Mexico rim. In southern Arkansas the Smackover reservoir facies

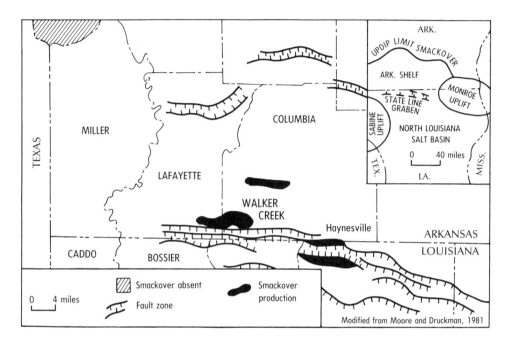

Fig. 7.6. Location and geologic setting of the Walker Creek field, southern Arkansas, U.S.A. Reprinted with permission, Gulf Coast Association Geological Societies.

consists of thick ooid grainstones deposited on a high-energy platform fronting an interior salt basin (Moore, 1984) (Fig. 7.6). Walker Creek is a major south Arkansas Smackover field with ultimate recovery in excess of 90 million barrels of oil. The Walker Creek reservoir consists of a relatively thick sequence of ooid, pellet, and algal-coated grainstones deposited over the crest of subtle salt structures close to the Smackover shelf margin (Brock and Moore, 1981) (Fig. 7.7). The reservoir consists of up to six separate porosity zones that exhibit independent pressure regimens that were originally thought to represent stacked, ooid bars separated by lagoonal muds (Chimene, 1976) (Fig.7.8). Brock and Moore (1981) suggested that the tight zones represented facies consisting of pellet, oncolite, and quartz-bearing grainstones that compacted easily during burial, while the porous intervals represented coarser grainstone facies dominated by large ooids and rhodolites that resisted burial compaction. Brock and Moore indicated that the stacked porosity zones represented a series of shoaling-upward cycles that developed over the sea floor expression of the subtle salt structures at Walker Creek.

A. Structure Map of the Top of the Smackover

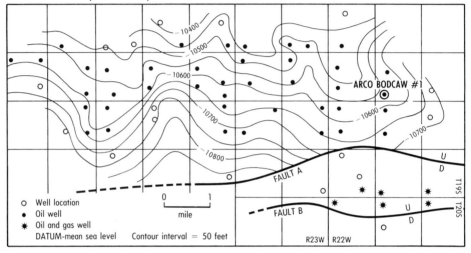

B. Net Porosity Isopach Map of the Upper Smackover

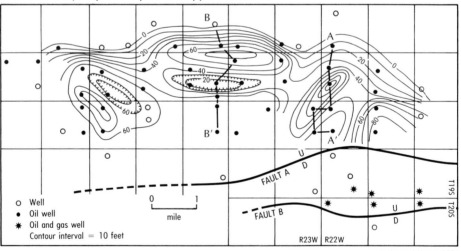

Becher and Moore, 1976

Fig. 7.7. Structure (A) and net porosity isopach (B) maps drawn on the Jurassic upper Smackover reservoir rocks at Walker Creek field southern Arkansas. Two cross section lines (A-A' and B-B') are shown on the net porosity isopach map (B). Reprinted with permission, Gulf Coast Association Geological Societies.

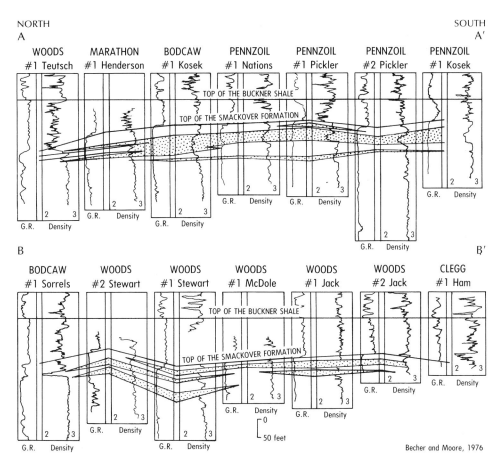

Fig. 7.8. Subsurface cross sections drawn across the Walker Creek field. Lines of cross section are shown on Figure 7.4. Stippled pattern indicates porous zones identified on density logs. Reprinted with permission, Gulf Coast Association Geological Societies.

After an intensive study of one of the Walker Creek cores, the Arco Bodcaw #1, Wagner and Mathews (1982) suggested that the porous zones represented the position of floating meteoric water lenses beneath Smackover islands developed over the salt structures at Walker Creek. Aragonite ooids were believed to have stabilized to calcite in the floating, meteoric phreatic zone. The sediments in the vadose zone above, and in the mixing and marine zone below, remained aragonite. Upon burial, the metastable sediments of the vadose and mixing zones were more susceptible to pressure solution than their stabilized counterparts in the main body of the lens, and consequently, lost their porosity during the burial process.

METEORIC DIAGENETIC ENVIRONMENTS

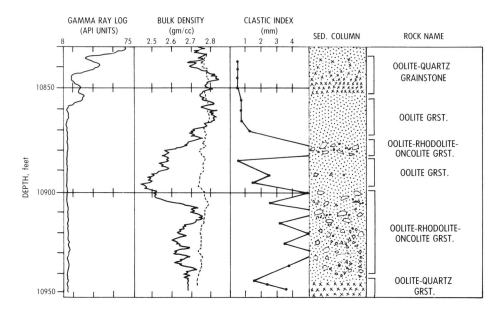

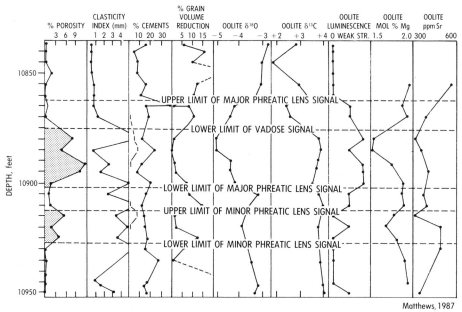

Fig. 7.9. Detailed log of the Arco Bodcaw #1 well from Walker Creek. Location of well is shown on Figure 7.7. (Top) Logs and lithologies for the upper 100 feet (30 m) of the well. (Bottom) Porosity in the Arco Bodcaw plotted against petrography and geochemistry. Interpreted distribution of phreatic and vadose signals are indicated. Used with permission, SEPM.

Wagner and Matthews based their interpretation on stable isotope and trace element compositional trends of grains in the Arco Bodcaw well (Fig. 7.9B) that closely parallel those outlined in the previous section as being characteristic of a local island model of meteoric diagenesis. The base of the vadose zone, which marks the position of the paleo-water table, was placed at a major carbon isotopic excursion toward more negative values, reflecting interaction with light vadose soil gas CO_2. In addition, the trace element compositional trends of the grains across the suspected water table are similar to those outlined by Wagner (1983) for Barbados Pleistocene vadose-to-phreatic transitions. Preferential grain volume reduction by compaction seems to coincide with the sequences that they would place either in the vadose zone or in the meteoric-marine mixing zone below (Fig. 7.9).

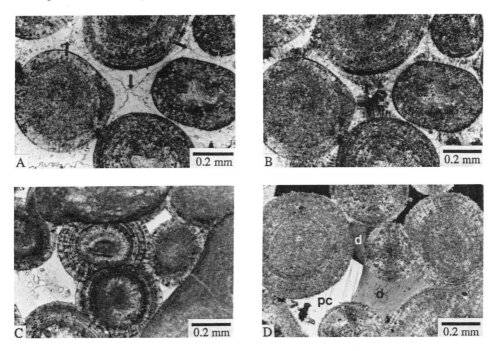

Fig. 7.10. (A) Early, precompaction, marine cement completely occluding porespace in Jurassic Smackover ooid grainstones. Note polygonal suture pattern (arrows). Arco Bodcaw #1, 10930' (3332 m) at Walker Creek, Columbia Co., Arkansas. Plain light. (B) Same as (A) under crossed polars. (C) Compaction-reduced intergranular porosity in Jurassic Smackover ooid grainstones at Walker Creek field, Columbia Co. Arkansas. Note the lack of early, precompaction cement. Arco Bodcaw #1, 10873' (3314 m). Plain light. (D) Compaction and cement-reduced intergranular porosity from the Smackover at Walker Creek field. Pore spaces are partially filled with postcompaction poikilotopic calcite (pc) and dolomite (d) cement. There is no early, precompaction cement present. Arco Bodcaw #1, 10892' (3321 m). Crossed polars.

The Wagner and Matthews (1982) island diagenetic model for Walker Creek was challenged by Moore and Brock (1982) on the basis of the lack of petrographic or textural evidence in core or thin section for subaerial exposure of the Smackover in the Arco Bodcaw, or any other core from Walker Creek. In addition, no obvious dissolution fabrics or significant precompaction meteoric cement was present in the zones believed to be under the influence of meteoric water. Wagner and Matthews had mistakenly suggested that early fibrous marine and late postcompaction calcite cement in the Arco Bodcaw (Moore and Druckman, 1981; Moore, 1985) was precipitated from meteoric waters (Fig. 7.10).

Subsequent work (Moore and others, 1986; Swirydczuk, 1988) indicates that the Smackover ooids at Walker Creek were originally calcite, or magnesian calcite, rather than aragonite (Figs.4.6A and B,6.7B). Therefore, one would not expect to observe either extensive dissolution fabrics or significant cementation associated with either calcite or magnesian calcite ooids undergoing diagenesis in a meteoric environment, particularly in a rather stagnant, floating meteoric lens. Finally, the presence of extensive marine cementation (as beach rock?) (Figs. 7.10A and B) associated with early salt-related structures at Walker Creek (Brock and Moore, 1981) certainly supports the idea of Smackover islands, and enhances the possibility of a floating meteoric lens at Walker Creek.

One major question remains. If the ooids were magnesian calcite, would one expect to retain the magnesian calcite in the vadose and marine zones into burial conditions deep enough so that pressure solution would preferentially destroy the porosity of the magnesian calcite sequences? Saller (1984a) indicates that all magnesian calcite has been removed from Pleistocene sequences at Enewetak, while aragonite has been removed from Pleistocene successions only where meteoric influence is associated with exposure surfaces. This data suggests that Mg-calcite stabilization should have taken place before significant burial into the subsurface at Walker Creek, casting some doubt on Wagner and Matthews' conclusion that meteoric water stabilization was the only control on porosity preservation at this site.

The use of the Wagner-Matthews (1982) model in the Smackover is complicated by the ubiquitous presence of the shoaling upward cycles in the upper Smackover recognized by Brock and Moore (1981). When these cycles are associated with early salt-related structures showing movement during deposition, it is reasonable to expect occasional emergence, island formation, and the development of a local meteoric lens, as Wagner and Matthews suggest for the Arco Bodcaw well at Walker Creek. Under these circumstances it may be impossible to determine whether porosity preservation is the result of preferential early stabilization under meteoric conditions, or preferential compaction related to grain type as controlled by energy at the site of deposition.

Therefore, when using the Wagner-Matthews model in the rock record to explain porosity preservation, its applicability must be fully evaluated in the context of a well-constrained geologic and sedimentologic setting, including a determination of the original mineralogy of the sediment affected (see also Humphrey and others, 1986).

Regional meteoric aquifer system

The gravity-driven regional aquifer system is perhaps one of the most important meteoric diagenetic environments relative to porosity modification. In this setting, large volumes of meteoric water can actively move through and interact with carbonate sediments of variable solubility, dissolving some carbonate phases, and precipitating others downstream as water chemistry responds to the addition of $CaCO_3$. This potential for large-scale transport of $CaCO_3$ down flow paths is a particularly important attribute of the meteoric aquifer diagenetic environment.

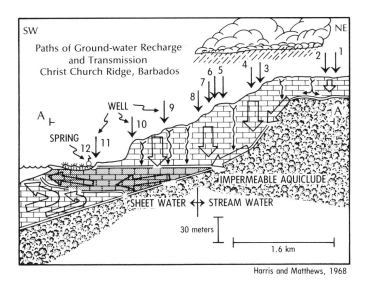

Fig. 7.11. *Hydrologic setting of the Christ Church Ridge, Barbados. Stippled area represents floating meteoric phreatic lens at the shoreline. Reprinted with permission from Science, v.160, p.78. Copyright (C), 1968 by the AAAS.*

The island of Barbados is the only modern, well-characterized, gravity-driven aquifer system of any size involving a young, mineralogically metastable carbonate terrain. Our knowledge of the Barbados system is based on a long-term research program at Brown University under the guidance of R. K .Matthews. The general Barbados setting is shown in Fig. 7.11. The main attribute of the system is an uplifted Pleistocene coral cap up to 90 m thick, resting on a tilted pre-Pleistocene siliciclastic aquaclude. Elevations of over 300 m are attained in the center of the island. Most of the recharge waters falling on the coral cap pass through the vadose zone in sinks and fissures directly into the phreatic zone, where gravity-driven ground water movement is by stream flow along the base of the coral cap. Where the stream drainage intersects sea level in the south, a 30-m-thick, floating, phreatic lens that extends some 1.6 km to the island coast is developed. Water movement through this lens is by sheet flow with fresh meteoric spring outflow along the coastline (Harris and Matthews, 1968).

Regional meteoric aquifer diagenetic model

The diagenetic drive for regional meteoric aquifers is generally furnished by the solubility contrast between aragonite and calcite, and the rapid movement of rather chemically simple water through the aquifer (Matthews, 1974). The aquifer is recharged with water undersaturated with respect to both aragonite and calcite. As long as aragonite is present, and the water is undersaturated with respect to calcite, solution of aragonite will be the dominant process (Fig. 7.12A). As $CaCO_3$ is added to the water by dissolution of aragonite, the water will ultimately reach saturation with respect to calcite and calcite precipitation will occur. Because of the calcite nucleation rate step, which is discussed in Chapter 6, and the movement of water through the aquifer, a net transfer of $CaCO_3$ down the water flow path results (Fig. 7.12A). Where the water is supersaturated with respect to calcite and undersaturated with respect to aragonite, simultaneous precipitation of calcite and dissolution of aragonite are possible (Fig. 7.12A, zone 2). As the water approaches saturation with respect to aragonite, microscale solution-precipitation may dominate and neomorphic grain stabilization is common (Fig. 7.12A, zone 3). With continued rock-water interaction down the flow path, the water ultimately becomes saturated with respect to both aragonite and calcite, and only calcite precipitation occurs (Fig. 7.12A, zones 3 and 4). When the meteoric water reaches the sea and begins to mix with marine waters, calcite cementation is shut down, and another round of dissolution is possible (Fig. 7.12A, zone 4) (Matthews, 1974).

Given enough time, the recharge area is stabilized to calcite, and calcite dissolution commences until equilibrium with the calcitized sequence is attained. Water emerging from this stabilized calcite terrain, while saturated with respect to calcite, is still undersaturated with respect to aragonite, and will dissolve any available aragonite down the flow path. The tendency, therefore, is for expansion of the stabilized calcite terrain down the flow path until all available aragonite is converted to calcite (Fig. 7.12B).

In order to establish a regional carbonate meteoric aquifer system, the following are necessary: 1) a widespread sedimentologic unit with hydrologic continuity

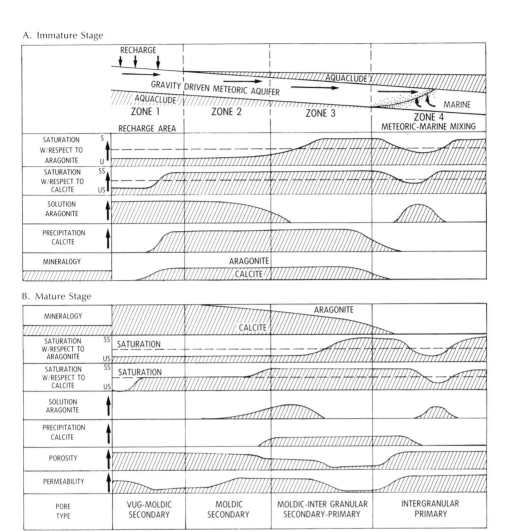

Fig. 7.12. Diagenetic model for a gravity-driven, confined, meteoric aquifer system. The hydrologic setting is shown at the top of the figure. The aquifer is divided into 4 zones, with the recharge area zone 1 and the area of meteoric-marine water mixing, zone 4. The model is divided into two stages: (A) Immature stage at the initiation of recharge. Aragonite is still present in the recharge area. (B) Mature stage. Aragonite has been destroyed by dissolution and only calcite is present in the recharge area.

(usually high-energy grainstones); 2) dip toward the sea to establish a system of gravity flow; and 3) a climate that will furnish more water than will be used by evaporation and transpiration. The high-energy terminal phase of major, shoaling-upward, shelf or platform cycles controlled by sea level, such as the Jurassic Smackover of the Gulf Coast, or the Mississippian of the continental interior of the United States, can well serve to set up major, regional, meteoric aquifer systems (Meyers, 1974; Moore, 1984).

Porosity development and predictability in regional meteoric aquifer environments

The transfer of significant volumes of $CaCO_3$ down the flow path during stabilization of unstable mineralogies in regional meteoric aquifer settings infers that total porosity volume may be significantly changed by solution or precipitation at any single site along the flow path during mineralogical stabilization. Furthermore, given a starting metastable mineralogy dominated by aragonite, the diagenetic facies outlined in the model above, including porosity characteristics, should always be developed in a predictable sequence along the flow path during the stabilization process. If stabilization is interrupted by sea level rise or subsidence, these distinctive diagenetic and porosity facies will be frozen and buried in the geologic record.

Fig. 7.12 B schematically illustrates the updip-to-downdip changes in the nature of carbonate pore systems that can be expected from such a system. In Zone 1, along the updip margin, moldic porosity is solution enlarged into vugs because calcite is undergoing dissolution. Permeability can be extremely low in areas of complete intergranular cementation. Grain neomorphism is generally not present. Zone 2 is characterized by ubiquitous moldic porosity and almost total occlusion of intergranular pores by calcite cement in the upflow areas of the zone. Calcite cements are generally not present in aragonite molds. Zone 3, where water is supersaturated with calcite and at equilibrium with aragonite, will generally be characterized by neomorphosed aragonite grains and abundant intergranular cement. The downflow portion of Zone 3 is probably the tightest porosity zone of the system. Intergranular calcite cement in Zone 4 becomes sparser in a downdip direction and porosity volume increases. Zone 4 is generally characterized by primary porosity preservation, little calcite cement, and common grain neomorphism.

Young metastable sedimentary sequences suffering an early meteoric diagenetic overprint in a regional meteoric aquifer environment, then, should exhibit a predictable diagenetic and porosity facies tract oriented normal to hydrologic flow direction. This diagenetic facies tract can be extremely useful in predicting the distribution and nature of porosity on a regional basis during hydrocarbon exploration and production operations (Moore, 1984).

Geochemical trends characteristic of a regional meteoric aquifer system

Any regional geochemical trends established in a meteoric aquifer system will generally parallel the meteoric flow paths. If the aquifer flows through a siliciclastic

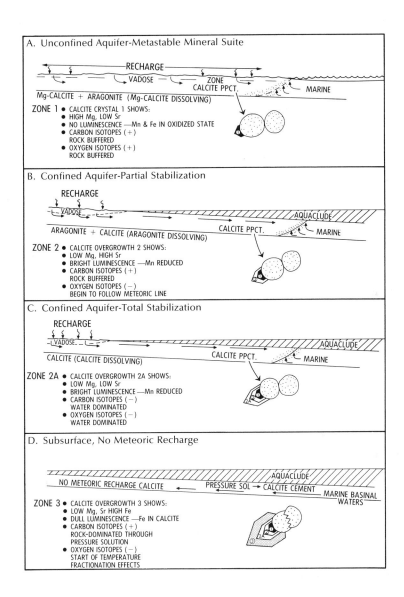

Fig. 7.13. Schematic diagram showing the geochemical trends characteristic of a regional meteoric aquifer system. There are four stages, A-D, representing the evolution of the aquifer system from an initial unconfined, to a final subsurface system cut off from meteoric recharge. Cements precipitated downflow from aquifer waters are shown diagrammatically on the right side of the diagram. Nonluminescent cement shown in black, brightly luminscent in white, and dully luminescent in a stippled pattern. Geochemical characteristics of each cement zone are shown on the left side of the diagram.

terrain before entering the carbonate system (not an unusual situation), carbonate cements may reflect the siliciclastic signature of precipitational waters by increased contents of radiogenic Sr, and the trace elements Fe and Mn in an updip direction. In an unconfined aquifer, the full extent of the aquifer should be oxygenated by recharge through the vadose zone. Calcite cements precipitated in this situation should show no luminescence because Fe^{+3} and Mn^{+4} are the favored valence states in oxygenated water, and Fe and Mn with these valences cannot be incorporated in the calcite lattice. If no manganese is present in the calcite, the calcite will exhibit no luminescence (Fig. 7.13A). If the aquifer is subsequently cut off from vadose recharge downflow and develops into a confined aquifer later, the water tends to become reduced, and Fe^{+2} and Mn^{+3} are the dominant valence states present, allowing Fe and Mn incorporation into calcite cements being precipitated as overgrowths on the initial nonluminescent cement. This second stage calcite overgrowth will tend to be brightly luminescent (Fig. 7.13B and C). If vadose recharge is maintained in the updip areas, a pH-Eh gradient may be established down aquifer flow, with nonluminescent cements continuing to be precipitated updip, while luminescent phases are precipitated downdip. As the aquifer system is buried, meteoric recharge is cut off and marine-related waters begin to dominate in the subsurface. Calcite cement overgrowths associated with pressure solution, precipitated from these subsurface fluids, are generally dully luminescent because of increasing Fe concentrations derived from subsurface fluids and grain-to-grain pressure solution. Fe tends to quench luminescence, hence as Fe concentration in the calcite increases, calcite luminescence tends to become duller. A calcite crystal precipitated at any one site, therefore, might well reflect the long-term pore fluid evolution of the aquifer by the complex cathodoluminescence zonation depicted in Fig. 7.13D. (Meyers, 1974; 1988; Meyers and Lohmann, 1985; Mussman and others, 1988; Niemann and Reed, 1988).

In addition, the trace element composition of individual calcite crystals should reflect the progressive mineralogic stabilization of the carbonate aquifer as it matures diagenetically through time. As Mg-calcites are recrystallized to calcites, the Mg-Ca ratio in the diagenetic water will increase, and increased Mg will be incorporated in the calcite cements downflow as a Mg spike (Fig. 7.13A). Aragonite will begin to be dissolved as soon as all Mg-calcite is converted to calcite, increasing the Sr-Ca ratio in the water, thus resulting in a Sr spike in the cements precipitated downflow. This Sr spike will occur in the calcite crystal after the earlier formed Mg spike (Fig. 7.13 B) (Benson and Matthews, 1971).

The stable isotopic composition of calcite cements precipitated in a regional meteoric aquifer system will generally follow Lohmann's (1988) meteoric water line with relatively large variations in carbon isotopic values and minimal variation in the oxygen isotopes (Fig. 6.4). Cement isotopic compositions may, however, exhibit some gross downflow regional trends, particularly as the aquifer matures diagenetically. During the early stages of stabilization, the isotopic signature will be dominated by the isotopic composition of the marine sediments being stabilized, so that both oxygen and carbon composition might be high (Fig. 7.13 A). As diagenesis progresses and the volume of metastable minerals decreases, meteoric phreatic water isotope signatures begin to dominate and both oxygen and carbon values tend to become lower (Fig. 7.13 B and C). As the sequence is buried into the subsurface, carbon isotopes will become rock-dominated again during pressure solution and become more positive. However, during burial, oxygen isotope values will tend to be more negative because of increasing temperature (Fig. 7.13 D).

These trends will usually be manifested within a single crystal. They will generally parallel the luminescent and trace element trends developed above, and reflect the progressive evolution of the meteoric phreatic waters as driven by the mineral stabilization process (Fig. 7.13D).

The Jurassic Smackover Formation, U.S. Gulf of Mexico: a case history of economic porosity evolution in a regional meteoric aquifer system

The Jurassic Smackover Formation produces copious quantities of oil and gas from an exploration-production trend some 100 km wide and 1000 km long around the margins of the Gulf of Mexico, extending from Texas to Florida (Fig. 7.14) (Moore, 1984). The Smackover was deposited on shallow marine shelves fronting a series of interior salt basins involved in the early opening history of the Gulf of Mexico. Smackover reservoir facies consist of widespread, blanket ooid grainstones that occur in the uppermost Smackover. These grainstones range from 30 to over 200 m thick. The upper Smackover ooid sands were deposited as high-energy platform facies during the terminal Oxfordian highstand (Fig. 7.14) (Moore, 1984; 1986). The upper Smackover grainstones pass into siliciclastics around the updip margins of the Smackover subcrop.

During the subsequent Kimmeridgian sea level rise, a rimmed shelf was developed and an extensive subaqueous lagoonal evaporite, termed the *Buckner*, was deposited across the eastern (Florida and Alabama) and western (Texas) limbs of the Jurassic subcrop (Fig. 7.14). In the central area (Louisiana and Mississippi), fine-grained siliciclastics, rather than evaporites, were deposited in the shelf lagoon overlying the upper Smackover grainstones. Extensive refluxion of evaporative waters from the Buckner lagoon resulted in the regional dolomitization of the upper Smackover in the east and west (see Chapter 5, this book), while the center portion of the subcrop remained undolomitized (Fig. 7.14). The following discussion is confined to the central portion of the Smackover subcrop in southern Arkansas and Louisiana (Fig. 7.14).

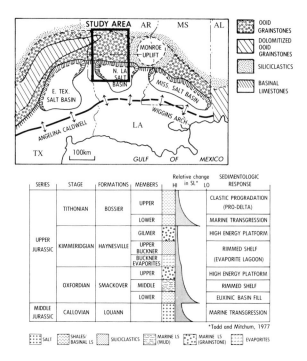

Fig. 7.14. (Top) Geologic setting of the Upper Jurassic Smackover Formation, southern Arkansas, northern Louisiana. In this area the Smackover Formation is a blanket ooid grainstone extending from the margin of the north Louisiana salt basin to the updip pinch-out of the Smackover. (Bottom) Stratigraphic-sedimentologic setting of the Upper Jurassic in the central Gulf Coast as related to sea level. Used with permission, SEPM.

In a regional study of the upper Smackover ooid grainstones of Arkansas and Louisiana, Moore and Druckman (1981) recognized three distinct diagenetic zones oriented parallel to depositional strike (Fig. 7.15). The northern diagenetic zone, extending updip to the upper Smackover's facies change from ooid grainstones into coarse quartz sands, was characterized by ubiquitous oomoldic-to-vuggy porosity and extensive intergranular calcite cements (Figs. 7.16 and 7.17A). Downdip, the transitional zone exhibited a gradient from oomoldic porosity with intergranular calcite cements in the north, to well-preserved ooids with intergranular calcite cements along the southern margins of the zone (Figs. 7.16, 7.17B and C). The southern zone consists of well-preserved ooids with no significant early cement, and with all porosity being represented by primary, preserved, intergranular pore types (Figs. 7.16 and 7.17D).

Moore and Druckman (1981) suggested that this well-defined regional diagenetic gradient was the result of a gravity-driven meteoric aquifer environment acting on an aragonite-dominated sediment shortly after deposition. Specific features in each zone

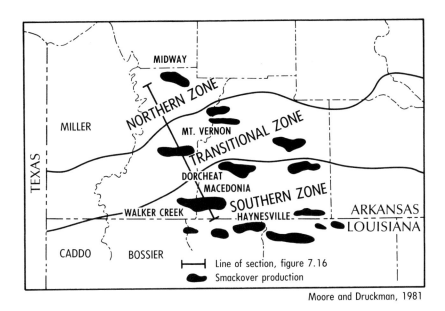

Fig. 7.15. Diagenetic zones developed in the upper Smackover Formation, southern Arkansas and northern Louisiana. Major oil fields are shown in black. Reprinted with permission of the American Association of Petroleum Geologists.

ZONES	NORTHERN	TRANSITIONAL	SOUTHERN
EARLY CEMENT VOLUME			
SECONDARY MOLDIC POROSITY			
PRIMARY INTERGRANULAR POROSITY			
PERMEABILITY			
POROSITY TYPES	VUG-OOMOLDIC	OOMOLDIC-INTERGRANULAR	INTERGRANULAR
INFERRED ORIGINAL MINERALOGY OF OOIDS	ARAGONITE	ARAGONITE CALCITE	CALCITE

Fig. 7.16. Schematic diagram showing the nature of the diagenetic gradients and the resultant porosity-permeability relationships developed in the upper Smackover as the result of a regional meteoric aquifer system shortly after deposition. The regional changes in the inferred original mineralogy of ooids in the Smackover are shown in the last line of the diagram (from the work of Moore and others, 1986). The line of section is shown in the last figure.

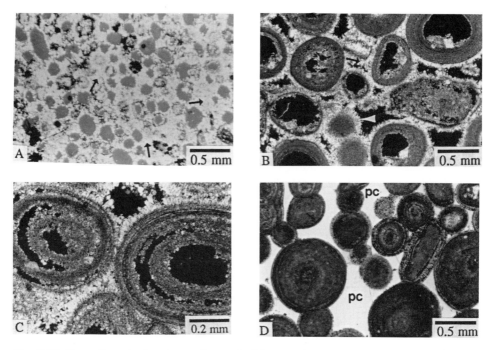

Fig. 7.17. Upper Jurassic Smackover diagenetic fabrics and porosity types in southern Arkansas developed as a result of the regional meteoric water system described by Moore and Druckman (1981) (Figure 7.16). (A) Oomoldic porosity with almost complete occlusion of intergranular porosity by early calcite cement in the northern diagenetic zone (arrows). Murphy I.P.C. #1, Miller Co. Arkansas, approximately 7500' (2287 m). Plain light. (B) Cement-reduced, intergranular porosity with oomoldic porosity. Calcite cement is well-developed, circumgranular crust characteristic of the transitional diagenetic zone (arrows). Murphy Giffco #1, Bowie Co. Texas, 7784' (2323 m). Crossed polars. (C) Close view of (B) showing characteristics of cement. Crossed polars. (D) Cement and compaction-reduced, intergranular porosity, characteristic of the southern diagenetic zone. Note that the ooids are well-preserved (inferred to be originally calcite) and that all cement (pc) is postcompaction. Arco Bodcaw #1 at Walker Creek field, 10873' (3315 m), Columbia Co. Arkansas. Plain light.

can be related to the basic regional aquifer model (Figs. 7.12 and 7.13), as follows:

1.) The oomoldic and vuggy porosity of the northern zone indicates that this sequence continued to receive meteoric water after sediment stabilization, enlarging some oomolds into vugs.

2.) The stable isotopic composition of the cements of the northern zone show suprisingly high values of both $\partial^{18}O$ and $\partial^{13}C$ (Moore and Druckman, 1981), suggesting that the stable isotopic composition of the cements was derived dominantly from marine ooids during aragonite stabilization, rather than exhibiting the light stable isotopic signature expected for meteoric waters.

3.) Pore-fill calcite cements of the northern and transitional zone consistently exhibit a distinct Sr spike (Moore and others, 1986), confirming that these cements derived their $CaCO_3$ from aragonite stabilization.

4.) These cements also exhibit a simple but consistent luminescent zonation, with an initial nonluminescent zone followed by bright luminescence (Moore and others, 1986). This pattern may well reflect a meteoric system initially open to oxygenated waters. Fe and Mn cannot be incorporated into the calcite lattice under oxygenated conditions; therefore, the initial calcite zones exhibit no luminescence. The second calcite zone's bright luminescence indicates that precipitational fluids later became reducing, allowing the incorporation of Fe and Mn into the crystal lattice. This luminescence pattern suggests that the open aquifer system evolved into a confined aquifer during the deposition of the overlying Buckner shales.

5.) Recrystallized aragonite ooids, showing some microsolution, are associated with calcite cements in the transitional zone. This relationship indicates that aquifer waters moving into the transitional zone, while saturated with respect to calcite, may have been close to equilibrium with respect to aragonite. The $CaCO_3$ necessary to bring the water into equilibrium with aragonite was derived from dissolution of aragonite in the northern zone and transported down-aquifer into the transitional zone.

6.) Ooids in the southern part of the transitional zone and in the southern zone have subsequently been identified as originally calcite ooids, rather than aragonite (see discussion, Chapter 6, Fig. 6.7, Moore and others, 1986; Swirydczuk,1988). In the southern part of the transitional zone, these calcite ooids furnished a ready nucleation site for the $CaCO_3$ derived from the stabilization of aragonite ooids to the north. This resulted in extensive calcite cementation and porosity occlusion along the southern margins of the transitional zone. In the southern zone proper, no significant early cementation or dissolution took place, and the ooid-dominated sequences were buried into the subsurface with all primary intergranular porosity intact. It is not certain whether the regional meteoric aquifer extended through the southern zone, because the sediments in this zone were dominated by calcite ooids, and interaction with meteoric waters may not have resulted in extensive cements or dissolution features.

The development of this regional aquifer system in a carbonate sequence dominated in its northern reaches by aragonite ooids resulted in the development of a regionally consistent, predictable, pattern of carbonate porosity—from vuggy, oomoldic, impermeable grainstones updip, to primary preserved, intergranular, porosity downdip (Fig. 7.16) (Moore, 1984). The best reservoir rock development occurs in the transitional zone, where original ooid mineralogy was still aragonite, and in the southern zone, where original ooid mineralogy was calcite. This pattern of porosity development, however, should also be expected in a sequence totally dominated by aragonite or a mixture of aragonite and calcite. Little cementation should be expected in the downdip portions of such a system where mixing with marine waters would be expected, since calcite cementation is not favored under the elevated Ca/Mg ratios encountered in marine fluids (Matthews, 1968; Folk, 1974).

Differential platform subsidence tilted upper Smackover blanket ooid grainstones toward the basin, thus creating the head necessary to establish the aquifer. Recharge was apparently through updip quartzose sands intercalated with upper Smackover ooid grainstones. This diagenetic pattern was later extended across the entire Smackover subcrop from Texas to Florida (Moore, 1984; Stewart, 1984; McGillis, 1984; Wilkinson, 1984; Meendsen and others, 1987). Reflux dolomitization to the east and west was concurrent with, or shortly after, the development of the upper Smackover meteoric aquifer (Moore and others, 1988).

Mississippian grainstones of southwestern New Mexico, U.S.A.: a case history of porosity destruction in a regional meteoric aquifer system

Upper Mississippian carbonates, from Kinderhookian to Chesterian in age, form a classic transgressive-regressive shoaling-upward sequence in the Sacramento and San Andres mountains, from Albuquerque, New Mexico to El Paso, Texas. The regressive leg of this cycle is represented by a relatively thick, crinoidal-lime grainstone blanket, similar to the Jurassic Smackover described above, and is termed, in part, the *Lake Valley Formation of Osagian Age* (Fig. 7.18). Meyers and his co-workers (Meyers, 1974, 1978, 1980,1988; Meyers and Lohmann, 1978; Meyers and others, 1982; Meyers and Lohmann, 1985) have established that the cementation and attendant porosity occlusion of these crinoidal sands were accomplished under the influence of a regional meteoric aquifer system during the early burial stages of the sequence.

In contrast to the Smackover described above, the Lake Valley grainstones consist almost exclusively of crinoid grains that were originally composed of 10 mole % $MgCO_3$ (Meyers and Lohmann, 1985). These grainstones have been cemented by a complex luminescent zoned calcite cement, consisting of four regionally correlatable luminescent zones (Meyers, 1974) (Fig. 7.19A). The earliest zones (1-3) are believed to be related to the regional meteoric water system. Meyers and Lohmann suggest that nonluminescent Zone 1 represents a nonconfined aquifer system developed shortly after deposition of the Lake Valley (Fig. 7.20A). Brightly luminescent Zone 2 is believed to have been precipitated after the aquifer was shut off from extensive vadose recharge (during a sea level highstand) (Fig. 7.20B), while nonluminescent Zone 3 represents a second stage of vadose recharge developed during a subsequent sea level rise (Meyers, 1988) (Fig. 7.20C). Zone 5 was believed to represent later post-Mississippian burial cementation (Meyers and Lohmann, 1985).

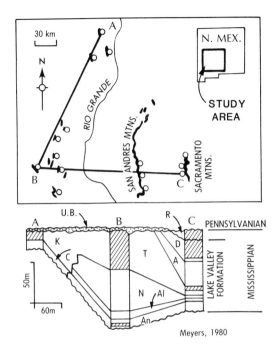

Fig. 7.18. Stratigraphic setting for the Mississippian Lake Valley Formation, southern New Mexico, U.S.A. Letters in lower diagram (cf. K) represent abbreviations of laterally equivalent formations in isolated outcrops along the Rio Grande river valley. Used with permission, SEPM.

The stable isotopic composition of these zones clearly reflects the progressive evolution of the regional meteoric aquifer water system during, and shortly after, stabilization of Lake Valley crinoidal sands (Fig. 7.21). The isotopic composition of Zone 1 is near the estimated original marine values for the Mississippian, indicating overwhelming rock buffering of the isotopic signal during initial mineral stabilization. Zones 2 and 3 occur along the meteoric water line, suggesting increasing dominance of the meteoric water signal as mineral stabilization proceeds.

All four cement zones comprise 26% of the volume of the grainstones and packstones of the Lake Valley (Meyers, 1974). If original porosity for these sediments is 40%, and micrite matrix is minimal, then compaction (both chemical and physical) is responsible for only 35% of the porosity loss (see also Meyers, 1980), while cementation is responsible for 65% porosity loss (Fig. 7.22). Of this 65%, approximately 50% of the cement was precipitated before significant burial (Zones 1-3), and dominantly in conjunction with the meteoric aquifer (Fig. 7.22). If compaction is assumed to have been accelerated by the Mississippian meteoric water system (as suggested by Meyers, 1980),

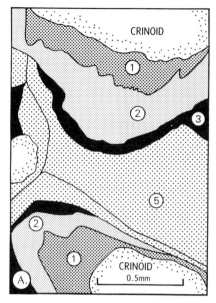
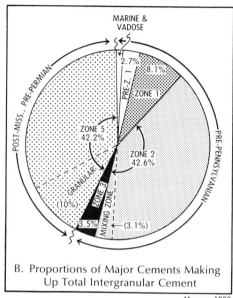

Fig. 7.19. (A) Schematic diagram showing luminescent zoned calcite cements of the Lake Valley Formation. Zone 3, shown in black is quenched. (B) Percentage of these zones found in total intergranular cement in the Lake Valley. Stippled patterns are the same as found in (B). Used with permission, SEPM.

then much of the 35% compaction documented for the sequence would have taken place in the Mississippian. Assuming this figure was 75%, the porosity of the Lake Valley at the time of burial would have been reduced to approximately 18% (Fig. 7.22). Preserved porosity (the above 18%) at the time of the post-Mississippian unconformity consists entirely of intergranular porosity and seems to have been distributed uniformly throughout the sequence. This porosity was lost by cement infill (Zone 5) and the remaining 25% was lost by compaction during subsequent burial of the sequence to depths near 1000 m.

The source of the Mississippian cement (Zones 1-3) is believed to be early grain-to-grain pressure solution (a few hundred meters burial) enhanced by the high solubility of the Mg-calcite crinoid grains in an active meteoric phreatic aquifer (Meyers, 1974; Meyers and Lohmann, 1985). In such a system, porosity occlusion would be particularly effective, because no secondary porosity would be developed and kinetic factors would be muted because of the presence of numerous calcite nucleation sites, favoring local precipitation of $CaCO_3$ gained from intergranular solution of the crinoid fragments. Because little $CaCO_3$ transport is necessary, the porosity occlusion rate due to cementation would be uniform down the flow path. In this situation, neither significant zones of

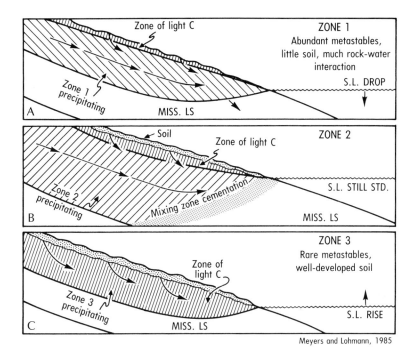

Fig. 7.20. Schematic diagram illustrating the inferred timing of calcite cement zones (1-3) of the Lake Valley with sea level fluctuations, based in part on isotopic composition of each zone. Zone 1 was precipitated during a sea level drop, zone 2 during a sea level stand, and zone 3 during an ensuing sea level rise. Used with permission, SEPM.

porosity enhancement nor porosity occlusion, such as are present in the Jurassic Smackover, would be developed.

The source of the post-Mississippian cements is believed to be primarily intraformational by pressure solution (although extraformational sources, also driven by pressure solution, cannot be ruled out), mainly in conjunction with stylolitization during the progressive burial of the unit down to a maximum burial depth of 1000 m (Meyers and Lohmann, 1985). Obviously, all porosity in this unit would have been destroyed before hydrocarbon maturation could have taken place.

In contrast, porosity within the Smackover at the time of burial into the subsurface ranged from 50%-0%, distributed unequally, but predictably, across the subcrop in response to the massive transport of $CaCO_3$ during the stabilization of aragonite ooids in the regional aquifer system. In the southern diagenetic zone, where Smackover ooid grainstones were calcite and there were no early cements, the sequence was buried into the subsurface with some 40% depositional porosity intact. If there had been a

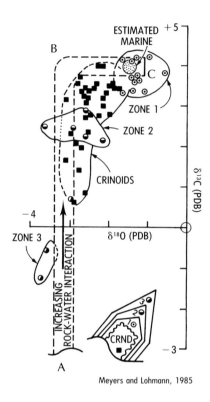

Fig. 7.21. Stable isotopic composition of calcite cement zones 1-3 of the Lake Valley. The stippled area is the estimated Mississippian marine stable isotopic composition, based on the composition of crinoids. Zone 1 cement clusters around this marine value indicating a rock-buffered system, probably during stabilization of unstable components. Zones 2 and 3 lie along the meteoric water line, with zone 3 reflecting the least amount of rock-water interaction, indicating precipitation after most stabilization was completed. Used with permission, SEPM.

comparable porosity loss by cementation and compaction, as seen in the Mississippian Lake Valley, it would be obvious that significant porosity (perhaps as much as 20%) would have been present in the Smackover at the time of hydrocarbon maturation and migration. Indeed, Moore and Druckman (1981) report an average of 10% late subsurface cement present in upper Smackover grainstones, and between 15 and 20% porosity in reservoirs of the sequence, at burial depths exceeding 3000 m.

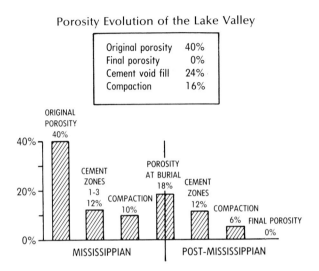

Fig. 7.22. Porosity evolution of the Lake Valley Formation. Initial porosity is assumed to be 40% with minimal micrite matrix. Final porosity is 0%

THE METEORIC DIAGENETIC ENVIRONMENT IN MATURE, MINERALOGICALLY STABLE SYSTEMS

Introduction

When a carbonate sequence has been mineralogically stabilized, buried, then exhumed in association with a subaerial unconformity, the stage is set for a second cycle of diagenesis (telogenesis of Choquette and Pray, 1970) that can again modify the porosity of the sequence. Both vadose and phreatic processes are active, and together are referred to as *karstification* (James and Choquette, 1988).The complex topic of karst development will not be examined in great detail (as in James and Choquette, 1988); however, the impact of karstification of mature, stable, diagenetic terrains on porosity evolution will be discussed in general terms.

Karst processes and products

Fig. 7.23 illustrates schematically the active processes, and some of the products produced during the development of a karst profile in a relatively humid climate. Dissolution, in both the vadose and the phreatic, is the dominant active process. This dissolution is driven by CO_2 derived from the atmosphere and soil gasses, and by the mixing of meteoric and marine waters at the shoreline or along the base of the meteoric lens (James and Choquette, 1984; James and Choquette, 1988). Solution cavities of all

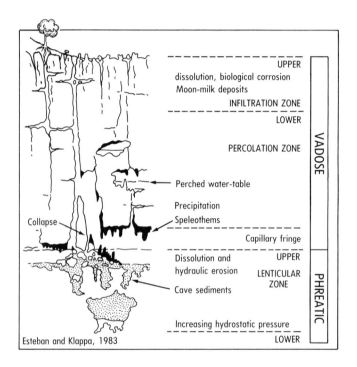

Fig. 7.23. Idealized karst profile illustrating some of the major processes and products one might expect to develop in a mineralogically mature carbonate terrain. Reprinted with permission of the American Association of Petroleum Geologists.

sizes, from tiny vugs to gigantic caverns, can be developed. Water movement in the vadose is often by conduit flow through fractures, which, with time, can be enlarged into vertically oriented pipes and sinks. Water movement in the phreatic zone is generally horizontal, and large, horizontally oriented cavity and cave systems are often developed (Craig, 1988; Ford, 1988). If the climate is humid enough and exposure long enough, a mature karst landscape featuring sinks, dolinen, solution towers, internal drainage, and considerable relief can develop (Esteban and Klappa, 1983; James and Choquette, 1988). Buried karst landscapes are often developed in association with important regional unconformities leading to significantly enhanced porosity and the development of major hydrocarbon reservoirs in the carbonate units below the unconformity.

The rapid and pervasive movement of water through soils associated with the vadose zone often carries soils and other sediments deep within the karst-related solution network to be deposited as internal sediments within caves and cavities. These sediments may either be siliciclastic (terra rosa), carbonate, or both. If soil-derived and siliciclastic, they can significantly occlude solution porosity at even the cave scale.

Finally, intense dissolution often results in solution-collapse features and the formation of extensive solution-collapse breccias that are often significantly more porous than the dissolved ground mass. These collapse breccias, commonly associated with regional unconformities, have served as receptor beds for extensive Paleozoic hydrothermal ore mineralization on a worldwide basis, as well as reservoir rocks for oil and gas (see earlier discussion of the Ellenburger in Chapter 5; Roehl and Choquette, 1985; Loucks and Anderson, 1985).

Solution, cementation and porosity evolution in a diagenetically mature system

In the regional meteoric aquifer model developed in the immature metastable carbonate terrains described in the preceding section, contrasts in mineral stability are the major drive for the diagenetic system, and furthermore, its porosity evolution. Dissolution and precipitation proceeded as long as the mineralogical and solubility differences were present, and regional transport of $CaCO_3$ to remote cementation sites was possible (Matthews, 1968; Matthews, 1974).

In the system considered in this section, however, no solubility contrast exists because the sequence is already stabilized. Thus, recharge waters undersaturated with respect to calcite will dissolve the calcite of the sequence until the waters reach equilibrium with the calcite of the rock mass and little further solution or precipitation can take place. The mature meteoric diagenetic system, therefore, is one driven primarily by solution, and one should not expect significant downstream cementation to be a major factor in porosity evolution (Matthews, 1968; Matthews, 1974; James and Choquette, 1984.).

Recharge waters will reach calcite equilibrium relatively quickly in a stabilized diagenetic terrain (Matthews, 1968). This will tend to constrain the depth to which porosity enhancement can be accomplished beneath regional subaerial unconformities.

Unconformity-related porosity development can be localized by favorable (more porous) facies in the carbonates beneath the unconformity that will tend to channelize the flow of meteoric water into and through the aquifer. In addition, fracture patterns associated with local structural features, such as anticlines and faults, can also localize karst-related solution porosity (Craig, 1988).

A number of regional meteoric aquifer systems have been described, such as the Mississippian Lake Valley (Meyers, 1974, 1988; see previous section, this chapter), and the Middle Ordovician and Mississippian of the Appalachians (Grover and Read, 1983; Mussman and others, 1988; Niemann and Read, 1988), where aquifer recharge is across a karsted terrain associated with regional unconformities. In each of these examples, significant early meteoric calcite cementation exhibiting regional cathodoluminescent zonation can be related to the regional aquifer system. In each case, however, the carbonate aquifer was young and immature, and was composed of metastable carbonate mineral components whose dissolution provided most of the $CaCO_3$ for the calcite cement.

The Cretaceous Golden Lane of Mexico (Boyd, 1963; Coogan and others, 1972); the Cretaceous Fateh field, Dubai, U.A.E. (Jordan and others, 1985); and the Yates field of west Texas, U.S.A. (Craig, 1988) are all giant fields, each with reserves in excess of 1 billion barrels of oil. All are examples of porosity developed in mature, stable, carbonate sequences related to karsting along regional unconformities. The Yates field will be discussed in detail in the next section.

Karst-related porosity in the Permian San Andres Formation at the Yates field, west Texas, Central basin platform, U.S.A.

Yates field is a giant oil field developed in Late Permian reservoirs of the San Andres, Queen, and Grayburg formations along the margins of the Central basin platform, fronting the Midland basin in west Texas (Fig. 7.24). To date, the Yates field has produced some 1.07 billion barrels of oil from a productive area covering over 32 sq mi (Shirley, 1987). A regional unconformity separates the San Andres from the Grayburg above (Fig. 7.24). At Yates, the San Andres exhibits two basic facies: a western facies consisting of an intertidal-lagoonal sequence of relatively impervious interbedded shales and dolomites; and an eastern facies characterized by porous dolomites that are believed to originally have been subtidal, shelf-margin grainstone sequences (Figs. 7.24 and 7.25). This west-to-east facies change divides the Yates field into western and eastern sectors, with most production coming from the more porous eastern sector (Fig. 7.25) (Craig, 1988). The Seven Rivers anhydrite sequence, which overlies the Permian carbonates, forms the seal for the field (Fig. 7.24).

Evidence from cores, logs, drilling records, and production characteristics, show that San Andres production is from a complex system of rather small caves developed

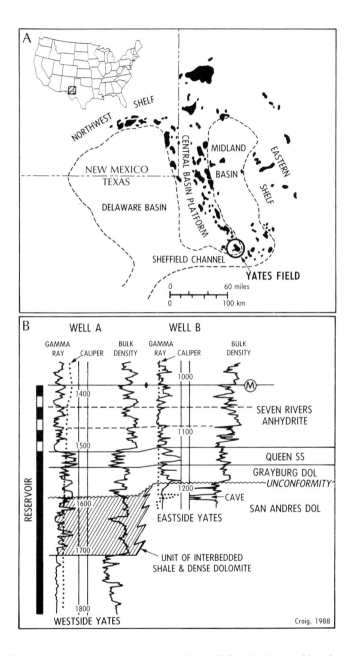

Fig. 7.24. (A) Geologic setting of the Yates field, west Texas, U.S.A. (B) Log profiles of two wells showing stratigraphic setting across the field, west to east. Note the unconformity at the top of the San Andres dolomite with caves. "M" is a marker used in the reconstruction of the paleogeography at Yates. Reprinted with permission from Paleokarst. Copyright (C) 1988, Springer-Verlag, New York.

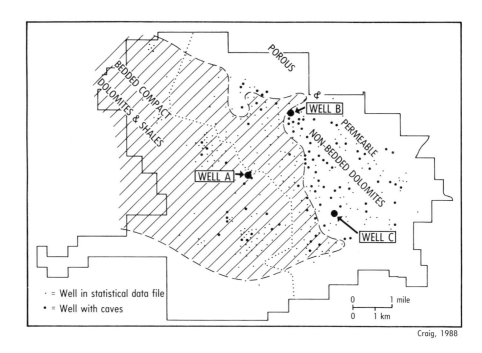

Fig. 7.25. Wells with caves at Yates field shown in relation to the paleogeographic distribution of major lithofacies exposed at the unconformity surface of the San Andres dolomite as shown in wells A and B in Figure 7.20. Reprinted with permission from Paleokarst. Copyright (C), 1988, Springer-Verlag, New York.

immediately below the regional unconformity at the top of the San Andres. The grainstones of the eastern sector were lithified and stabilized to limestone prior to karstification, and apparently were dolomitized after karstification by evaporative reflux from the Seven Rivers evaporative sequence above (Craig, 1988).

The distribution of caves in the San Andres is controlled in part by original facies, with most caves occurring in the more porous eastern sector of the field (Fig. 7.25). Some structural control of cave formation by joints and faults has been documented (Craig, 1988).

The restoration of the paleotopography on top of the San Andres unconformity seems to illuminate a classic karst surface complete with sinkholes, karst towers, and a chain of possible islands along the eastern sector of the field (Fig. 7.26). Craig (1988) weaves a persuasive argument in support of using an island-based, floating, freshwater lens to form the cave complex in the San Andres. However, it is questionable whether an

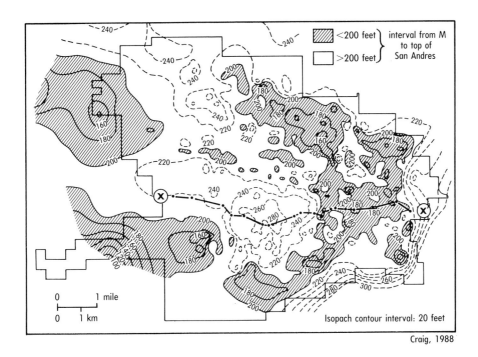

Fig. 7.26. Paleotopography on the San Andres unconformity expressed in terms of the isopach interval from the Seven Rivers "M" marker to the top of the San Andres. Thins represent topographic highs and thicks are topographic lows. Hachured areas are San Andres islands which were created when the unconformity surface was drained or flooded to the level of 200 feet (61 m) below the "M" marker, that is, to the elevation of the Late Permian (Guadalupian) sea. Reprinted with permission from Paleokarst. Copyright (C), 1988, Springer-Verlag, New York.

island meteoric lens is a diagenetic system active enough to have formed the extensive dissolution complex documented at Yates (see discussion earlier in this chapter). While the cave system obviously marks a paleowater table (Fig. 7.27), meteoric-marine mixing zone dissolution is possible where an active Permian regional meteoric aquifer intersects the coast at the edge of the platform. A similar situation has been described in the Yucatan Peninsula of Mexico by Back and others (1979; see discussion earlier in this chapter). At Xel Ha, large volumes of meteoric water discharge into, and mix with, marine waters of the Caribbean. Extensive dissolution, including cave formation, mark the meteoric-marine mixing zone. The unusual morphology of the Xel Ha coastline is the result of extensive cave collapse in the mixing zone.

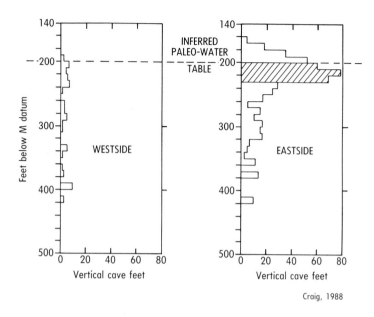

Fig. 7.27. Vertical cave feet (feet of cave) as a function of depth below the "M" marker. Cave abundance in eastside Yates is focused in a relatively thin zone of 30 feet (9 m) which is thought to represent a Permian paleo-water table coinciding with a Permian sea level position (-200 feet below the "M" marker. Reprinted with permission from Paleokarst. Copyright (C), 1988, Springer-Verlag, New York.

SUMMARY

The meteoric vadose diagenetic environment is unique because it is a two-phase system—air and water. This system is reflected in the unequal distributional patterns and unusual morphology of the cements precipitated in the environment. Calcite cements precipitated in the vadose zone tend to be concentrated at grain contacts, and exhibit microstalactitic and meniscus fabrics.

Diagenesis associated with mineral stabilization is generally slow in the vadose zone. Therefore, it is possible that a metastable sequence (aragonite dominated, but perhaps not Mg-calcite) that remains in the vadose zone will retain its unstable mineralogy and significant depositional porosity upon its burial into the subsurface. Subsequently, the metastable sequence might well lose its porosity more quickly under overburden pressure during burial than will a unit containing stabilized calcite because of the higher susceptibility of the metastable mineral phases to grain-to-grain pressure solution.

SUMMARY

The meteoric phreatic diagenetic environment is the most important meteoric environment relative to porosity modification because of the large volume of water available for dissolution and precipitation. Two basic models can be recognized: the local floating meteoric lens, and the regional meteoric aquifer system.

The impact of the local floating meteoric lens on porosity modification is limited because of the relatively small volume of meteoric water available, and the constraints of water flux through the system. Most diagenesis (solution and precipitation) is centered around the water table, and cementation-dissolution is usually balanced with no resulting loss or gain in porosity. Mineral stabilization, however, can be accomplished rather quickly. Those sequences that have encountered a floating meteoric lens are usually stabilized prior to burial into the subsurface and have the potential to retain preserved depositional porosity to significant burial depths.

In contrast, the gravity-driven meteoric aquifer system developed in metastable sediments shows the potential for large-scale regional transport of $CaCO_3$ derived from dissolution, leading to the development of significant secondary porosity in an upflow direction and the occlusion of porosity by calcite cementation in a downflow direction. The progressive stabilization of a carbonate sequence in an active aquifer system generally leads to the development of a well-defined diagenetic gradient in a carbonate unit. This unit parallels aquifer flow directions, and allows the mapping, and ultimately the prediction, of porosity types and porosity quality on a regional scale. These characteristics firmly establish the regional meteoric aquifer system as the most important diagenetic setting in the meteoric realm.

When a regional meteoric aquifer system is developed in a diagenetically mature sequence, such as in association with an unconformity, dissolution is the dominant diagenetic process and significant secondary porosity development can be the result. The characteristics of the affected sequence, such as remnant porosity and fractures, can significantly control the pattern of dissolution. Dissolution effects are limited by the rapidity of the rock-water equilibration in a mineralogically stable system. Several giant oil fields are associated with karstic dissolution along regional unconformities.

Chapter 8

DOLOMITIZATION ASSOCIATED WITH METEORIC AND MIXED METEORIC AND MARINE WATERS

INTRODUCTION

One of the most popular and widely used dolomitization models is the mixed meteoric-marine water model, more commonly referred to as the *mixing* or *Dorag* model (Badiozamani, 1973). This model is generally applied to ancient dolomitized sequences that are believed to have been under the influence of mixed meteoric-marine waters, such as at the terminus of regional meteoric aquifers, where meteoric flow mixes with marine waters at the coast, or at the transition of meteoric to marine waters below and adjacent to floating meteoric water lenses (Land, 1980, 1986). Dolomites attributed to the mixing environment are important oil and gas reservoirs (Roehl and Choquette, 1985).

Another meteoric-diagenetic setting in which dolomite is accumulating involves evaporated, continental, alkali ground waters in coastal lakes and lagoons of the Coorong region of south Australia (von der Borch and others, 1975). The Coorong model has seldom been applied to ancient sequences, and may have neither the wide applicability nor the economic importance of the mixing model (Morrow, 1982b).

In the following paragraphs, each of these two models will be discussed. Emphasis will be on the validity and constraints of the mixing model, applicability of the Coorong model to the geologic record, and the ultimate porosity potential for each model.

Meteoric-marine mixing, or Dorag model of dolomitization

Fig. 6.3A illustrates the thermodynamic basis for the mixing model of dolomitization. When a meteoric water saturated with respect to calcite and undersaturated with respect to dolomite is mixed with a marine water strongly supersaturated with respect to both phases, intermediate mixtures (because calcite saturation does not covary in a linear fashion with changes in salinity) are undersaturated with respect to calcite, and strongly supersaturated with respect to dolomite. Hence, dolomitization is believed to be favored in those intermediate mixtures of meteoric and marine water (Badiozamani, 1973).

The model was formulated and published almost simultaneously by Badiozamani (1973), who based his *Dorag* model on Middle Ordovician sequences of Wisconsin, and by Land (1973a and b), who based his version of the mixed model on the dolomitization of Holocene and Pleistocene reef sequences on the north coast of Jamaica. Both Badiozamani and Land drew heavily on the report of Hanshaw and others (1971) concerning dolomitization in the Floridian meteoric aquifer, and Runnels' (1969) work on mixing of natural waters.

This model has been widely applied to ancient dolomitized sequences of all ages (Zenger and Dunham, 1980). It has had particular appeal because of the inevitability of meteoric influence on shallow marine carbonate sequences, and the fact that these meteoric waters replace and mix with original marine pore fluids in a variety of hydrologic situations (Fig. 6.9). In each instance in which an isolated, ancient floating meteoric water lens is identified in the record, dolomitization is possible in the mixing zone beneath. For every regional meteoric aquifer system, there is the potential for dolomitization in the mixing zone at the coastal terminus of the aquifer, or in the mixing zone beneath the aquifer. Prograding shoreline sequences usually support a meteoric lens. The mixing zone beneath, and laterally adjacent to, the lens will migrate with the shoreline and can ultimately affect a wide swath of the carbonate progradational sequence. The model has been used to explain local patchy dolomitization (Choquette and Steinen, 1985; see below), as well as regional platform dolomitization (Loucks and Budd, 1981).

The recognition of mixed water dolomites in the geologic record is generally based on the following: 1) their association with petrographic fabrics and textures that can be related to the meteoric diagenetic environment, such as moldic porosity, early circumgranular-crust, equant calcite cements, and associated meteoric vadose fabrics; 2) occurrence within a geologic setting compatible with meteoric diagenetic environments; 3) use of the isotopic and trace element composition of the dolomite itself as a critical line of evidence because of the strong contrast between the isotopic and trace element composition of marine, meteoric, and evaporative waters (see chapters 3 and 6). In practice, in those sequences that have been totally dolomitized, destroying hard evidence for meteoric conditions, dolomite geochemistry is usually used as the prime interpretive criterion.

Lohmann (1988) suggests that mixing zone dolomites should fall in a stable isotopic population parallel to the meteoric calcite line, but offset toward higher $\partial^{18}O$ values because of dolomite fractionation (Fig. 6.4). Dilution by meteoric waters should result in reduced levels of Sr and Na in mixed zone dolomites relative to dolomites forming in seawater or evaporated seawater (Land and others, 1975; Land, 1985).

Concerns relative to the validity of the Dorag, or mixing model of dolomitization, and its application to ancient rock sequences

For a long time there has been uncertainty over the validity of the Dorag model. Carpenter (1976), and Machel and Mountjoy (1986) have expressed skepticism concerning the thermodynamic foundations of the model, concluding that mixing seawater and meteoric water should actually inhibit dolomitization. Hardie (1987) points out that Badiozamani (1973) used the solubility constant for partially ordered dolomite (10^{-17}) in his meteoric-marine water saturation calculations. Hardie contends that the solubility constant for disordered dolomite ($10^{-16.52}$) should be used because most modern dolomites forming on the surface today are disordered. Use of the disordered dolomite solubility constant to calculate dolomite saturation in a mixture of meteoric and seawater may result in a much smaller dolomite window (between 30-40% seawater, versus 5-50%) (Fig. 6.3B).

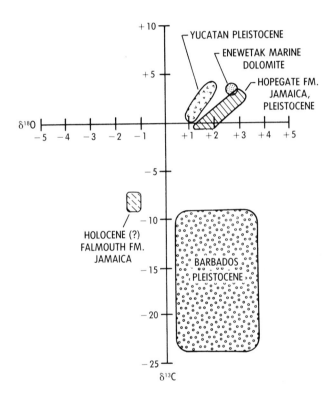

Fig. 8.1. Stable isotopic composition of Yucatan Pleistocene dolomites compared to other dolomite occurrences. Data sources: Jamaica Holocene and Pleistocene (Land, 1973a and b); Yucatan Pleistocene (Ward and Halley, 1985); Enewetak marine dolomite (Saller, 1984b); Barbados Pleistocene (Humphrey, 1988). Used with permission of SEPM.

Perhaps the most serious concern about the model is the fact that there are few unequivocal examples of mixing zone dolomitization in action today. Land (1973b) presents a compelling case for a working mixed meteoric-marine water dolomitization model at Discovery Bay, Jamaica. At Discovery Bay, 120,000-year-old Pleistocene Falmouth reef limestones are seemingly being dolomitized in a meteoric-marine mixing zone presently established within the Falmouth reef limestones. An analysis of ^{14}C dates, and the patterns of diagenetic alteration within the Falmouth relative to the present position of the water table and the mixing zone, provide strong support for a present-day operational mixing zone diagenetic system in Jamaica. The stable isotopic composition of these dolomites (Fig. 8.1), particularly the low $\partial^{13}C$ values, certainly suggests significant meteoric water influence during formation. Subsequently, Hardie (1987) has expressed some doubts (without sufficient justification in the opinion of this author) regarding the validity of Land's conclusions concerning the Falmouth dolomites. In 1980, Margaritz and others also described an example of mixing zone dolomitization in a modern hydrologic setting in Israel.

In the years following Badiozamani's work in the Paleozoic and Land's description of the Falmouth model, numerous other researchers have failed to find dolomites in meteoric-marine water mixing zones associated with present beaches (Grand Cayman, Moore, 1973; St. Croix, 1977), or in modern island-floating meteoric lenses of various sizes and hydrologic characteristics (Joulters Cay, Halley and Harris, 1979; Schooner Cay, Budd, 1984; Enewetak, Saller, 1984a; Bermuda, Plummer and others, 1976; Barbados, Matthews, 1974). Gebelein (1977), however, reported dolomitization beneath palm hammocks that supported meteoric lenses on the tidal flats of Andros Island. Subsequent investigations, however, conclude that no significant potential for dolomitization existed in these mixing zones (Gebelein, and others, 1980).

Some dolomite similar to that documented by Land (1973a) for the 300,000-yr-old Pleistocene Hope Gate Formation in Jamaica, and the Pliocene Seroe Domi Formation of Bonaire (Sibley, 1980), has been found in Pleistocene rocks of the Yucatan in Mexico (Ward and Halley, 1985). The Yucatan dolomites are found in 122,000-year-old Pleistocene carbonates cropping out along the coast, near the regional mixing zone described by Back and others (1979).

Initially, it was hoped that these dolomites represented another example of an active, modern, mixing-zone dolomitization system similar to that found by Land (1973b), and operating in the Falmouth Formation in Jamaica. Ward and Halley determined, however, that these dolomites were formed in the Pleistocene, and are totally unrelated to the present meteoric-marine mixing zone. Dolomitization seems to have occurred during a period of eustatic sea level fall after the 122,000 BP sea level high stand (Fig.8.2). As sea level began to drop, and meteoric water began to mix with marine

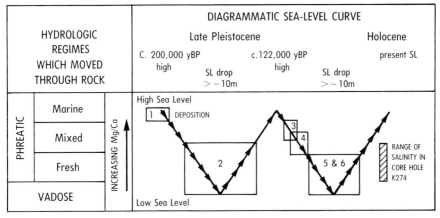

Fig. 8.2. Diagenetic events affecting Pleistocene carbonates in the Yucatan related to pore-water regimes as controlled by changes in sea level. The sea-level curve is not to scale in time or space; these rocks spent most of their history in the vadose zone. Used with permission of SEPM.

waters, dolomitization of grains and matrix ensued. As meteoric water influence increased, alternating zones of calcite and dolomite cements were precipitated in a complex zonal arrangement, as was reported by Land for the Hope Gate in Jamaica, and by Sibley for the Pliocene-Pleistocene in Bonaire (Fig. 8.2). At the point when meteoric water apparently began to dominate, calcite cements were precipitated (Fig. 8.2).

The stable isotopic composition of these dolomites (Fig. 8.1) suggests that they formed in waters ranging in composition from 75%-100% seawater. While these results certainly show some meteoric water involvement, the dolomitization process in the Yucatan seems to have been dominated by marine waters. The stable isotopic composition of the Yucatan dolomites is close to the values reported by Land for the Hope Gate dolomites in Jamaica, suggesting that the Hope Gate dolomites were also formed in marine-dominated waters.

Humphrey (1988) has recently described an interesting occurrence of dolomite in a 200,000-year-old, raised reef sequence on the island of Barbados that has significance relative to the question of mixing-zone dolomitization. Humphrey interprets these dolomites as mixing-zone dolomites based on petrography, geologic setting, and geochemistry. The Barbados dolomites are petrographically similar to the Yucatan and Jamaica occurrences described above. The Barbados dolomite cements, and their Yucatan counterparts, were precipitated in complex intercalations of calcite and dolomite, in a diagenetic framework dominated by meteoric-related fabrics. The most compelling evidence for a mixing-zone origin, however, rests in the very low $\partial^{13}C$ compositions of the dolomite (-10 to -20 permil, PDB) (Fig. 8.1). Humphrey believes that given the geologic setting, these light carbon isotopic compositions could only have been the result of the influence of soil gas CO_2. The wide variation of carbon isotopes and the consistent small variation of oxygen isotopic composition exhibited by the Barbados dolomites display trends similar to those reported by Land (1973b) for the Falmouth mixing zone dolomites in Jamaica (Fig. 8.1), and follow closely the isotopic signature for mixed-zone dolomites predicted by Lohmann (1988). Humphrey also suggests that these light carbon values constrain dolomitization to a mixture consisting of 95% meteoric water and only 5% marine water. Finally, Humphrey makes a compelling case (based on the well-defined geologic framework available for the Barbados Pleistocene) for these dolomites to have formed in approximately 5000 years during a falling sea level event.

The Barbados dolomites described above seem to be one of the most persuasive examples of dolomitization in a mixing zone setting described to date. However, dolomitization does not appear to be occurring in the present extensive mixing zone along the Barbados shoreline (Matthews, 1974), nor has significant dolomite been reported from any of the other raised Pleistocene reef tracts on Barbados. Nevertheless, with further study, Barbados may present a remarkable opportunity to extend the understanding of the mixed water diagenetic environment and dolomitization processes in this complex setting.

At this point, then, it can be said that dolomites seem to have formed in mixing zone settings in the past, and to a very limited extent, in some mixing zones today. These dolomites, however, do not necessarily obey the mixing model, as developed by Badiozamani (1973) (Fig. 6.3), because the dolomitizing fluids may be dominated either by marine, or almost pure meteoric waters, rather than the intermediate compositions predicted by the model. The dolomitization events seem to coincide with falling sea levels, rather than high stands or low stands. The fact that they may have formed over a very short time during active sea level movement belies the conventional wisdom that "the tremendous variation in the position of mixing zones in the last million or so years has apparently not been conducive to extensive dolomitization in many places" (Land, 1985, p. 120). Land (1980) and Hardie (1987) have clearly demonstrated that the stable isotopic and trace element geochemistry of ancient dolomites cannot adequately differentiate between dolomites forming in a mixed marine-meteoric water system from those forming under hydrothermal or marine-dominated systems. In those cases where dolomitizing fluids are dominated by meteoric vadose waters, the light carbon isotopic signature may indeed be diagnostic.

The problems involved in the recognition of mixing zone dolomites are compounded when faced with the very real possibility of recrystallization and chemical reequilibration of earlier-formed metastable dolomites with later diagenetic fluids, as was discussed in Chapters 3 and 5 (see also Land, 1980; Land, 1986; Hardie, 1987; Moore and others, 1988).

These uncertainties should certainly give the carbonate geologist pause when using the mixing model to explain the occurrence of an ancient dolomite, particularly if the interpretation is based dominantly on geochemical evidence, and other plausible explanations are available. Hardie's (1987) concerns summarize this problem: "All in all, we have been too quick to accept uncritically the mixing- zone concept which, on close scrutiny, proves to have very weak underpinnings. Rather than being hailed as a major breakthrough in the dolomite problem, the concept should be treated as another working hypothesis to be explored and tested alongside other working hypotheses of dolomitization, providing we can first establish the necessary diagnostic criteria by which to differentiate mixing-zone dolomites from dolomites of other origins".

The following discussion of a Mississippian reservoir in the Illinois basin presents a case history of dolomitization where setting, geologic history, and dolomite geochemistry are compatible with a mixed marine-meteoric water origin, and reasonable alternative dolomitization models cannot be applied.

Mississippian North Bridgeport field, Illinois basin, U.S.A.: mixed water dolomite reservoirs

The North Bridgeport field is a shallow (490 m, 1600 ft), relatively small accumulation that was discovered in the Mississippian Ste. Genevieve Formation in 1909 in the northeastern sector of the Illinois basin, along the LaSalle arch (Fig.8.3). The field has undergone a series of water flood projects since 1959, and has produced some 3.1 million barrels of 39° API gravity oil from a productive area of 1100 acres through March 1983. Ultimate recovery is estimated at 4.0 million barrels (Choquette and Steinen, 1985).

The North Bridgeport reservoir is developed in Ste. Genevieve ooid grainstones and dolomitized lime mudstones and wackestones. The ooid shoals, localized along the crest of the LaSalle arch, occur as elongate tidal bars and channels cutting the arch at a high angle (Fig. 8.4). They are believed to be analogous to the ooid shoals forming along the northwestern margin of the present Great Bahama Platform. The dolomitized

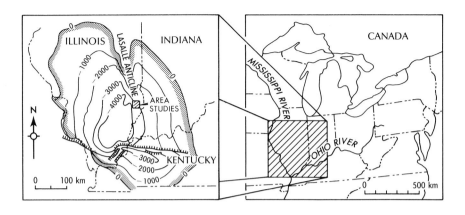

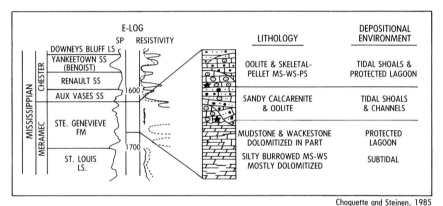

Fig. 8.3. Location maps and generalized Ste. Genevieve lithologic column. Map at left shows regional structure on top of the New Albany Shale (Upper Devonian-Lower Mississippian). Lithologic column is a conceptualized version of a cored sequence in a Bridgeport field well. Used with permission of SEPM.

mudstones and wackestones probably represent protected lagoonal environments associated with the tidal channels and shoals (Fig. 8.3) (Choquette and Steinen, 1985).

The ooid and dolomite reservoirs are generally associated with one another, with the dolomitized sequence usually appearing beneath, and/or laterally adjacent to, the ooid reservoirs and allied sandy calcarenite channel sequences (Fig. 8.4). Table 8.1 and Fig. 8.5 outline the differences in the pore systems supported by these two reservoir types. The ooid grainstones contain cement-reduced intergranular porosity averaging some 12%, with 275 md permeability. These rocks have no secondary porosity development. The dolomite reservoirs show an average 27% porosity and 12 md permeability developed in intercrystalline and moldic pore types. The molds represent dissolution of

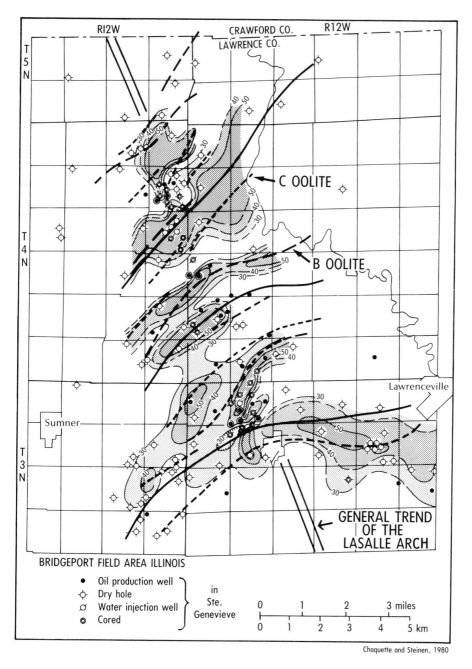

Fig. 8.4. Map of percent dolomite with carbonate-sand trends superimposed. Solid lines are channel-fill calcarenite trends, long and short dashed lines are B and C oolite trends respectively. Used with permission of SEPM.

various bioclasts and scattered ooids. Comparative porosity-permeability plots clearly illustrate the high porosity-low permeability characteristics of the dolomite sequences versus the low porosity-high permeability attributes of the ooid grainstone reservoirs (Fig. 8.5) (Choquette and Steinen, 1985).

Table 8.1. Comparison of some reservoir properties in oolite and dolomite reservoirs[1], Bridgeport field. Used with permission of SEPM.

	Oolite	Dolomite
Porosity types	Interparticle (BP) Intraparticle (WP)	Microintercrystal (mBC) Moldic (MO)
Inferred origin	Primary, reduced by cementation	Dissolution of $CaCO_3$
Pore size range (apparent)	$10^{-2} - 10^0$ mm	$10^{-4} - 10^{-2}$ mm (BC) $10^{-1} - 10^0$ mm (MO)
Porosity average range n	12% 2-22% 117	27% 13-40% 90
Permeability average range n	250 md 0.1-9500 md 115	12 md 0.7-130 md 87

[1] Porosity types follow Choquette and Pray (1970). Porosity and single-point gas permeability were determined by standard core analysis of ¾-inch (1.9-cm) diameter plugs drilled from cores.

Choquette and Steinen, 1985

The ooid grainstones, while not possessing secondary porosity, do show some evidence of subaerial exposure and meteoric water influence based on the occurrence of vadose meniscus and microstalactitic cements in the upper parts of the sand bodies. Meteoric-phreatic cements, however, are not common in the ooid grainstones. The common moldic porosity present in association with the dolomites is believed to represent significant meteoric water influence early in the diagenetic history of the Ste. Genevieve (Choquette and Steinen, 1980).

Choquette and Steinen (1980) suggested that the dolomitization of the lagoonal mudstones took place in a meteoric-marine mixing zone formed along the margins of the ooid shoals and sandy calcarenite channels. The meteoric waters were believed to be distributed by a gravity-driven regional meteoric aquifer system that used the porous channel and shoal systems as a distributional network (Fig. 8.6). Evaporative waters were not invoked because of the absence of any evidence for supratidal-sabkha conditions, or the presence of stratigraphically associated evaporative sequences

Stable isotope and trace element composition of the dolomite and associated calcite cements was used (Choquette and Steinen, 1980) to support a mixing zone origin for the dolomites (Fig. 8.7). The Ste. Genevieve dolomites have isotopic compositions

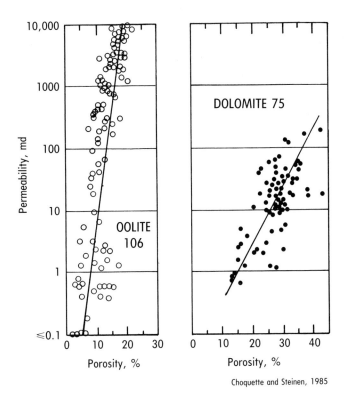

Fig. 8.5. *Semi-log plots of percent porosity vs air permeability in an oolite reservoir (left) and a dolomite reservoir (right) at Bridgeport field. Used with permission of SEPM.*

almost identical to those of the Hope Gate and the Yucatan dolomites described above. Therefore, these data suggest that if the Ste. Genevieve dolomites were actually formed in the mixing zone, the water composition may have been dominated by marine waters rather than meteoric waters, making meteoric water influence almost impossible to detect. The presence of intricate zoning in the dolomites, however, seems to preclude recrystallization and chemical reequilibration of an earlier, more unstable phase.

As illustrated in Choquette and Steinen's (1980) photomicrographs, the Mississippian ooids at North Bridgeport field are exceptionally well preserved, suggesting that they may originally have been composed of calcite. Sandberg (1983) indicates that the Mississippian was a time of calcite seas, implying that ooids precipitated during this interval should consist of calcite. If, indeed, a major meteoric water aquifer had been developed in these grainstones updip, the water would have quickly come to equilibrium with respect to calcite and little carbonate cement would be expected to precipitate as the water moved through the shoals. But where these waters began to mix with marine

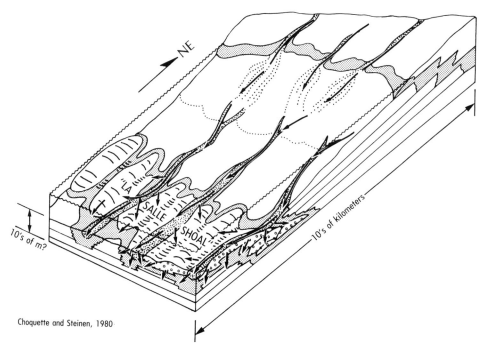

Fig. 8.6. Suggested hydrologic model for dolomitization of lime-mud sediments in upper Ste. Genevieve Limestone. Mixed meteoric-marine groundwaters fed from a recharge area northeast of the LaSalle paleoshoal, gained access to the lime muds via carbonate-sand conduits. Used with permission of SEPM.

waters, a zone of calcite dissolution would be expected, as is seen in the dolomite-associated sequences below and adjacent to the shoals (Matthews, 1974; Choquette and Steinen, 1985).

Geological evidence would seem to favor mixing-zone dolomitization, while geochemical evidence certainly does not seem to preclude the use of the mixed water model for the dolomite reservoirs at North Bridgeport.

Dolomitization by continental waters, Coorong Lagoon, south Australia

Modern dolomites are forming in the coastal plain of southeastern Australia in evaporative, ephemeral lakes fed by a regional, seaward-flowing, unconfined aquifer. The lakes are formed in depressions between Pleistocene ridges where the groundwaters emerge from the aquifer in spring lines near the sea (Fig. 8.8). Cores taken from a seaward-to-landward transect of ephemeral lakes and lagoons show that increasing amounts of dolomite occur in the more landward lakes (Fig. 8.9) (von der Borch and others, 1975).

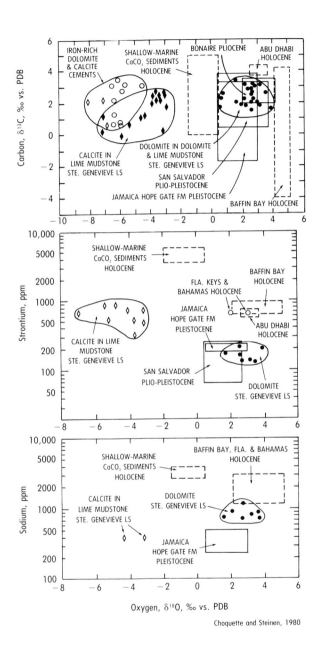

Fig. 8.7. Comparison of the stable isotopic and trace element composition of Ste. Genevieve dolomites and calcite cements with dolomites from various settings. Sources of data may be found in Choquette and Steinen (1980). Used with permission of SEPM.

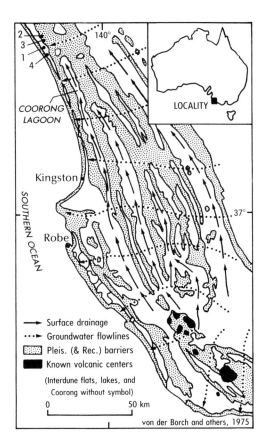

Fig. 8.8. Locality and physiography of the Coorong Lagoon southeastern Australia. Surface drainage shown in solid lines with arrows, groundwater flowlines in dotted lines with arrows. Originally published in Geology, 1975, v.3, pp. 283-285. Reprinted with permission.

Waters sampled from coastal versus inland sites show that water from the inland sites is meteoric, with low Mg/Ca ratios, but with high CO_3^{-2} compositions. On the other hand, water from the seaward sites shows elevated Mg/Ca ratios and low CO_3^{-2} contents (von der Borch and others, 1975). The dolomites in the remote lakes are well-ordered, very fine crystalline, and form a distinctive white mud layer. The sparse dolomites forming in the more marine lakes and lagoons are generally calcium-rich and distinctly disordered. Von der Borch and others (1975) suggest that the disordered dolomites in the lagoons and near lagoonal lakes formed from marine-to-mixed meteoric marine waters, while the ordered dolomites of the remote lakes were formed from meteoric waters. They also suggest that meteoric water leaching from surrounding volcanic terrains furnished much of the magnesium necessary for dolomitization. Their findings seem to support

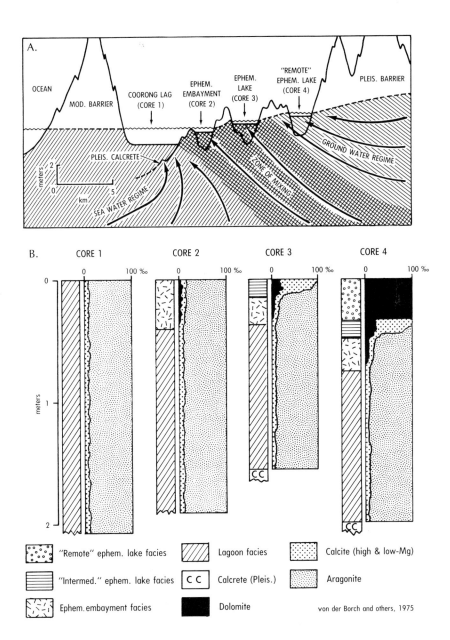

Fig. 8.9. (A) Cross section across the Coorong lagoon showing hydrologic setting, and location of cores shown in B. (B) Series of cores taken from Coorong Lagoon inland to a major Pleistocene barrier. Environments shown on the left, and mineralogy of each core to the right. Dolomite shown in black. Originally published in Geology, 1975, v.3, pp. 283-285. Reprinted with permission.

Morrow's (1982a) and Lohmann's (1988) contention that elevated CO_3^{-2} levels enhance dolomitization (see Fig. 6.2).

The dolomites observed in the Coorong region of south Australia (although an interesting occurrence of dolomite forming from continental, meteoric waters) seem to have only limited applicability as a model for the dolomitization of ancient rock sequences. Indeed, because of their stability at the time of formation and their extremely fine crystal size, it would seem that dolomites formed in this manner would tend to form reservoir seals rather than act as hosts for oil and gas.

SUMMARY

The mixing zone at the transition between marine and meteoric waters is an important diagenetic setting. While significant dissolution has been documented in this zone, perhaps the most important diagenetic characteristic is its perceived potential for dolomitization, with subsequent enhancement of porosity. The mixed, or Dorag dolomitization model, has been applied widely to the geologic record in a variety of geologic settings.

There is growing concern, however, that the Dorag model was originally based on equivocal thermodynamic concepts and inadequate modern analogues. Extensive subsequent work around the world has revealed only a few modern analogues for the model.

Recent investigations of Pleistocene examples of inferred mixed-zone dolomitization in Yucatan and Barbados do provide us with some new, interesting information on the setting of mixed-zone dolomitization. These studies indicate that mixed-zone dolomitization may take place rapidly, under conditions of falling sea level. In addition, dolomitization may take place in either meteoric or marine-dominated fluids, rather than the intermediate water mixtures predicted by the Badiozamani (1973) model. These data imply that isotopic composition of dolomites cannot adequately differentiate between mixed-water dolomites and marine, evaporative dolomites. The light carbon isotopic signature of vadose meteoric-dominated, mixed-water dolomites, however, seems to be a solid criterion for their recognition.

Because of these considerations, and the potential for early formed dolomite to recrystallize, it is believed that modern geological researchers place an inordinate amount of confidence in geochemical and stable isotopic criteria for the recognition of ancient mixed-zone dolomites. The Dorag, or mixed model, then, should be used with great care until adequate criteria can be developed for its recognition, and the model tested.

Finally, dolomitization by continental waters, such as at the Coorong Lagoon in southern Australia, while interesting and important to a fundamental understanding of the controls of dolomitization, may have limited applicability to ancient rock sequences.

Chapter 9

BURIAL DIAGENETIC ENVIRONMENT

INTRODUCTION

Early surficial porosity modifications, such as dissolution, cementation, and dolomitization, discussed in previous chapters, are generally accomplished in an infinitesimally small geological time frame. In fact, carbonate sequences spend most of their geologic history, measured in tens to hundreds of millions of years in the subsurface, in a diagenetic environment quite different from the environments encountered by newly formed carbonate sediments during the early stages of lithification and burial.

The nature of this alien environment, and the diagenetic processes that affect carbonate rock sequences have received increased attention during the '80s, as exploratory drilling has expanded into progressively deeper frontiers, as new and improved analytical techniques have become available, and as the numbers of researchers interested in all aspects of diagenesis have increased. Several excellent review papers and the work of a number of investigators have focused on the burial regimen, and have clearly demonstrated its importance in casting the ultimate evolutionary pathways of carbonate porosity (Bathurst, 1980; Moore and Druckman, 1981; Moore, 1985; Scholle and Halley, 1985; Bathurst, 1986; Halley, 1987; Choquette and James, 1987). Indeed, so much interest has been aroused that the importance of burial diagenetic processes in questions of carbonate porosity evolution may be overstated.

In this chapter the present state of knowledge of this important, but poorly known, diagenetic regimen will be presented, and a reasonable assessment of the fate of carbonate porosity during the burial process will be developed.

THE BURIAL SETTING

Introduction

In all previously discussed diagenetic environments, the primary emphasis was on surficial conditions, where temperatures remained near surface levels, pressure was not a consideration, and the rocks contained enough pore space to enable the free movement of diagenetic fluids in response to tides, waves, gravity, and evaporation. The starting materials were generally metastable, and pore fluids evolved rapidly by mixing or by chemical exchange during stabilization. The system often had access to significant, localized reservoirs of CO_2 that drove diagenetic processes at a relatively rapid clip. The major processes affecting porosity were dissolution and cementation.

The burial diagenetic environment starts when sedimentary sequences are buried beneath the reach of surface-related processes. It includes the mesogenetic or deeper burial regime of Choquette and Pray (1970), and may extend into the zone of low-grade metamorphism in the deeper reaches of some sedimentary basins (Fig. 9.1). While extending into the shallow subsurface, regional confined meteoric aquifers are treated as surficial environments because of surface meteoric recharge.

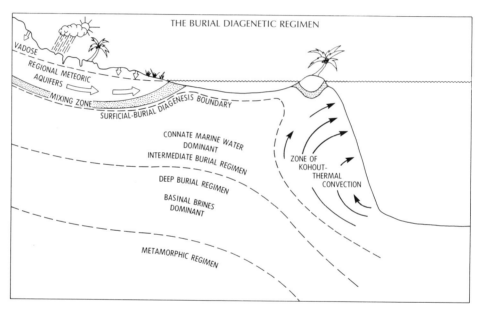

Fig. 9.1. Schematic diagram showing the relationship of surficial diagenetic environments to the several zones of the burial diagenetic regimen.

In this environment, diagenetic fluids are cut off from free exchange with chemically active gases of the atmosphere (most particularly oxygen and CO_2) temperature and pressure progressively increase, and the exchange of diagenetic fluids diminishes because of continuous porosity reduction. Pore fluids undergo a slow compositional evolution driven by rock-water interaction and mixing of basin-derived waters. The major process affecting porosity is compaction.

In the following paragraphs we will discuss some of the most important parameters of the burial setting, such as pressure, temperature, and water chemistry in order to set the stage for consideration of burial diagenetic processes and products.

Pressure

There are three types of pressure that influence sedimentary sequences during burial: hydrostatic, lithostatic, and directed pressure. Hydrostatic pressure is transmitted only through the water column, as represented by the sediment's pore system. Lithostatic pressure is transmitted through the rock framework, while directed pressure is related to tectonic stresses (Collins, 1975; Bathurst, 1975, 1980, 1986; Choquette and James, 1987). Fig. 9.2A illustrates the ranges of pressures that may be expected in a sedimentary basin setting. Hydrostatic pressure will vary, depending on fluid salinity and temperature. Most ambient hydrostatic pressures found during drilling will fall in the shaded area shown on Fig. 9.2A. The net pressure on a sedimentary particle in the subsurface is found by subtracting the hydrostatic pressure from the lithostatic pressure. The application of this net pressure on the rock mass results in strain, which is relieved by dissolution, and is the force that drives chemical compaction, or pressure solution.

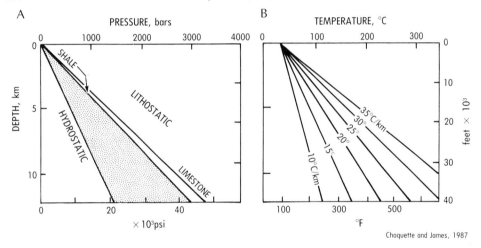

Fig. 9.2. Graphs showing the general ranges of pressure and temperature in the burial diagenetic regimen. (A) Static pressure variations. The position of the hydrostatic pressure curve varies depending on the concentration and density of pore waters. Most pore-fluid pressures in the subsurface would plot on curves in the stippled area. (B) Temperature ranges encountered in the subsurface assuming different geothermal gradients. Reprinted with permission of the Geological Association of Canada.

During burial, if interstitial water is trapped in the pore spaces while the sediment is undergoing compaction, hydrostatic pressure can increase dramatically and even approach ambient lithostatic pressure, resulting in very low net pressures at rock and grain interfaces. This condition is called *overpressuring*, or *geopressure* (Choquette and James, 1987). These low net pressures, acting at grain contacts, may slow and even stop physical and chemical compaction. In this case, the high fluid pressures act as an intergranular buttress, resulting in exceptionally high porosities under relatively deep burial conditions (Scholle and others, 1983) (see, for example, the North Sea chalk case history later in this chapter).

Geopressure may occur under conditions of rapid burial where aquacludes, such as evaporites, and shales, or even thin, cemented layers, such as marine hardgrounds or paleo-water tables, are present in the section. Geopressured zones tend to isolate pore fluids from surrounding diagenetic waters, preventing fluid and ion transfer and slowing or preventing diagenetic processes such as cementation (Feazel and Schatzinger, 1985). Hydrofracturing associated with geopressured zones may enhance porosity (as in the case of the Monterey Formation in California, as described by Roehl and Weinbrandt in 1985), but CO_2 degassing associated with the release of pressure during fracturing can lead to massive cementation (Woronick and Land, 1985).

Temperature

The progressive increase of temperature with depth into the subsurface is termed the *geothermal gradient*. Fig. 9.2B illustrates the range of temperatures that might be expected in the subsurface, assuming different geothermal gradients. Each sedimentary basin has a different thermal history, and therefore, geothermal gradient, as driven by burial history, sediment type, tectonic setting, and hydrology.

Taken alone, increasing temperature speeds chemical reactions and the rate of ionic diffusion. Increasing temperature, however, will decrease the solubility of carbonates because of the influence of CO_2 on the carbonate system. These relationships favor dolomitization, and the precipitation of calcite cements in the subsurface (Bathurst, 1986). The elevated temperatures found under burial conditions strongly influence the oxygen isotopic composition of subsurface calcite cements and dolomites because of the strong temperature dependency of oxygen fractionation between solid carbonate phases and water (Chapter 3; Anderson and Arthur, 1983).

Increasing both temperature and pressure during burial triggers a series of mineral reactions and phase changes that may release water and ions that can become involved in carbonate diagenetic processes in the subsurface. For example, the conversion of gypsum to anhydrite at about 1000 m releases significant water that may become involved in solution, cementation, or dolomitization (Kendall, 1984); and the conversion of smectite to illite commencing at some 2000 m and 60° C releases water of crystallization as well as cations such as Mg, which could be used in dolomitization (Boles and Franks, 1979; Wanless, 1979).

The application of elevated temperatures over extended time periods to sedimentary organic material leads to the catagenesis of the organic compounds and the formation of oil and gas (Tissot and Welte, 1978). Fig. 9.3 outlines the general framework of organic diagenesis relative to vitrinite reflectance, a measure of thermal maturity. Thermal maturity is a function of temperature and the time interval over which that temperature has been applied, and is generally independent of lithostatic pressure (Tissot and Welte, 1978).

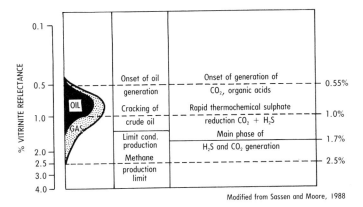

Fig. 9.3. *Framework of hydrocarbon maturation, destruction, and organic diagenesis as a function of vitrinite reflectance. Reprinted with permission of the American Association of Petroleum Geologists.*

The by-products of the later stages of organic catagenesis can dramatically influence deep burial carbonate diagenesis. As seen in Fig. 9.3, at the onset of oil generation and at a vitrinite reflectance of about .55%, CO_2 and organic acids are produced. At a vitrinite reflectance of 1.0%, rapid thermochemical sulfate reduction commences and CO_2 and H_2S are generated, reaching a peak at a vitrinite reflectance of about 1.7% (Sassen and Moore, 1988). These organic diagenetic processes can lead to the formation of aggressive subsurface fluids that may effect carbonate dissolution and produce secondary porosity (Schmidt and McDonald, 1979; Surdam and others, 1984). The $CaCO_3$ furnished to the pore fluids as a result of this dissolution phase is then available for precipitation as calcite cement or dolomite elsewhere in the system (Moore, 1985; Sassen and Moore, 1988).

Deep burial pore fluids

Our knowledge of subsurface fluids is garnered from samples of oil field waters gathered during initial testing for, or later production of, oil and gas. Fig. 9.4 illustrates the relative abundance of oil field waters of different salinities. Some 74% of these waters are classed as saline water or brines (10,000-100,000 ppm dissolved salts) and over 50%

of these waters are more saline than seawater. Most of the subsurface waters that have been analyzed are CaCl waters with major cations in the following proportions: Na>Ca>Mg (Collins, 1975). Those subsurface waters that are classified as brines (>100,000 ppm dissolved solids) are believed to be related to evaporites, originating either as evaporite interstitial fluids (Carpenter, 1978) or as the result of subsurface evaporite dissolution (Land and Prezbindowski, 1981). Those subsurface fluids with salinities less than sea water are believed to be mixtures of meteoric, marine, and perhaps, basinal fluids (Land and Prezbindowski, 1981; Stoessell and Moore, 1983).

In detail, the composition of subsurface fluids is complex and varies widely within and between basins because of mixing of chemically dissimilar waters, and continuous rock-water interaction (dissolution as well as precipitation) during burial (Collins, 1975; Carpenter, 1978; Land and Prezbindowski, 1981). Most subsurface waters, however, have very low Mg/Ca (Fig. 6.1), and tend to exhibit progressively higher oxygen isotopic values with depth, because of temperature fractionation effects (^{16}O is preferentially incorporated in the solid phase at elevated temperatures, resulting in higher concentra-

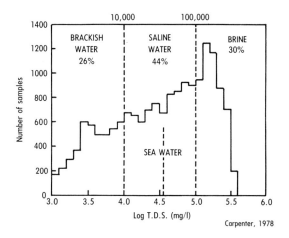

Fig. 9.4. Histogram illustrating the relative abundance of oil-field waters of differing salinities. Reprinted with permission of the Oklahoma Geological Survey.

tions of ^{18}O in the fluid phase). Land and Prezbindowski (1981) report $\partial^{18}O$ values for Cretaceous formation waters in excess of +20 ‰ (SMOW). See Table 3.1 for the mean composition of a Jurassic brine from the Gulf Coast of the U.S.

Hydrology of subsurface fluids

Most of our knowledge of hydrology concerns gravity-driven systems, such as were treated in the section on meteoric diagenetic environments. Deep basin hydrology, however, has assumed a position of importance, as investigators have taken a more integrated approach to basin studies, and as our exploration frontiers have deepened. Several important hydrologic factors can be identified that may play a significant role in burial diagenesis with an impact on subsequent porosity evolution.

During the compaction phase of burial, while significant porosity is still available, regional, thermally driven, connate pore fluid circulation is possible, and large-scale diagenetic events, such as cementation and dolomitization, may take place (Simms, 1984).

The up-section movement of highly saline basinal waters into, and through, shallow water carbonate shelf sequences is common around the margins of basins, as sedimentary basin sequences are progressively buried into the subsurface. Vertical brine migration generally follows fault and fracture zones until porous shelf sequences are encountered where aquacludes may allow lateral brine migration (Land and Prezbindowski, 1981; Moore and Druckman, 1981; Woronick and Land, 1985).

As porosity is destroyed during burial by compaction and cementation, the volume of water available for diagenesis is progressively diminished. As water volume decreases, its flux through the rock is slowed, and the diagenetic system changes from an open-water dominated system to a closed-rock-dominated system (Prezbindowski, 1985).

The diagenetic processes and products generated in the progressively harsher conditions of the burial environment will now be considered.

COMPACTION

Introduction

The one certainty in any discussion of carbonate porosity is that the major porosity trend encountered during the burial history of carbonate rock sequences seems to be one of overall porosity destruction (Choquette and Pray, 1970; Scholle and Halley, 1985; Choquette and James, 1987). The two principal mechanisms for porosity destruction are cementation and compaction.

Compaction includes mechanical compaction, dewatering, and chemical compaction, as exemplified by pressure solution along stylolites and between grains. Porosity destruction by compaction dominates those sedimentary sequences buried with marine pore fluids, such as pelagic oozes, rapidly subsiding shelf-margin sequences, and, in some cases, mud-dominated shelf-lagoon complexes (Scholle and Halley, 1985; Choquette and James, 1987). Chemical compaction is believed to be a significant source for later, porosity-occluding, subsurface cements.

Mechanical compaction and dewatering

As might be expected, mechanical compaction and dewatering is particularly important in mud-dominated sediments. Most of the work on compaction has centered on studies of pelagic oozes and chalks encountered during oil drilling activities in the North Sea, on the deep sea drilling program, and on the classic chalk outcrops of Great Britain and Europe (Scholle, 1971; Neugebauer, 1973; Schlanger and Douglas, 1974; Matter and others, 1975; Scholle, 1977; Garrison, 1981; Scholle and others, 1983).

Pelagic oozes commonly have a total porosity of up to 80%, consisting of 45% intragranular porosity and 35% intergranular porosity (Garrison, 1981). This intergranular porosity is close to that observed by Graton and Fraser (1935) for packed spheres used in their early porosity experiments, suggesting that most pelagic oozes have the initial porosity characteristics of a grain-supported sediment.

Matter and others (1975) observed a 10% porosity loss for these oozes in the first 50 m of burial. This porosity loss is approximately the same as that observed by Graton and Fraser (1935) in their packing experiments between unstable and stable packing geometries. Therefore, the initial gravitational porosity loss suffered by pelagic oozes is basically a mechanical rearrangement to a more stable grain geometry which takes place at about 70% porosity. At this point, the breakage of thin-walled foraminifera caused by loading may decrease the porosity further to some 60% (Fig. 9.5).

Mechanical compaction of muddy shelf sequences may take a path different from pelagic oozes because: the shape of the shelf mud particles are elongate needles versus the more equant shapes for the pelagic oozes (Enos and Sawatsky, 1981); there is a general lack of large volumes of intragranular porosity in the shelf muds; and significant (up to 40%) tightly-bound water sheaths surrounding the elongate particles of shelf muds commonly occur. Enos and Sawatsky (1981) indicated that the elongate aragonite needles were bipolar, thus attracting and binding water molecules. Under these conditions, the dewatering phase is probably the most important early compaction event in muddy shelf sequences, and would lead to the highest loss of porosity, perhaps down to 40% or less, before a stable grain framework is established (Fig. 9.5).

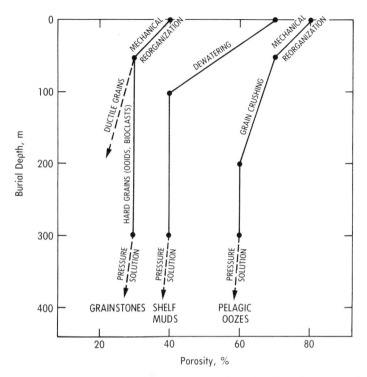

Fig. 9.5. Schematic diagram illustrating the porosity evolution of shelf grainstones and mud versus pelagic oozes during the first 400 m of burial. Curves based on data from Neugebauer (1973), Scholle (1977), Garrison (1981), and Enos and Sawatsky (1981).

The compaction experiments of Shinn and Robbin (1983), using modern, mud-rich sediments from many of the same environments and localities as sampled by Enos and Sawatsky (1981), seem to confirm the importance of the dewatering phase in shelf sediment compaction. They report a decrease of porosity from 70-40% under a load of 933 psi (approximately equivalent to 933 ft, or 298 m of burial) for mud-dominated sediment cores from Florida Bay. This porosity loss was not accompanied by significant grain breakage, or detectable pressure dissolution of the fine aragonitic matrix. This dramatic porosity drop was noted within the first 24 hours of the experiment. No further significant porosity decline was observed, regardless of the duration of time that the pressure was maintained. These results indicate that this initial porosity loss was dominantly a result of dewatering, and that an apparently stable grain framework is reached for this type of sediment at approximately 40% porosity under burial conditions of 100 m or less (Shinn and Robbin, 1983).

Grain-supported sediments, such as carbonate sands, will have a slightly different compactional history than that of the mud-dominated shelf sediments discussed above, and will more nearly parallel the early burial history of pelagic oozes. The first loss of porosity is associated with mechanical grain rearrangement to the most stable packing geometry, which, if the grains are spheres such as ooids, would involve a loss of up to 10% intergranular porosity (Graton and Fraser, 1935). The next phase of compaction for grain-supported sediments involves the mechanical failure of the grains themselves, either by fracturing, or by plastic deformation, both leading to the loss of porosity (Coogan, 1970; Bhattacharyya and Friedman, 1979) (Figs. 9.5 and 9.6).

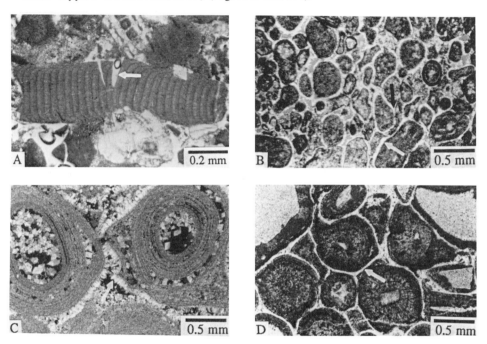

Fig. 9.6. (A) Halimeda fractured by burial compaction (arrow) under 762 m (2500') of limestone at Enewetak Atoll, western Pacific. Sample taken from Miocene section in well F-1. Plain light. (B) Compacted Jurassic Smackover Formation peloids under 9000' (2744 m) of burial, southern Arkansas. The peloids were coated by early isopachous cement that is also involved in compaction (arrow). Plain light. (C) Compaction of Jurassic Smackover Formation ooids that have undergone dissolution during stabilization. Early isopachous calcite cements are also involved in compaction (arrow). Murphy Giffco #1, 7788' (2374 m) Bowie County, Texas. Crossed polars. (D) Strongly compacted Jurassic Smackover Formation calcite ooids. Again note the early isopachous cement that was also involved in compaction (arrow). Phillips Flurry A-1, 19894' (6065 m) southern Mississippi, U.S.A.

The experimental work of Fruth and others (1966), however, suggests that while the mechanical failure during burial of hard, well-organized carbonate grains such as ooids and many bioclasts can be easily documented, the actual loss of porosity because of this failure may be less than 10% at confining pressures equivalent to some 2500 m overburden. Other grain types, such as fecal pellets and oncoids, which are soft or ductile at the time of deposition, or bioclasts that have very thin shell walls or have been partially dissolved during early diagenesis, may be particularly susceptible to mechanical compaction, failing by plastic deformation, or crushing, with an attendant loss of significant porosity (Fig. 9.6) (Rittenhouse, 1971; Enos and Sawatsky, 1981; Shinn and Robbin, 1983; Moore and Brock, 1982; see Walker Creek case history, Chapter 7)

Mechanical compaction effects in grain-supported sediments can be moderated by early cementation either in the marine environment or in the meteoric diagenetic environment, where intergranular cements tend to shield the enclosed grains from increasing overburden pressure during burial (Moore and Druckman, 1981; Shinn, 1983; Purser, 1978). In some cases, if the grains were originally soft, even early cementation may not be sufficient to stop grain failure and attendant porosity loss (Fig. 9.6).

Chemical compaction

Once a carbonate sediment has been mechanically compacted, and a stable grain framework has been established (regardless of whether the grains are mud-sized or sand-sized), continued burial will increase the elastic strain at individual grain contacts. Increased strain leads to an increased chemical potential, reflected as an increase in solubility at the grain contact, and ultimately resulting in point dissolution at the contact. Away from the grain contact, elastic strain is less, the relative solubility of calcium carbonate reduced, and ions diffusing away from the site of dissolution will tend to precipitate on the unstrained grain surfaces (Bathurst, 1975).

Pressure solution is manifested in a variety of scales from the microscopic, sutured, grain contacts generated during grain-to-grain pressure solution, to the classic, macroscopic, high-amplitude stylolites that generally cut across most textural elements of the rock along with a number of intermediate forms, such as horsetails, solution seams, and others (Figs. 9.7 and 9.8), (Bathurst, 1975; Wanless, 1979; Bathurst, 1984; Koepnick, 1984).

In grainstones, initial pressure solution is concentrated at grain contacts, resulting in microscopic sutured contacts with progressive grain interpenetration (Figs. 9.8A and B). During this stage, increasing lithostatic load is accommodated by individual grain solution. Ultimately, however, the rock will begin to act as a unit, and lithostatic load will then be accommodated by macroscopic stratal solution features, such as stylolites (Figs 9.8C thru F). It is unclear whether mud-dominated limestones undergo a similar compactional evolution. The dual process of pressure solution and cementation, described above, is commonly referred to as *chemical compaction* (Lloyd, 1977; Bathurst, 1984; Scholle and Halley, 1985; Choquette and James, 1987).

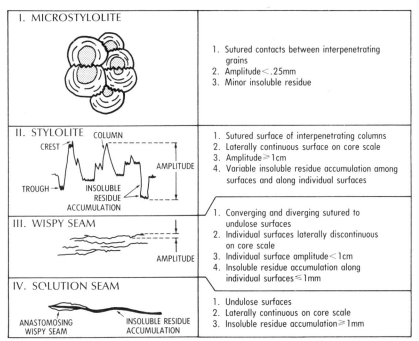

Fig. 9.7. Types and characteristics of pressure solution features encountered in the subsurface. Reprinted with permission of the Abu Dhabi National Reservoir Research Foundation.

The pressure solution process (ignoring cementation) involves the reduction of the bulk volume of the rock with a resultant loss in porosity (Rittenhouse, 1971). For example, a sediment composed of spheres, such as an oolite, with orthorhombic packing that loses 40% of its bulk volume to pressure solution will suffer a 44% reduction in its original porosity.

Pressure solution, however, is undoubtedly coupled with cementation, derived from the products of solution (Bathurst, 1975; Coogan and Manus, 1975; Bathurst, 1984; Scholle and Halley, 1985). Rittenhouse (1971) calculated the volumes of cement that could be derived from grain-to-grain pressure solution (Rittenhouse's work was based on siliciclastic sands, but the principles apply equally to carbonates) using a variety of packing geometries and textural arrangements.

Using Rittenhouse's (1971) method and his orthorhombic packing case, and assuming an original porosity of 40% and a bulk volume decrease of 30% by pressure

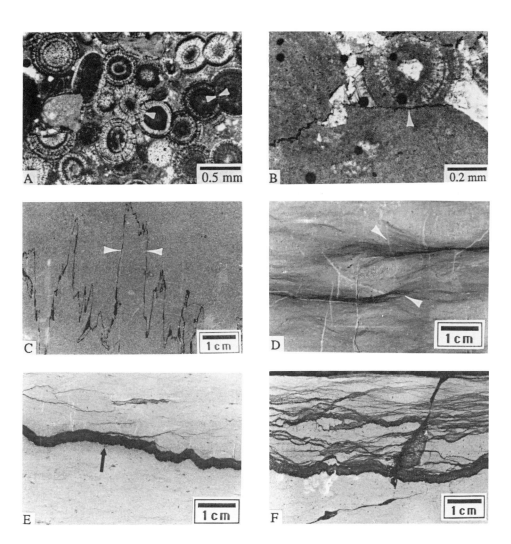

Fig. 9.8 (A) Strongly compacted Jurassic Smackover Formation ooid grainstone showing grain-to-grain pressure solution (arrows). I.P.C.#1, 9742' (2970 m), Columbia Co. Arkansas, U.S.A. (B) Grain-to-grain pressure solution between ooid and larger algally coated grains (arrow), Jurassic Smackover Formation, Arco Bodcaw #1, Walker Creek field, 10906' (3325 m), Columbia Co., Arkansas. (C) High amplitude stylolites (arrows) in Jurassic Smackover Formation, Getty #1 Reddoch 17-15, 13548' (4130 m), Clarke Co., Mississippi. (D) Wispy seam stylolites with "horse tails" (arrows) in the Jurassic Smackover Formation. LL and E Shaw Everett, 13657' (4164 m), Clarke Co., Mississippi. (E) Solution seam with insoluble residue accumulation (arrow) in the Jurassic Smackover. Same well as (D), 13603' (4147 m). Anastomosing wispy seams in the Jurassic Smackover Formation. Same well as (D), 13635' (4156 m).

solution, a decrease of 36% of the original porosity down to 25% can be calculated. With this 14% drop in original porosity, the potential production of only about 4% cement can be expected, assuming that the cementation process is 100% efficient (Fig. 9.9, case 1). We can, therefore, only reduce our original porosity down to approximately 21% by

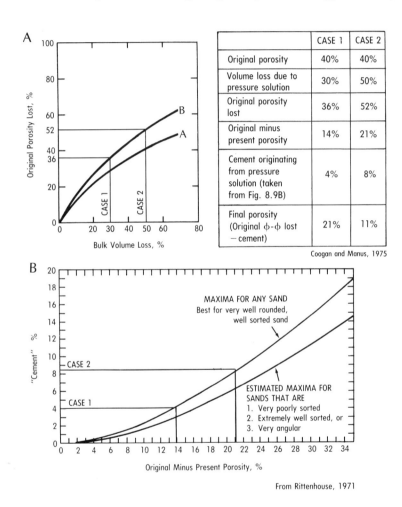

Fig. 9.9. *Volume of cement, and porosity loss that may be attributed to pressure solution. Starting porosity in both cases is 40%. In case 1, 30% of the total volume of the rock is lost by pressure solution, and the final porosity is 21%, while in case 2, 50% of the total volume of the rock is lost by pressure solution and the final porosity is 11%. (A) Original porosity lost plotted vs bulk volume loss. Curve B is orthorhombic packing. (B) Original minus present porosity plotted vs % cement. This is the curve used to estimate volume of cement that can be gained from pressure solution. Reprinted with permission of the American Association of Petroleum Geologists.*

chemical compaction, assuming a bulk volume decrease of 30% by pressure solution (Fig. 9.9, case 1). The 30% volume decrease used in the calculations above has been cited by several authors as common in the geological record (Coogan and Manus, 1975; Bathurst, 1975, 1984). The 30% figure was originally derived by measuring the accumulated amplitudes of megascopic stylolites; it ignores, among other things, volume lost by grain interpenetration, solution seams, and mechanical compaction (Bathurst, 1984). If the estimate of bulk volume lost during burial is increased to 50% (Fig. 9.9, case 2) to accommodate the possible effects of grain-to-grain pressure solution and mechanical compaction, the original porosity can be reduced down to approximately 11%. Even in this more extreme case, only 8% cement is produced as a result of chemical compaction (Fig. 9.9, case 2).

This exercise suggests that under some circumstances the actual volume reduction attributed to chemical compaction can be underestimated, while the volume of cement that can result from pressure solution processes can often be overestimated. It should be emphasized that the estimates of cement produced by pressure solution using the Rittenhouse technique assumes a 100% efficiency, a level seldom attained in nature.

Factors affecting the efficiency of chemical compaction

Burial depth with attendant lithostatic load is one of the most critical factors determining the onset and ultimate efficiency of chemical compaction. Dunnington (1967), in his careful analysis of compaction effects on carbonate reservoirs, indicated that 600 m was a minimum burial depth for the onset of pressure solution. Buxton and Sibley (1981) deduced from stratigraphic reconstructions, that chemical compaction took place under some 1500 m burial in the Devonian of the Michigan Basin. Neugebauer (1973, 1974) stated that while minor solution takes place in the first 200 m of burial, significant pressure solution does not commence in chalks until at least 1000 m of burial. Saller (1984a) observed no pressure solution in uncemented grainstones in cores taken at Enewetak Atoll until some 800 m of burial. Meyers and Hill (1983), however, believed that pressure solution in Mississippian grainstones started after only tens to hundreds of meters of burial (see discussion of these grainstones in Chapter 7). In addition, Bathurst (1975) cited numerous occurrences of shallow burial onset of pressure solution. These occurrences clearly indicate that sweeping generalizations concerning minimum burial needed for initiation of chemical compaction cannot be made because of the many factors, as noted in the following paragraphs, that may enhance or retard the process (Table 9.1).

The original mineralogy of the sediment undergoing compaction is a very important control over the nature of its early chemical compactional history. A

Table 9.1

FACTORS AFFECTING CHEMICAL COMPACTION

FACTORS ENHANCING CHEMICAL COMPACTION	FACTORS RETARDING CHEMICAL COMPACTION
1. Metastable (i.e.: aragonite) mineralogy at the time of burial (Wagner and Matthews, 1982)	1. Stable (i.e.: calcite, dolomite) mineralogy at the time of burial (Moshier, 1987)
2. Magnesium poor meteoric waters in pores (Neugebauer, 1973)	2. Oil in pores (Dunnington, 1967; Feazel and Schatzinger, 1985)
3. Insolubles such as clays, quartz, and organics (Weyl, 1959)	3. Elevated pore pressures as found in geopressured zones (Harper and Shaw, 1974)
4. Tectonic stress (Bathurst, 1975)	4. Organic boundstones (Playford, 1980)

mud-bearing sediment composed of metastable mineral components (aragonite) should be much more susceptible to early chemical compaction and chemically related porosity loss than the stable (less soluble) calcite mineralogy characteristic of deep marine pelagic oozes and chalks (Scholle, 1977). If the sediments are buried with marine pore waters, the solubility of the particles in contact will determine the timing of the onset of pressure solution, and the metastable minerals such as aragonite should begin to dissolve well before calcite or dolomite (see discussion on the Jurassic Walker Creek case history detailed in Chapter 7; Wagner and Matthews, 1982; Scholle and Halley, 1985).

While most shallow marine, mud-bearing carbonate sequences do lose their porosity early in their burial history (Choquette and Pray, 1970; Bathurst, 1975; Wilson 1975; Moshier, 1987), mud matrix microporosity is common and very important in certain Mesozoic reservoirs, particularly in the Middle East (Wilson, 1980). Some researchers relate this microporosity to dissolution associated with unconformity exposure (Harris and Frost, 1984). Moshier (1987), however, makes a compelling case for an original calcite mineralogy for the former Cretaceous shallow marine muds, and implies that the microporosity is preserved depositional porosity because of the mineralogical stability of the original mud that allows the sequence to follow the characteristic porosity evolution of chalks and pelagic oozes. Moshier's thesis is consistent with Sandberg's view (1983) that the Cretaceous was a calcite sea in which most abiotic components (including mud-sized material precipitated as whitings across shallow marine shelves) would be formed as calcite.

The chemistry of pore fluids in contact with a sedimentary sequence during chemical compaction may have a dramatic impact on the rate of porosity loss during burial. Neugebauer (1973, 1974), in his experimental work, and Scholle (1977), in his field and subsurface studies, both indicate that meteoric waters (Mg poor) enhance the process of chemical compaction, as compared to marine pore fluids. On the other hand, the early introduction of hydrocarbons into porespace can totally stop solution transfer and hence chemical compaction (Dunnington, 1967; Feazel and Schatzinger, 1985).

Pore fluid pressures can dramatically affect the rate of chemical compaction. In over-pressured settings, where hydrostatic pressures begin to approach lithostatic load, the stress at grain contacts, and therefore, the strain on the mineral, is reduced, and pressure solution is slowed or stopped. This effect has been demonstrated in the North Sea (Harper and Shaw, 1974; Scholle, 1977) where chalks exhibit porosities of some 40% at 3000 m burial in zones of over-pressuring. Using Scholle and others' (1983) burial curve for chalks (Fig. 9.12), this porosity would be expected in chalks encountered at 1000 m of burial or less.

Insolubles, such as clay and organic material, seem to enhance chemical compaction in fine-grained carbonate sequences (Choquette and James, 1987). Weyl (1959) and Oldershaw and Scoffin (1967) observed the preferential formation of stylolites in clay-rich sequences, while Sassen and others (1987) related stylolite intensity to fine-grained, organic-rich, source beds in the Jurassic of the Gulf of Mexico.

Any factor that increases stress at the point of contact between carbonate grains will tend to enhance chemical compaction. Tectonic stress, such as folding or faulting, concentrates stress into well-defined zones, leading to tectonically controlled concentration of pressure-solution-related phenomena such as stylolites (Bathurst, 1984). Miran (1977) described stylolitization, and massive cementation triggered by tectonism involving the Cretaceous chalk of England.

The temporal subsidence and thermal history of a basin should be expected to affect the efficiency of the chemical compaction process. In siliciclastics, the effect of thermal regimen on porosity-depth relationships has long been known. Loucks and others (1979) documented a greater loss of sandstone porosity with depth in south Texas, than in Louisiana on the U.S. Gulf Coast, because of the much higher geothermal gradient of south Texas, resulting in more efficient chemical compaction in that region (Fig. 1.12). Lockridge and Scholle (1978) could detect no difference in the porosity-depth curves for chalks between North America and the North Sea, even though the geothermal gradients in the North Sea are 1.5 to 2 times higher than encountered in North America. Schmoker (1984), however, has generated a compelling case for subsurface porosity loss (presumably driven by chemical compaction), depending exponentially on temperature and linearly on time, and hence operating as a function of thermal maturity of the basin of deposition. This important topic will be examined further in the summary discussion of carbonate porosity evolution later in this chapter.

Finally, boundstones, will commonly, as a facies, show little evidence of compaction of any type, because of the presence of an integral biologic framework and intense cementation at the time of deposition. The porosities of reef-related lagoonal, shelf, and slope sequences, however, are significantly reduced by mechanical as well as chemical compaction. Shelf margin relief relative to adjacent shelf and periplatform facies is believed to have been enhanced in the Devonian sequences of the Canning Basin by the resistance of the well-cemented shelf-margin reefs to compaction, while adjacent muddy and grainy facies show abundant stylolitization (Playford, 1980). Wong and Oldershaw (1981) demonstrated a similar facies control over chemical compaction in the Devonian Kaybob reef, Swan Hills Formation, Alberta (see Fig. 9.23 for location).

  The North Sea Ekofisk field: a case history of porosity preservation in chalks

Ekofisk field is a giant oil field developed in Upper Cretaceous and Lower Tertiary chalks in the Norwegian and Danish sectors of the central graben of the North Sea Basin (Fig. 9.10). Discovered in 1969, it had produced 715 million barrels of oil until 1983, with

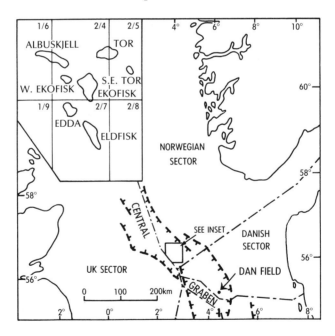

Fig. 9.10. Location of Ekofisk field and other major Mesozoic chalk fields in the Central Graben area of the North Sea. From Scholle and others, 1983. Reprinted with permission of the American Association of Petroleum Geologists.

an ultimate recovery estimated at 1.2 billions of barrels of oil out of an estimated 5.4 billions of barrels of oil in place (Feazel and others, 1985). The chalk fields of the central graben are anticlinal structures developed as a result of Permian salt movement. The reservoirs are sealed by overpressured Late Tertiary shales (Fig. 9.11). At present, the reservoirs are encountered between 2930 and 3080 m.

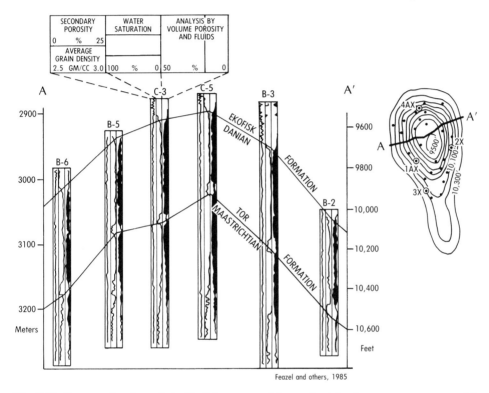

Fig. 9.11. Cross section through the Ekofisk structure. Map to the right shows structure on top of the Ekofisk Formation. Black area to the right of one set of curves shows amount of oil-filled porosity. Reprinted with permission from Carbonate Petroleum Reservoirs. Copyright (C), 1985, Springer-Verlag, New York.

The reservoirs are Danian and Maastrichtian chalks deposited slowly in the central graben area in relatively deep waters as a pelagic rain, and rapidly as mass flow deposits originating along the margins of the graben (Scholle and others, 1983; Taylor and Lapré, 1987). Salt-related structures were present within the graben during the Late Cretaceous and Early Tertiary and may have affected the deposition of the chalks (Feazel and others, 1985). Oil production is from preserved primary porosity that ranges from 25 to 42%. Permeabilities are low, averaging about 1 md. Production tests indicate effective permeabilities of over 12 md as a result of the extensive fracturing common to most of the chalk fields of the central graben. The reservoirs are sourced from the underlying Jurassic Kimmeridgian shales (Scholle and others, 1983).

The biggest surprise concerning production from the Ekofisk field is the extraordinary level of primary porosity encountered in chalks buried to depths of some 3000 m (Fig.9.11). Based on his studies of European and North American chalk sequences, Scholle (1977) suggests that the original porosity of most chalks should be completely destroyed by mechanical and chemical compaction by this depth (Fig. 9.12).

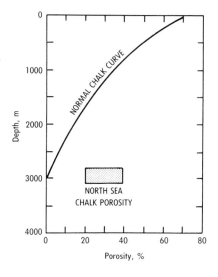

From Scholle and others, 1983

Fig. 9.12. Range of North Sea chalk porosity encountered at a depth of 3000 m compared to an average chalk burial-porosity curve. Reprinted with permission of the American Association of Petroleum Geologists.

Most geologists at Ekofisk agree that reservoir overpressuring is one of the major contributory factors to porosity preservation in the greater Ekofisk area (Scholle, 1977; Van den Bark and Thomas, 1981; Feazel and others, 1985; Taylor and Lapré, 1987). Fig. 9.13 is a generalized depth-pressure plot for a typical Ekofisk well. At 3050 m, the average depth of Ekofisk reservoirs, normal pore pressures should be approximately 4300 psi and lithostatic pressure 9000 psi. However, pore fluid pressures of over 7000 psi have been reported at that depth, reducing the net lithostatic load responsible for compaction to a level of approximately 1900 psi. This pressure is equivalent to about 1000 m burial depth, and using Scholle and others' (1983) chalk burial curve (Fig. 9.12) one would expect to encounter porosities in the range found at Ekofisk.

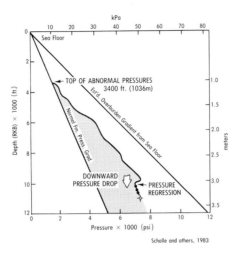

Fig. 9.13. Generalized pressure-depth plot for a typical Ekofisk well in the Central Graben of the North Sea compared to expected formation pressures. Abnormal pressures are encountered in most of the lower Tertiary and Upper Cretaceous section including the chalk reservoirs at 3 to 3.5 km depth. Reprinted with permission of the American Association of Petroleum Geologists.

The early migration of oil into the reservoirs may be another important contributory factor affecting porosity preservation at Ekofisk. The geothermal gradients coupled with the burial history of the Jurassic source rocks point toward early maturation and migration in the region (Harper, 1971; Van den Bark and Thomas, 1981). As oil displaces water in the pores, the chemical compaction process will shut down because solution and precipitation of carbonate require a film of water.

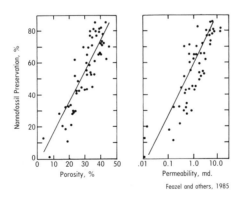

Fig. 9.14. Correlation of nannofossil preservation and rock properties as measured from core samples in the Ekofisk field. Reprinted with permission from Carbonate Petroleum Reservoirs. Copyright (C), 1985, Springer-Verlag, New York.

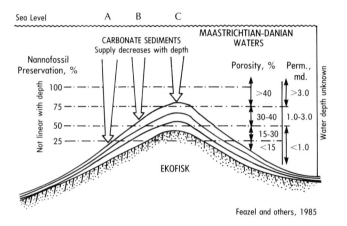

Fig. 9.15. *Depositional model for chalks, central North Sea reservoirs. Relationships illustrated from cores taken at wells A, B, C. Reprinted with permission from Carbonate Petroleum Reservoirs. Copyrght (C) 1985, Springer-Verlag, New York.*

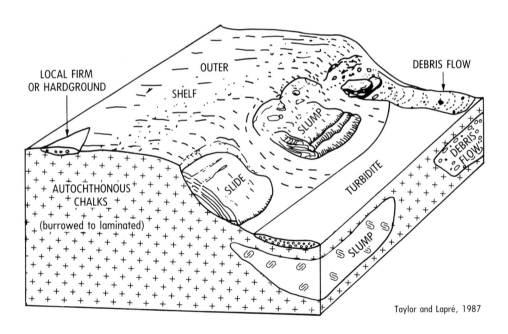

Fig. 9.16. *Depositional model for chalks emphasizing slope failure and transport of autochthonous chalks in the moderately deep marine environment of the Central Graben area of the North Sea. Reprinted with permission from Petroleum Geology of Northwest Europe. Copyright (C) 1986, Petroleum Geology 86 Limited.*

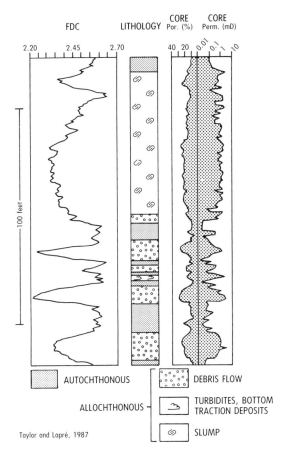

Fig. 9.17. Schematic of a core log of a well characteristic of the Ekofisk field in the Central Graben showing relationship between lithology and reservoir characteristics. Note that highest porosity and permeability coincides with allochthonous sequences. Reprinted with permission from Petroleum Geology of Northwest Europe. Copyright (C) 1986, Petroleum Geology 86 Limited.

Finally, a number of researchers (Feazel and others, 1985; Taylor and Lapré, 1987) contend that there is also an early sedimentological control on porosity in the Ekofisk field. Feazel and others (1985) have noted that there seems to be a direct relationship between the state of preservation of the nannofossils comprising the chalk and the porosity and permeability of the chalk (the best preservation being associated with the highest porosity) (Fig. 9.14). The model that they invoke is pelagic sedimentation in deep water over sea-floor topography formed by penecontemporaneous Permian salt movement (Fig. 9.15). Pelagic oozes deposited on the crest of salt structures would be above the lysocline (zone of rapid calcite dissolution, located above the calcite compensation depth), and would undergo no dissolution. At this site, the rate of sedimentation would be relatively rapid and there would be little chance for marine cementation and the formation of hardgrounds. The oozes deposited in deeper water down the flanks of the structures, however, would suffer progressively greater dissolution, the rate of actual sedimentation would drop, and the likelihood of marine cementation would increase. The crest of the salt structures would then support a greater chalk thickness, with no early porosity-occluding marine cements.

Taylor and Lapré (1987), on the other hand, believe that extensive mass-movement of the chalks destroyed early sea floor cements, and deposited poorly packed sediments rapidly across future reservoir areas. These rapidly deposited allochthonous chalks were more porous than the slowly deposited *in situ* chalks because they were not subjected to as intensive bioturbation and resulting dewatering during deposition (Fig. 9.16). This porosity contrast was retained into the subsurface. During subsequent burial, chemical compaction reduced the porosity of both chalk types equally until compaction was shut down by hydrocarbon migration into the reservoirs at about 1000 m burial depth.

Fig. 9.17 illustrates a typical section through Cretaceous chalks in the Ekofisk area showing the close relationship between high porosity/permeability and interpreted allochthonous chalk facies. The Taylor-Lapré model seems to imply that salt movement took place after deposition of most of the productive chalk section. The Feazel and others model, however, demands significant sea floor relief associated with salt movement during deposition of the productive chalks. Feazel and Farrall (1988; Feazel personal communication) have since indicated that this depositional model was probably flawed, and that hydrocarbon migration in the structure, shutting down chemical compaction, could explain most of the features observed at Ekofisk.

The exceptional production characteristics of Ekofisk reservoirs (the discovery well had an IP of over 10,000 barrels of oil per day, Feazel and others, 1985) are the result of high porosity, thick productive intervals (over 150 m), low oil viscosity (36° API gravity), high GOR (1547:1, Feazel and others, 1985), and intensive fracturing, which enhances their rather low permeabilities.

BURIAL CEMENTATION

The problem of source of $CaCO_3$ for burial cements

The passive introduction of cement into pore space requires a ready source of $CaCO_3$ during the precipitational event. In surficial diagenetic environments, large volumes of cement are precipitated into pore spaces prior to significant burial (James and Choquette, 1984). The $CaCO_3$ needed for these cements is provided directly from supersaturated marine and evaporated marine waters, or indirectly, with the $CaCO_3$ coming from the dissolution of metastable carbonate phases in various meteoric and mixing zone waters. In each case, the rate of precipitation and the ultimate volume of cement emplaced is influenced strongly by the rate of fluid flux through the pore system.

In the subsurface, while the volume of cement is not as large as encountered in pre-burial environments, burial cements have been documented from a number of sedimentary basins (Meyers, 1974; Moore, 1985; Niemann and Read, 1988). Under burial conditions, however, supersaturated surface marine waters are not available, a ready reservoir of atmospheric and soil gas CO_2 is not available, and most subsurface fluids are in equilibrium with $CaCO_3$. Under these conditions, neither solution nor precipitation tends to take place (Matthews, 1974; Choquette and James, 1987). If metastable mineral grains survive early surficial diagenesis and are buried into the subsurface, a solubility contrast that must be satisfied by dissolution and precipitation is created (see Chapter 7). Mineralogical stabilization in the subsurface tends to takes place by microsolution and precipitation because subsurface diagenetic fluids move slowly and are so near equilibrium with the surrounding rock. In this case, then, solution and precipitation take place in a closed system and little $CaCO_3$ leaves the grain to be made available for cementation. While increasing temperature does favor cementation in the subsurface (Bathurst, 1986), an adequate source of $CaCO_3$ to form the cement is an obvious problem (Bathurst, 1975).

Pressure solution, active during chemical compaction, is favored as a $CaCO_3$ source for the bulk of the burial cements encountered in the record (Meyers, 1974; Bathurst, 1975; Scholle and Halley, 1985; Choquette and James, 1987). In the discussion of chemical compaction earlier in this chapter, however, it was pointed out that an enormous amount of section must be removed by pressure solution to form a relatively small volume of cement. It seems obvious that the $CaCO_3$ provided by pressure solution must be supplemented either by $CaCO_3$ gleaned from late dissolution of limestones and dolomites by aggressive fluids formed during the diagenesis of organic material under elevated temperatures in the subsurface, or by importing it from adjacent subsurface carbonate units also undergoing chemical compaction (Bathurst, 1975; Schmidt and McDonald, 1979; Moore and Druckman, 1981; Moore, 1985; Sassen and Moore, 1988).

The following sections will briefly outline those criteria useful for the recognition of subsurface cements, then evaluate the impact of subsurface cementation on reservoir porosity evolution.

Petrography of burial cements

The following discussion will only consider those cements precipitated in the subsurface from waters that are isolated from surficial influences. Regional artesian meteoric aquifer systems were considered in the previous chapter and will not be discussed here (Fig. 9.1)

In general, subsurface cements are coarse, commonly poikilotopic, dully luminescent, and not as inclined to luminescent zoning as their surficial counterparts (Moore, 1985; Choquette and James, 1987). These burial cements are commonly either calcite, or saddle dolomite (Figs. 9.18B-F).

While burial cements will invariably exhibit two-phase fluid inclusions, the presence of these inclusions in a cement is not a valid criterion for late subsurface origin. Moore and Druckman (1981) noted that Upper Jurassic ooid grainstones were the host for early, precompaction meteoric cements that contained common two-phase fluid inclusions, presumably as a result of physical reequilibration in the subsurface (see also the discussion of fluid inclusions in Chapter 3, this book)

Perhaps the most reliable criterion for the late subsurface origin for a cement is the determination that the cement clearly postdates burial phenomena such as fractures; compacted and fractured grains; stylolites; and the products of the thermal degradation of hydrocarbons, such as pyrobitumen (Moore, 1985; Choquette and James, 1987) (see Chapter 3, this book) (Fig. 9.18).

Geochemistry of burial cements

Subsurface cements are precipitated from fluids that are in the process of evolving, chemically, as a result of continuous rock-water interaction in the burial diagenetic environment over extremely long periods of time. These rock-water interactions are not confined specifically to carbonates, but commonly include related basinal siliciclastics and evaporites. Rock-water interaction with such a mineralogically diverse rock mass often results in the formation of progressively more concentrated brines rich in metals such as Fe, Mn, Pb and Zn (Collins,1975; Carpenter, 1978; Land and Prezbindowski, 1981).

Burial cements, then, commonly are Fe and Mn rich, with late saddle dolomites often containing over 5 wt.% Fe. Lead and zinc mineralization is often associated with these late Fe-rich cements and dolomites (Carpenter, 1978; Prezbindowski, 1985; Moore, 1985). Sr concentrations, however, are commonly quite low, even though many subsurface brines exhibit high Sr/Ca ratios (Collins, 1975; Moore, 1985). Moore and Druckman (1981), and Moore (1985) indicate that the slow rate of precipitation that

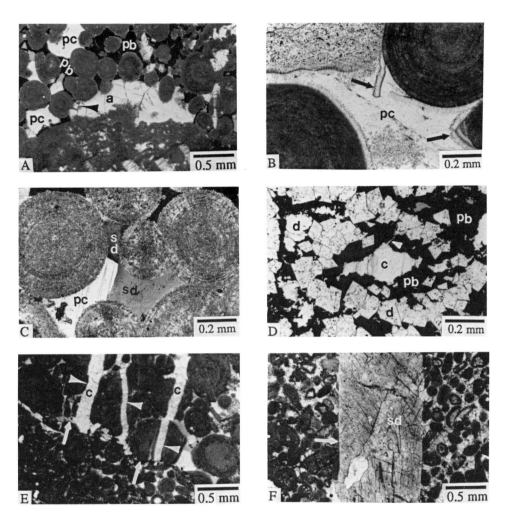

Fig. 9.18. A. Jurassic Smackover with intergranular pores filled with bitumen (pb). Late anhydrite (a) filling pores and replacing grains (arrow). poikilotopic calcite (pc) is late postcompaction. Getty Masonite 18-8, 13,613' (4150 m) Clarke Co., Miss. Plain light. B. Postcompaction poikilotopic calcite cement (pc). Note spalled, early, circumgranular calcite cement (arrow) incorporated in later cement. Same well as (A), 13,550' (4131 m). Plain light. C. Postcompaction poikilotopic calcite cement (pc) and saddle dolomite (sd) filling pore space in Jurassic Smackover. Arco Bodcaw #1, 10,910' (3326 m). Crossed polars. D. Pervasive dolomite (d) in Jurassic Smackover. Porosity filled with bitumen (pb). Mold filled with post-bitumen calcite (c). Phillips Flurry A-1, 19,894' (6065 m), Stone Co. Miss. Sample furnished by E. Heydari. Plain light. E. Calcite (c) filled tension gash fractures (small arrows) associated with stylolites (large arrows). Smackover Formation, Getty #1 Reddoch, Clarke Co., Mississippi, 13,692" (4174 m). Sample furnished by E. Heydari. Plain light. F. Large fracture (arrow) filled with saddle dolomite (sd) in Jurassic Smackover, Evers and Rhodes #1, 11,647' (3551 m), east Texas. Plain light.

characterizes subsurface calcite and dolomite cements may be responsible for the apparent, very small Sr distribution coefficients indicated for these cements (see Chapter 3, this book for a discussion of distribution coefficients).

The oxygen isotopic compositions of burial calcite and dolomite cements are clearly affected by the strongly temperature-dependant fractionation of oxygen isotopes between cement and fluid (Anderson and Arthur, 1983; Chapter 3, this book). As temperature increases during burial, the calcites precipitated should exhibit progressively lower oxygen isotopic values (Moore and Druckman, 1981; Moore, 1985;

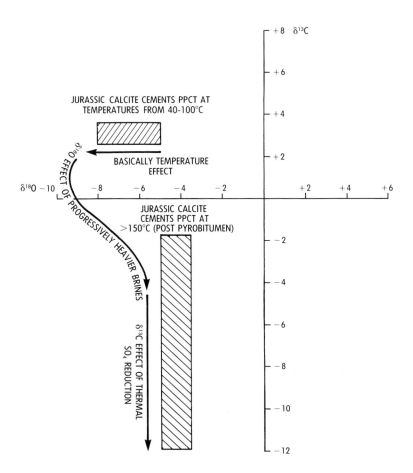

Fig. 9.19. Stable isotopic composition of burial calcite cements showing the effects of increasing temperatures, increasingly heavier brines, and thermal sulfate reduction on the oxygen and carbon isotopic composition. Data on 40 to 100°C calcites, Moore (1985); post pyrobitumen calcites, Heydari and Moore (1988).

Choquette and James, 1987). However this same fractionation process during subsurface rock-water interaction leads to progressively higher $\partial^{18}O$ values in subsurface fluids (Land and Prezbindowski, 1981). The progressive decrease of ^{18}O values in the cement, therefore, may ultimately be buffered, and even reversed, by the increasingly higher $\partial^{18}O$ values in the pore fluid through time (Heydari and Moore, 1988). Compare the stable isotopic compositions of the two Jurassic burial calcite cement populations shown on (Fig. 9.19). The early, low temperature population shows the normal temperature-dependent trend toward lower oxygen isotopic values, while the later, higher temperature population exhibits higher $\partial^{18}O$ values as a result of being precipitated from basinal fluids with an oxygen isotopic composition perhaps as high as $+15°/_{oo}$(PDB) (Heydari and Moore, 1988).

The carbon isotopic composition of most subsurface cements is generally rock-buffered as a result of chemical compaction, so will show little variation, and commonly have high $\partial^{13}C$ values. However, there is usually a trend toward slightly depleted ^{13}C compositions in those cements precipitated shortly before hydrocarbon migration. This trend is probably the result of the presence of increasing volumes of ^{12}C as a by-product of hydrocarbon maturation (Moore, 1985) (Fig. 9.17).

Very late cementation events, particularly those involved with late sulfate reduction associated with thermal degradation of hydrocarbons at temperatures in excess of 150° C, can exhibit very light carbon isotopic compositions (Heydari and others, 1988). Fig. 9.19 shows the general stable isotopic trends commonly associated with late subsurface cements.

Sr isotopes can be quite useful in the study of late burial cements. During the chemical compaction phase of burial diagenesis, before pore fluids have evolved by mixing with basinal brines, the Sr isotopic composition of burial cements will generally represent the Sr isotopic composition of the compacting limestones and dolomites, and therefore will reflect the Sr isotopic composition of the marine water in which the limestones and dolomites formed. Basinal fluids commonly have the opportunity to interact with siliciclastic, clay-rich sequences and hence are often enriched with respect to ^{87}Sr. As these pore fluids arrive (generally, shortly before hydrocarbon migration) and mix with marine connate waters in carbonate shelfal sequences, the $^{87}Sr/^{86}Sr$ ratio of the pore fluid, as well as the cements precipitated from the pore fluid, increases. The $^{87}Sr/^{86}Sr$ ratio in the burial cement may be a measure of the relative proportions of basinal brine and connate water present at the time of cement precipitation, and hence may reflect the stage of pore fluid evolution at the time of the cementation event (See Fig. 9.20) (Moore, 1985; Heydari and Moore, 1988).

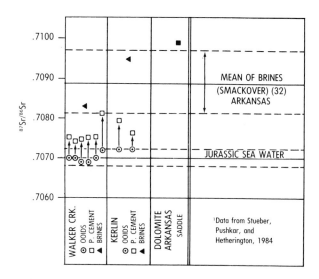

Fig. 9.20. Strontium isotopic composition of carbonate components and formation waters from oil fields in southern Arkansas, U.S.A. Jurassic seawater value of .7070 is taken from the curve of Burke and others, 1983. Reprinted with permission of the American Association of Petroleum Geologists.

Impact of late subsurface cementation on reservoir porosity

Compared to pressure solution, postcompaction subsurface cements, excluding regional meteoric aquifer systems, are probably of minor importance in reducing the total pore volume available for hydrocarbons in a subsurface reservoir. Several observations support this conclusion. The actual volume of late, postcompaction cements documented in hydrocarbon reservoirs is relatively small (Moore, 1985; Scholle and Halley, 1985; Prezbindowski, 1985). These data support Rittenhouse's conclusion (1971) that a large volume reduction by pressure solution yields a comparatively small volume of cement. This small yield is compounded by the fact that hydrocarbon migration will generally commence, reservoirs will generally be filled, and reservoir diagenesis terminated by the presence of hydrocarbons, long before chemical compaction (the main source for cementation materials) is completed. Even though cementation will continue as long as chemical compaction and subsurface dissolution provide a carbonate source, these later cements will only impact that porosity not filled with oil. Therefore, they are economically unimportant.

SUBSURFACE DISSOLUTION

There is abundant evidence that aggressive pore fluids, derived as a result of hydrocarbon maturation and hydrocarbon thermal degradation, exist, and affect rock sequences in the subsurface (Schmidt and McDonald, 1979; Surdam and others, 1984). Moore and Druckman (1981) and Druckman and Moore (1985) documented the enhancement of preserved intergranular pore systems in a number of Upper Jurassic carbonate reservoirs in southern Arkansas in the central Gulf Coast of the United States. Elliott (1982) reported similar porosity enhancement in Mississippian limestones in the Williston Basin, as did Davies and Krouse (1975), in Paleozoic sequences of the Alberta sedimentary basin. They called upon the decarboxilation (loss of -COOH group) of organic material during the maturation process to provide the CO_2 and organic acids necessary to significantly expand the pores by dissolution (Fig. 9.3).

Porosity enhancement related to hydrocarbon maturation seems to take place just prior to the arrival of hydrocarbons in reservoirs; consequently, it is of economic concern. Pore types are generally solution-enlarged intergranular, evolving into vugs; moldic porosity is seldom seen. Late stage pores are generally rounded and can be confused with meniscus vadose pores. They cut all textural elements, including demonstrable late subsurface cements and may enlarge stylolites (Fig. 9.21).

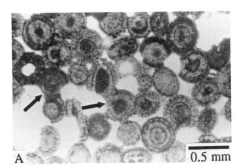 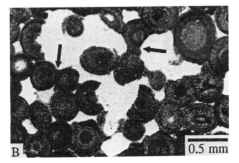

Fig. 9.21. Late subsurface dissolution fabrics of Jurassic Smackover Formation ooid grainstone reservoir rocks at Walker Creek field, Columbia Co., Arkansas. Note the rounded pores that mimic vadose pore geometries. Both samples from Arco Bodcaw #1 at 10896' (3322 m) and 10894' (3321 m) respectively. Both samples plain light.

The thermal degradation of hydrocarbons after emplacement in the reservoir (at temperatures in excess of 150°C, Fig. 9.3) leads to the production of CO_2, H_2S, methane, and solid pyrobitumen (Tissot and Welte, 1978; Sassen and Moore, 1988). While the CO_2

and H_2S may combine with water to trigger porosity enhancement in the growing gas cap by dissolution, the concurrent precipitation of solid pyrobitumen can seriously degrade the porosity and permeability of the reservoir (Fig. 9.18D). The presence of SO_4 acts as a catalyst in the breakdown of methane and can accelerate the production of aggressive pore fluids (Sassen and Moore, 1988).

At this point, there is not enough data to fully evaluate the importance of subsurface dissolution and the impact of late organic diagenetic processes on the final stages of porosity evolution in carbonate rock sequences. However, there is some certainty that porosity is created in the deep subsurface; dissolution provides carbonate for further cementation; and pyrobitumen formation during thermal destruction of hydrocarbons is an important, but seldom-considered porosity destructive process.

SUBSURFACE DOLOMITIZATION

Introduction

Late subsurface dolomitization models have periodically been used to explain massive dolomitization events, such as the dolomitization of platform margins, and reefs in the Alberta sedimentary basin of western Canada (Illing, 1959; Griffin, 1965; Mattes and Mountjoy, 1980). While the elevated temperatures of the subsurface tend to favor dolomitization, a major constraint on late burial dolomitization is the source and delivery of Mg to the site of dolomitization (Morrow, 1982a; Land, 1985).

There are three principal sources in the subsurface of the large volumes of Mg needed for dolomitization: bittern salts (polyhalite and carnalite); Mg derived from clay mineral transformations; and interstitial water (either connate marine water, or connate evaporative brines) (Land, 1985).

While bittern salts are a significant source of Mg, they are generally rare in the geologic column and their volumes are normally not adequate to source the Mg requirements of the pervasive dolomitization events with which we are concerned.

Massive expulsion of Mg during shale burial and diagenesis with subsequent dolomitization of adjacent carbonate-bearing units has not yet been documented. Indeed, Hower and others (1976) demonstrated that calcium loss of Gulf Coast shales exceeds magnesium loss by at least a factor of six. In addition, the formation of chlorite from smectite during burial is an efficient internal Mg sink often overlooked when budgeting Mg during burial diagenesis. These observations suggest that subsurface clay transformations do not provide a reliable and adequate Mg source for pervasive dolomitization

(Morrow, 1983; Land, 1985).

Connate evaporative brines could provide more than enough Mg to allow pervasive dolomitization on a regional scale. These brines, however, tend to be released early in the compactional history of an evaporite, and therefore, may not be available for dolomitization in the deeper subsurface (Land, 1985; see Stoessell and Moore, 1983 for a different view).

It would seem then, that only connate marine waters expelled during compaction can provide adequate Mg for dolomitization during intermediate and later stages of burial diagenesis. Large volumes of marine water (at least 650 pore volumes) must be provided, because a single pore volume of marine water can precipitate so little dolomite (Land, 1985). Therefore, dolomitization by marine connate waters can only pervasively dolomitize significant volumes of limestone if the waters are hydrologically focused into the limestone aquifer during compaction expulsion (Land, 1985; Machel, 1985; Garven, 1985).

Finally, while small volumes of burial dolomite are obviously associated with pressure solution and stylolitization (Wanless, 1979; Moore and others, 1988), dolomitization on a large scale during chemical compaction (by remobilization of previously formed dolomite) to form regional pervasive dolomites has not yet been documented (Land, 1985).

Petrography and geochemistry

Three general burial dolomite types are commonly seen: 1) scattered, coarse, euhedral dolomite rhombs with strong, often complex luminescent zonation, generally associated with stylolites and pressure solution (Fig. 9.8B); 2) pervasive, coarsely crystalline dolomite that may exhibit fabric selectivity and well-developed porosity, or may form dense interlocking mosaics (Fig. 9.18D); 3) saddle dolomite, which is common, as mentioned above, and generally occurs as a very late pore-fill cement (Figs. 9.18C and F). Cathodoluminescence in saddle dolomites tends to be subdued.

As in subsurface calcite cements, burial dolomites, particularly late saddle dolomites, contain abundant two-phase fluid inclusions. These inclusions generally show homogenization temperatures in excess of 100°C, and depressed melting temperatures indicating high salinity fluids (Anderson, 1985; Moore and others, 1988).

The geochemistry of burial dolomites parallels that of the subsurface calcite cements discussed earlier. In general, pervasive burial dolomites contain low Sr concentrations, range from calcium-rich to near stoichiometric, but are generally well-ordered (Mattes and Mountjoy, 1980; Anderson, 1985). As indicated earlier, late saddle dolomite cements are commonly Fe-rich, while earlier-formed, pervasive burial dolomites are iron poor (Mattes and Mountjoy, 1980; Anderson, 1985).

The stable isotopic composition of pervasive burial dolomites generally shows moderately low $\partial^{18}O$ values reflecting elevated temperatures and high $\partial^{13}C$ values, indicating buffering by rock carbon during the dolomitization event (Fig. 9.22). The later saddle dolomite cements generally show somewhat lower $\partial^{18}O$ and $\partial^{13}C$ values, reflecting higher temperatures of formation and the input of light carbon from thermal diagenesis of organic matter (Fig. 9.22) (Moore, 1985; Moore and others, 1988).

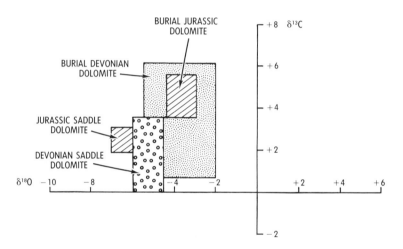

Fig. 9.22. Stable isotopic composition of various subsurface dolomites. Data from: Jurassic saddle and burial dolomite, Moore and others, 1988; Devonian burial and saddle dolomite from Anderson, 1985.

The strontium isotopic compositions of subsurface dolomites, like their calcite cement counterparts, tend to show $^{87}Sr/^{86}Sr$ ratios above sea water values, indicating that the dolomitizing fluids had interacted with siliciclastics (presumably basinal shales) prior to their arrival at the dolomitization site. The later the dolomitization event, the higher the $^{87}Sr/^{86}Sr$ ratio in the dolomite tends to become. Therefore, late saddle dolomites usually show significantly higher $^{87}Sr^{86}Sr$ ratios than earlier pervasive phases (Anderson, 1985; Moore and others, 1988) (Fig. 9.20).

Impact of burial dolomitization on reservoir porosity

The scattered, burial dolomites associated with pressure solution generally have little effect on reservoir porosity. However, pervasive subsurface dolomites can be economically important in certain situations. If the dolomitization event only partially dolomitizes a sequence, and the remaining $CaCO_3$ is not removed by dissolution, the tendency is for no enhancement of porosity (Murray, 1960; Anderson, 1985). If a muddy sequence is totally dolomitized in a non-fabric-selective manner, an interlocking dolomite mosaic may form, with actual porosity destruction (Mattes and Mountjoy, 1980). Pervasive burial dolomitization of porous grainstones, or other porous facies, however, will generally lead to porosity enhancement, and porosity preservation under deep burial conditions. Fabric selective dolomitization, accompanied with $CaCO_3$ dissolution, can also lead to the formation of exceptionally favorable reservoir characteristics (Anderson, 1985).

Upper Devonian dolomitized sequences of Alberta, Canada: a case history of burial dolomitization

Upper Devonian Frasnian age sequences of western Canada, including some of Canada's most important oil reservoirs, such as the Leduc reefs, are commonly pervasively dolomitized (Fig. 9.23). Illing (1959) indicated that Frasnian pervasive dolomitization was a subsurface event, with dolomitizing fluids being derived from compacting basinal and interreefal shales. Subsequent work (Mattes and Mountjoy, 1980; Anderson, 1985; Machel, 1986) has only served to support a subsurface origin for the main body of pervasive upper Devonian dolomitization.

The dolomitization of Frasnian Nisku reef oil reservoirs in the Bigoray, Pembina, and Brazeau areas, some 150 km southwest of Edmonton in western Canada (Figs. 9.23 and 9.24), has recently been intensely studied by Machel (1985, 1986) and Anderson (1985), making these occurrences perhaps the best documented, economically important burial dolomites to date.

Nisku carbonates and shales were deposited during a major Upper Frasnian regressive cycle that followed the initial Lower Frasnian transgressive cycle responsible for the sequential backstepping deposition of the underlying Swan Hills and Leduc reef complexes (Fig. 9.24). The Nisku Formation was part of an extensive carbonate shelf system that surrounded a major shale basin termed the Winterburn (Fig. 9.24). Nisku reefs were deposited offshelf on the slope into the Winterburn shale basin and were ultimately encased in basinal shales (Stoakes, 1979; Anderson, 1985).

Fig. 9.25 is a generalized map of the Nisku facies in the Bigoray, Pembina, and Brazeau areas, indicating the relative location of Nisku reefs. Reefs in the Bigoray area are limestone, while reefs in the Pembina and Brazeau areas are at least partially dolomitized. Many reefs in these trends show preferential dolomitization in their southwestern quadrants, indicating a definite dolomitization gradient from southwest to northeast (Machel, 1986). Fig. 9.26 is a schematic paleogeographic reconstruction of Nisku environments developed by Anderson (1985), representing the major Nisku

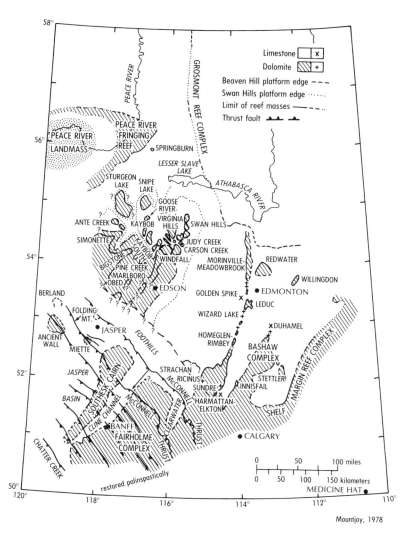

Fig. 9.23. Distribution of Upper Devonian reef complexes and platforms in southern Alberta showing those which are dolomitized. Reprinted with permission of the Canadian Society of Petroleum Geologists.

reef-building episode. There is no evidence that Nisku reefs were exposed to subaerial conditions during deposition. They were subsequently encased in shale and buried into the subsurface without significant meteoric water influence (Machel, 1985; Anderson, 1985).

The burial history of the Nisku is shown in Fig. 9.27. Rapid burial to 1000 m occurred in the Mississippian. At this point in the burial history of the Nisku, the

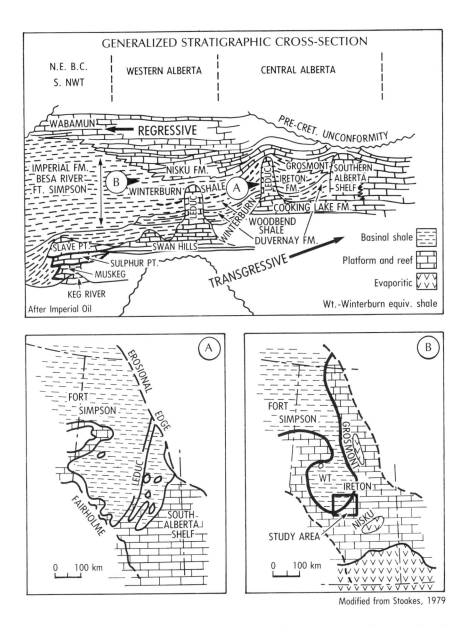

Fig. 9.24. Stratigraphic (upper diagram) and paleogeographic setting (lower diagram) of the Upper Devonian Leduc and Nisku formations. Reprinted with permission of the author.

compactional expulsion of connate waters was at a maximum. In the Late Paleozoic to Early Jurassic, the craton was progressively tilted to the west, extensive erosion of post-Nisku sequences ensued, and a gravity-driven hydrologic system may have been

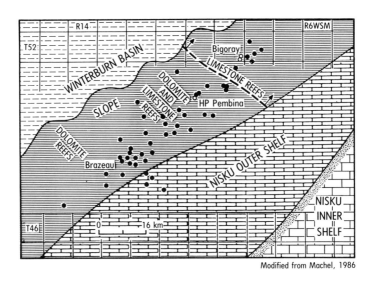

Fig. 9.25. Location map of the Nisku slope reef trends, with positions of basin, outer shelf and inner shelf indicated. Note that reefs in the Bigoray area are all limestone, those in the Pembina area limestone and dolomite, while reefs in the Brazeau area are all dolomite. Township and range surveys are shown. Reprinted with permission from, Reef Diagenesis. Copyright (C), 1986, Springer-Verlag, Berlin.

established in the basin. It was during this period that the pervasive dolomitization of the Nisku was accomplished (Anderson, 1985). Subsequently, in the Early Cretaceous, the Nisku was deeply buried under a major clastic wedge over 2 km thick. Hydrocarbon migration commenced in the Late Cretaceous, Early Tertiary (Anderson, 1985).

The dominant type of dolomitization is matrix selective, with larger, less permeable calcite constituents unreplaced, and commonly removed by solution. Major porosity types are biomolds, solution-enhanced molds, and vugs (Machel, 1986). Without calcite solution or fabric-selective dolomitization, an interlocking dolomite mosaic is commonly formed and porosity significantly diminished. A major subsurface calcite dissolution event, believed to be related to and overlapping with dolomitization, is responsible for the favorable reservoir characteristics of the Nisku reefs.

Geologic setting, burial history, trace element geochemistry, and stable isotopic composition all suggest that dolomitizing fluids were basically derived from marine waters (Machel, 1985; Anderson, 1985). Values of $^{87}Sr/^{86}Sr$ over Devonian seawater levels in pervasive dolomite indicate that these marine waters had interacted with the surrounding shales prior to dolomitization (Anderson, 1985). Oxygen isotopic composition of dolomites indicates formation at temperatures between 45 and 55°C, at a burial depth between 300 to 1000 m (Machel, 1985; Anderson, 1985).

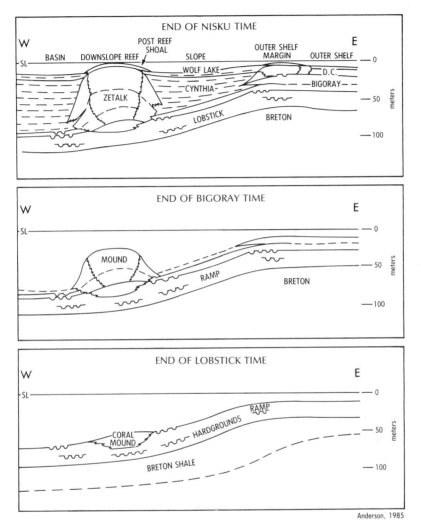

Fig. 9.26. Schematic diagram showing the development of Nisku slope reefs through time. Cross section west to east from the Winterburn shale basin to the Nisku shelf. Reprinted with permission of the author.

Machel (1985, 1986) calls upon connate marine waters from compacting basinal shales during intermediate burial to dolomitize the Nisku reefs. He appeals to the focused flow of compactional fluids through porous carbonate conduits to overcome the serious mass flow constraints of the system. In addition, Machel (1985) uses reef-related faulting to deflect and channel flow into the southwestern portions of the Nisku slope reef complex. Machel's model basically supports the concepts developed by Illing (1959) for the dolomitization of the underlying Leduc reefs.

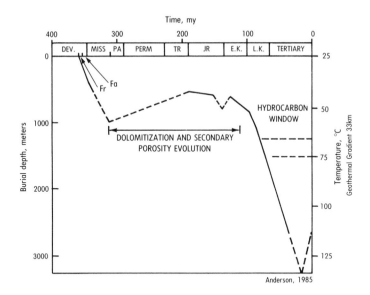

Fig. 9.27. Plot of the burial history of the Nisku reefs, uncorrected for compaction. Reprinted with permission of the author.

Anderson's (1985) mass balance calculations suggest that it would have taken over 400 ft of shale with 40% porosity spread over 3000 km² to have furnished the Mg in connate interstitial water to dolomitize the Nisku reef complex. He noted that the Nisku reef complex was small compared to other Devonian reef complexes in Canada. When he considered further the volume of all of the dolomitized carbonate reefs and the platform margins in the basin, he concluded that the Machel-Illing compactional dolomitization model was untenable. Anderson surmises that additional Mg-bearing fluids above the volume of connate marine interstitial waters trapped at deposition were required.

Anderson (1985) calls upon the development of an active gravity-driven meteoric hydrologic system in the shallower parts of the basin during the Late Paleozoic associated with erosion during the tilting of the craton to the west (Fig. 9.24). He indicates that this shallow meteoric system may have affected deeper basinal flow by thermal Kohout-convection, much in the style of the Florida peninsula. Flow paths followed more permeable carbonate conduits, as conceived by Illing (1959), and would have lead to distinct dolomitization gradients, as seen in the Nisku complex.

The Nisku dolomites most certainly originated in the subsurface from modified marine waters. The method by which sufficient Mg was delivered to the dolomitization site is uncertain at this point because of our relative ignorance concerning the details of evolving basin hydrology during the compactional phase of burial. This case history clearly indicates the importance of basin hydrology in the evolution of carbonate porosity in the subsurface diagenetic environment.

THE ROLE OF EARLY, SURFICIAL DEPOSITIONAL AND DIAGENETIC PROCESSES VERSUS BURIAL PROCESSES IN SHAPING ULTIMATE POROSITY EVOLUTION

"Much (and perhaps most) cementation and formation of secondary porosity (except fractures) in carbonates occurs at relatively shallow depths in one of four major diagenetic environments: the vadose zone, meteoric phreatic zone, mixing zone, and marine phreatic zone" (Longman, 1980, p.461). Scholle and Halley (1985, p. 309) responded to this statement of the conventional wisdom of the early 1980s by stating: "... there generally is little or no porosity loss in the zone of near-surface water circulation (that is, in the vadose, meteoric-phreatic, or mixing zone[s]). Thus, the transition from very porous carbonate sediments to well-cemented, low-porosity rocks is a dominantly subsurface process."

From what is known of carbonate diagenesis and carbonate porosity evolution, as briefly outlined in the preceding chapters of this book, both these viewpoints are obviously extreme and certainly off target. The Scholle and Halley position will be considered first.

Scholle and Halley (1985) use the lack of porosity change during mineralogical stabilization of oolites in Florida and the Bahamas to support their position of no porosity generation or destruction in surficial diagenetic environments (Fig. 9.28). Their analysis is misleading because they have disregarded the reality of massive marine cementation in reef and shelf margin sequences (Chapter 5). They disregard also the potential for $CaCO_3$ transport in more active, regional meteoric aquifer systems, where secondary porosity can be generated in one area by dissolution, and primary porosity can be destroyed far downstream by precipitation of calcite cements (Chapter 7). Instead, they use gross total porosity, disregarding geographic and spatial distribution of secondary porosity development, and porosity occlusion by cementation. While Scholle and Halley's points concerning the sourcing of subsurface cements by pressure solution are valid (see the preceding discussion), their assessment of the volume and importance of these compaction-related cements does not seem to be supported by actual observations of workers dealing with subsurface diagenetic studies (Prezbindowski, 1985; Moore, 1985; Woronick and Land, 1985; Chapter 9). On the other hand, the geologic record is

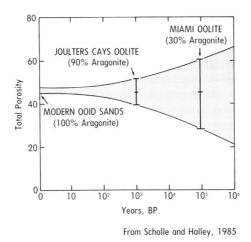

Fig. 9.28. Diagram illustrating lack of porosity change during mineralogical stabilization of oolites in Florida and the Bahamas. Hachured area represents the range of porosity, while tick mark represents the mean porosity value. Used with permission of the SEPM.

replete with examples of total porosity destruction by early marine and meteoric phreatic cementation (Chapters 5 and 7 this book, Roehl and Choquette, 1985; Harris and Schneiderman, 1985).

In short, Scholle and Halley seem to have simply overstated their case, perhaps to make a point concerning the importance of burial diagenesis. In reality, the situation is that depositional fabrics and textures combined with early surficial diagenetic processes centered on dissolution, cementation, and dolomitization are the major factors controlling the ultimate *distribution of economic porosity and permeability* in the subsurface.

In turn, Longman (1980) and others in the late 1970s obviously tended to ignore important burial diagenetic processes, such as compaction and cementation, that tend to *reduce porosity volume and quality* during the burial history of a sequence. However, as was pointed out in the discussion above, hydrocarbon migration and reservoir filling generally take place well before the completion of the compaction process, resulting in termination of compaction and cementation processes at the reservoir site. The fact that compaction and cementation may continue to reduce porosity in rocks surrounding the reservoir, after hydrocarbon migration is complete, is economically irrelevant.

PREDICTING CHANGES IN POROSITY WITH DEPTH

In 1977, Scholle published a series of curves illustrating the porosity-depth trends for chalks in North America and Europe (Fig. 9.12). These curves illustrated clearly the progressive loss of porosity with depth occasioned by closed system chemical compaction in these fine-grained calcite sediments. Obviously, this type of curve would be very helpful in future chalk exploration efforts (Lockridge and Scholle, 1978). In 1982 Schmoker and Halley presented a depth-porosity curve for south Florida subsurface limestones derived from log analysis of selected wells in the south Florida basin (Fig. 9.29). The smooth depth versus porosity curve was derived from a least-squares

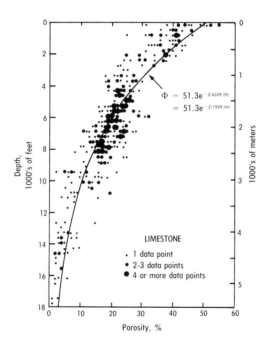

Fig. 9.29. Porosity-depth values and exponential representation of composite data for limestones of the South Florida basin. Reprinted with permission of the American Association of Petroleum Geologists.

exponential fit of a rather scattered data swarm.

Halley and Schmoker (1983), and later Scholle and Halley (1985), concluded that these porosity trends were basically the result of chemical compaction during burial in a semi-closed system, paralleling the processes controlling porosity changes with depth in chalks. They concluded that near-surface diagenetic processes did not significantly affect ultimate porosity evolution.

When the spread of the actual data swarms for depth-porosity relationships in shallow marine limestones (Fig. 9.29) is examined, it is obvious that this view is oversimplified. However, as discussed in the preceding chapters, there are other, important diagenetic processes that may actively destroy, as well as generate, porosity

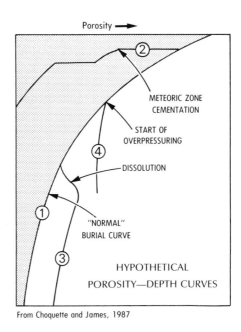

From Choquette and James, 1987

Fig. 9.30. Hypothetical porosity-depth curves in response to early diagenetic overprint, geopressured situations, and organic-related late diagenesis. (1) Normal burial curve based on chalk curves. (2) Porosity destruction and development related to early meteoric zone diagenesis. (3) Porosity development related to late stage dissolution, as a result of hydrocarbon maturation and destruction. (4) Porosity preservation as a result of overpressuring. Reprinted with permission of the Geological Association of Canada.

throughout the burial history of a sequence (see Fig. 9.30 for a theoretical burial curve that incorporates many of these processes).

Schmoker (1984) increased the utility of these types of plots as well as the understanding of the role of early diagenesis in porosity evolution when he convincingly demonstrated the influence of thermal maturity, that is, the temperature-time history of

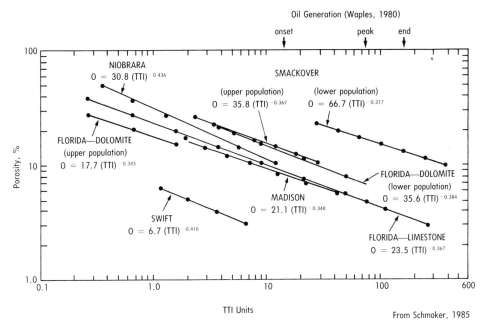

Fig. 9.31. *Porosity versus Lopatin's time-temperature index of thermal maturity (TTI) for various carbonate units. Reprinted with permission of the American Association of Petroleum Geologists.*

a rock, over the ultimate paths of carbonate porosity evolution.

Taking the porosity-depth relationships of a number of carbonate sequences from separate basins with different thermal histories, Schmoker plotted porosity versus calculated thermal maturity based on Lopatin models of the individual basins on a logarithmic base (Fig. 9.31). The result was a series of distinct, but subparallel, straight line plots related by a power function in the form $\phi = a(TTI)^b$. This relationship strongly supports the concept that subsurface processes of porosity loss are functions of time-temperature history. Coefficients of the power functions relating porosity and thermal maturity for Schmoker's data sets are compiled in Table 9.2. Schmoker indicated that the multiplying coefficient (a) represents the net effect on porosity of all petrologic parameters and obviously includes early diagenetic effects such as cementation, secondary porosity, and dolomitization. This exponent is widely variable in response to variations in original depositional fabrics and subsequent diagenetic overprint (Table 9.2). The exponent (b), however, is nearly constant (Table 8.2) and results in the subparallel plots of porosity populations shown in Fig. 9.31. Schmoker believes that exponent (b) represents the rate-limiting step in chemical compaction, that is, diffusive solution film transport from the site of solution to site of precipitation—and hence is essentially independent of the rock matrix. It is the positive temperature influence over rates of diffusion that links thermal history to porosity destruction in the subsurface.

Table 9.2. Reprinted with permission of the American Association of Petroleum Geologists.

COEFFICIENTS OF POWER FUNCTION
($\Theta = a\,[TTI]^b$) RELATING
POROSITY AND THERMAL MATURITY

	Multiplying coefficient a	Exponent b
Florida		
limestone	23.5	−0.367
dolomite (upper)	17.7	−0.345
dolomite (lower)	35.6	−0.384
Niobrara	30.8	−0.436
Swift	6.7	−0.410
Madison	21.1	−0.348
Smackover		
(upper)	35.8	−0.367
(lower)	66.7	−0.317
	Avg. =	0.372 ±0.038

The Schmoker approach seems to adequately accommodate the dramatic initial fabric and textural differences within a carbonate sequence that may result from depositional processes and local or subregional intensive early diagenesis. These diagenetic overprints would be reflected as separate, but parallel, lines on a porosity-TTI plot, such as the two Smackover populations shown on Fig. 9.31.

What then, is the real significance of depth-porosity curves, and how should they be viewed? If constructed carefully (see Brown, 1984), TTI-porosity plots can help define, and most importantly, predict the average or typical porosity of a given potential reservoir within a basin's depth-thermal maturity framework. This average then becomes a standard or reference porosity against which subsurface porosity-diagenetic studies can be compared in order to isolate those extraordinary conditions that may lead to new drilling frontiers and discoveries, such as depositional chalk porosity preservation in the North Sea by geopressures (see discussion of Ekofisk earlier in Chapter 9).

SUMMARY

Mechanical and chemical compaction are the dominant porosity-modifying agents in the burial diagenetic environment. Environmental factors such as sediment texture (presence or absence of mud and grain size), the relative compactibility of grain types, and the presence or absence of bound water may affect the early mechanical compaction history of carbonate sediments. Early cementation, organic framework development, reservoir overpressuring, pores filling with oil, and pervasive dolomitization all tend to retard the onset and efficiency of chemical compaction. Aggressive pore fluids combined with active hydrologic systems, the presence of metastable mineral phases, and admixtures of siliciclastics all tend to accelerate the chemical compaction process.

In comparison to compaction, cementation has a relatively minor impact on porosity evolution in the subsurface. The major source of $CaCO_3$ for cementation is believed to be pressure solution. However, the volume of cement that can be generated from pressure solution is limited, and porosity in reservoirs is generally filled with oil before chemical compaction is complete.

Additional $CaCO_3$ for cementation may be gained from carbonate dissolution by aggressive fluids associated with thermally driven, organic diagenesis. While carbonate reservoir enhancement by subsurface dissolution has been documented, its actual economic importance has not yet been fully evaluated.

Geologic setting, petrography, and geochemistry must be used together to determine the burial origin of passive cements and replacement dolomites. Burial dolomitization is promoted by the elevated temperatures encountered in the subsurface. However, the major constraints on the formation of regional, pervasive, subsurface dolomite are a reliable source of Mg, and a Mg delivery system to the point of dolomitization. Marine waters derived from basinal compaction, focused and driven through porous carbonate conduits, such as reefs and shelf-margin, reef-sand shoal complexes, have long been employed to explain the pervasive subsurface dolomites of the Upper Devonian of western Canada. Recent, intensive work on dolomitization in the Miette reef and Nisku slope reefs has confirmed that the dolomitizing fluids were indeed of marine origin. However, mass balance considerations suggest that compactional waters alone were insufficient to form the volume of dolomite present in the Upper Devonian of western Canada. A thermal, convective, Kohout-style model tied to an active meteoric water system formed during the Upper Paleozoic and Lower Mesozoic has been called on to provide the additional Mg needed to form the volume of dolomite present in the Upper Devonian of western Canada. Porosity development in these burial dolomites is generally dependent on a concurrent limestone dissolution event.

Early depositional and surficial diagenetic processes generally tend to control the distribution of porosity in the subsurface, while burial diagenetic processes frequently tend to reduce porosity volume and quality during the burial history of a sequence.

Finally, porosity-thermal maturity plots may be an important tool in the future for predicting porosity.

REFERENCES

Achauer, C. A., 1977. Contrasts in cementation, dissolution and porosity development between two Lower Cretaceous reefs of Texas. In: R. G. Loucks and D. G. Bebout (Eds.), Cretaceous Carbonates of Texas and Mexico. Bureau of Econ. Geol., University of Texas/Austin, Texas, 89: 127-135.

Adams, J. E. and Rhodes, M. L., 1960. Dolomitization by seepage refluxion. Am. Assoc. Petrol. Geol. Bull., 44: 1912-1920.

Aharon, P., Kolodny, Y. and Sass, E., 1977. Recent hot brine dolomitization in the "solar lake", Gulf of Elat: isotopic, chemical, and mineralogical study. J. Geol., 85: 27-48.

Aharon, P., Socki, R. A. and Chan, L., 1987. Dolomitization of atolls by sea water convection flow: test of a hypothesis at Niue, South Pacific. J. Geol., 95: 187-203.

Aïssaoui, D. M., Buigues, D. and Purser, B. H., 1986. Model of reef diagenesis: Mururoa Atoll, French Polynesia. In: J. H. Schroeder and B. H. Purser (Eds.), Reef Diagenesis. Springer-Verlag/Berlin, pp. 27-52.

Allen, J. R. and Matthews, R. K., 1982. Isotope signatures associated with early meteoric diagenesis. Sedimentology, 29: 797-817.

Anderson, J. H., 1985. Depositional facies and carbonate diagenesis of the downslope reefs in the Nisku Formation (U. Devonian), central Alberta, Canada. Dissertation. The University of Texas/ Austin, Texas, 393 pp.

Anderson, T. F. and Arthur, M. A., 1983. Stable isotopes of oxygen and carbon and their application to sedimentologic and paleoenvironmental problems. In: M. A. Arthur (Ed.), Stable Isotopes in Sedimentary Geology. SEPM Short Course No. 10, pp. 1.1-1.151.

Asquith, G. B., 1979. Subsurface Carbonate Depositional Models: A Concise Review. The Petroleum Publ. Co./ Tulsa, Okla., 121 pp.

Assereto, R. and Folk, R. L., 1980. Diagenetic fabrics of aragonite, calcite, and dolomite in an ancient peritidal-spelean environment: Triassic Calcare Rosso, Lombardia, Italy. J. Sediment. Petrol., 50: 371-394.

Back, W., Hanshaw, B. B., Plye, T. E., Plummer, L. N. and Weidie, A. E., 1979. Geochemical significance of groundwater discharge and carbonate solution to the formation of Caleta Xel Ha, Quintana Roo, Mexico. Water Resources Res., 15: 1521-1535.

Badiozamani, K., 1973. The dorag dolomitization model-application to the Middle Ordovician of Wisconsin. J. Sediment. Petrol., 43: 965-984.

Ball, M. M., 1967. Carbonate sand bodies of Florida and the Bahamas. J. Sediment. Petrol., 37: 556-596.

Bandoian, C. A. and Murray, R. C., 1974. Plio-Pleistocene carbonate rocks of Bonaire, Netherlands Antilles. Geol. Soc. Am. Bull., 85: 1243-1252.

Banner, J. L., 1986. Petrologic and Geochemical Constraints on the Origin of Regionally Extensive Dolomites of the Mississippian Burlington and Keokuk Fms., Iowa, Illinois and Missouri. Dissertation. State University of New York/Stony Brook, New York, 368 pp.

Banner, J. L., Hanson, G. N. and Meyers, W. J., 1988. Determination of initial Sr - isotopic compositions of dolostones from the Burlington-Keokuk Formation (Mississippian): constraints from cathodoluminescence, glauconite paragenesis and analytical methods. J. Sediment. Petrol., 58: 673-687.

Banner, J. L., Hanson, G. N. and Meyers, W. J. (to be published), 1989. Water-rock interaction history of regionally extensive dolomites of the Burlington-Keokuk Formation (Mississippian): isotopic evidence. In: P. A. Baker and V. Shukla (Eds.), Sedimentology and Geochemistry of Dolostones. SEPM Spec. Pub. No. 43.

Barnes, V. E., Cloud, Jr., P. E., Dixon, L. P., Folk, R. L., Jonas, E. C., Palmer, A. R. and Tynan, E. J., 1959. Stratigraphy of the pre-Simpson Paleozoic subsurface rocks of Texas and southeast New Mexico. Bureau of Econ. Geol., University of Texas/Austin, Texas, 1,2: 836 pp.

Bates, R. L., 1942. Lateral gradation in the Seven Rivers Formation, Rocky Arroyo, Eddy County, New Mexico. Am. Assoc. Petrol. Geol. Bull., 26: 80-99.

Bathurst, R. G. C., 1959. The cavernous structure of some Mississippian stromatactis reefs in Lancashire, England. J. Geol., 67: 506-521.

Bathurst, R. G. C., 1966. Boring algae, micrite envelopes, and lithification of molluscan biosparites. Geol. Jour., 5: 15-23.

Bathurst, R. G. C., 1971. Carbonate Sediments and Their Diagenesis. Developments in Sedimentology 12. (First Edition) Elsevier/Amsterdam, 658 pp.

Bathurst, R. G. C., 1974. Marine diagenesis of shallow water calcium carbonate sediments. F. A. Donath, F. G. Stehli and G. W. Wetherill (Eds.), Ann. Rev. Earth Planetary Sci., pp. 257-274.

Bathurst, R. G. C., 1975. Carbonate sediments and their diagenesis. Developments in Sedimentology 12. (Second Edition) Elsevier/Amsterdam, 658 pp.

Bathurst, R. G. C., 1980. Deep crustal diagenesis in limestones. Revista del Instituto de Investigaciones Geologicas: Universidad de Barcelona, 34: 89-100.

Bathurst, R. G. C., 1982. Genesis of stromatactis cavities between submarine crusts in Paleozoic carbonate mud buildups. Jour. Geol. Soc. London, 139: 165-181.

Bathurst, R. G. C., 1984. The integration of pressure-solution with mechanical compaction and cementation. In: Stylolite and Associated Phenomena Relevance to Hydrocarbon Reservoirs. Abu Dhabi National Reserv. Res. Found. Spec. Pub., pp. 41-55.

Bathurst, R. G. C., 1986. Carbonate diagenesis and reservoir development: conservation, destruction and creation of pores. In: J. E. Warme and K. W. Shanley (Eds.), Carbonate Depositional Environments, Modern and Ancient. Part 5: Diagenesis I. Colorado School of Mines Press/Golden, Colorado. Colo. School of Mines Quarterly, 81: 1-25.

Beales, F. W., 1956. Conditions of deposition of Palliser (Devonian) limestone of southwestern Alberta. Am. Assoc. Petrol. Geol. Bull., 40: 848-870.

Bebout, D. G. and Loucks, R. G., 1974. Stuart City Trend, Lower Cretaceous, South Texas (A Carbonate Shelf-Margin Model of Hydrocarbon Exploration). Bureau of Econ. Geol., University of Texas, Austin, Texas, 78: 1-80.

Becher, J. W. and Moore, C. H., 1976. The Walker Creek field, a Smackover diagenetic trap. Trans. Gulf Coast Assoc. Geol. Socs. 26: 34-35.

Behrens, E. W. and Land, L. S., 1972. Subtidal Holocene dolomite, Baffin Bay, Texas. J. Sediment. Petrol., 42: 155-161.

Benson, L. V., 1974. Transformation of a polyphase sedimentary assemblage into a single phase rock: a chemical approach. J. Sediment. Petrol., 44: 123-135.

Benson, L. V. and Matthews, R. K., 1971. Electron microprobe studies of magnesium distribution in carbonate cements and recrystallized skeletal grainstones from the Pleistocene of Barbados, West Indies. J. Sediment. Petrol., 41: 1018-1025.

Berner, R. A., 1965. Dolomitization of the Mid-Pacific atolls. Science, 147: 1297-1299.

Berner, R. A., 1971. Principles of Chemical Sedimentology. McGraw-Hill/New York, 240 pp.

Berner, R. A., Berner, E. K. and Keir, R. S., 1976. Aragonite dissolution on the Bermuda pedestal: its depth and geochemical significance. Earth Planet. Sci. Lett., 30: 169-178.

Bhattacharyya, Ajit and Friedman, G. M., 1979. Experimental compaction of ooids and lime mud and its implication for lithification during burial. J. Sediment. Petrol., 29: 1279-1286.

Blatt, H., Middleton, G. and Murray, R. 1972. Origin of Sedimentary Rocks. Prentice-Hall/New Jersey, 634 pp.

Blount, D. and Moore, C. H., 1969. Depositional and non-depositional carbonate breccias, Chiantla Quadrangle Guatemala. Geol. Soc. Am. Bull., 80: 429-442.

Boles, J. R. and Franks, G., 1979. Clay diagenesis in Wilcox sandstone of southeast Texas: implications of smectite diagenesis on sandstone cementation. J. Sediment. Petrol., 49: 55-70.

Bourrouilh, F., 1972. Diagenese recifale: calcitisation et dolomitisation leur repartition horizontale dans un atoll souleve Ile Lifou, Territoire de la Nouvelle Caledonie. Cah. ORSTOM, ser. Geol., 4: 121-148.

Boyd, D. R., 1963. Geology of the Golden Lane trend and related fields of the Tampico embayment. In: Geology of Peregrina Canyon and Sierra de El Abra, Mexico. Corpus Christi Geol. Soc. Am. Field Trip Guidebook, pp. 49-56.

Brand, U. and Veizer, J., 1980. Chemical diagenesis of a multicomponent carbonate system - 1. Trace elements. J. Sediment. Petrol., 50: 1219-1236.

Brock, F. C. and Moore, C. H., 1981. Walker Creek revisited: a reinterpretation of the diagenesis of the Smackover Formation of Walker Creek field, Arkansas. Trans. Gulf Coast Assoc. Geol. Socs., 31: 49-58.

Bromley, R. G., 1978. Hardground diagenesis. In: R. W. Fairbridge and J. Bourgeois (Eds.), The Encyclopedia of Sedimentology. Dowden, Hutchinson and Ross/ Stroudsburg, Pennsylvania, pp. 397-400.

Brown, A., 1984. Empirical relation between carbonate porosity and thermal maturity: an approach to regional porosity prediction: Discussion. Am. Assoc. Petrol. Geol. Bull., 69: 2024-2028.

Budd, D. A., 1984. Freshwater Diagenesis of Holocene Ooid Sands, Schooner Cays, Bahamas. Dissertation. The University of Texas/Austin, Texas, 491 pp.

Budd, D. A. and Perkins, R. P., 1980. Bathymetric zonation and Paleoecological significance of microborings in Puerto Rican shelf and slope sediments. J. Sediment. Petrol. 50: 881-903

Buddemeier, R. W. and Holladay, G., 1977. Atoll hydrology: island groundwater characteristics and their relationship to diagenesis. Proc. 3rd. Inter. Coral reef Sym., Miami, Fla., 2: 167-173.

Burke, W. H., Denison, R. E., Hetherington, E. A., Koepnick, R. B., Nelson, H. F. and Otto, J. B., 1982. Variation of seawater $^{87}Sr/^{86}Sr$ throughout Phanerozoic time. Geology, 10: 516-519.

Burri, P., Du Dresnay, R. and Wagner, C. W., 1973. Tepee structures and associated diagenetic features in intertidal carbonate sands (Lower Jurassic, Morocco). Sediment. Geol., 9: 221-228.

Bush, P., 1973. Some aspects of diagenetic history of the sabkha in Abu Dhabi, Persian Gulf. In: B. H. Purser (Ed.), The Persian Gulf - Holocene Carbonate Sedimentation and Diagenesis in a Shallow Epicontinental Sea. Springer-Verlag/Berlin, pp. 393-407.

Butler, G. P., 1969. Modern evaporite deposition and geochemistry of coexisting brines, the sabkha, Trucial Coast, Arabian Gulf. J. Sediment. Petrol., 39: 70-89.

Buxton, T. M. and Sibley, D. F., 1981. Pressure solution features in a shallow buried limestone. J. Sediment. Petrol., 51: 19-26.

Carballo, J. D., Land, L. S. and Miser, D. E., 1987. Holocene dolomitization of supratidal sediments by active tidal pumping, Sugarloaf Key, Florida. J. Sediment. Petrol., 57: 153-165.

Carozzi, A. V. and Textoris, D. A., 1967. Paleozoic Carbonate Microfacies of the Eastern Stable Interior (U.S.A.). Intern. Sed. Petrograph. Ser. 11, Brill/Leiden, 41 pp.

Carpenter, A. B., 1976. Dorag dolomitization model by K. Badiozamani - a discussion. J. Sediment. Petrol., 46: 258-261.

Carpenter, A. B., 1978. Origin and chemical evolution of brines in sedimentary basins. Okla. Geol. Surv. Circ. 79, 60-77.

Carrasco-V., B., 1977. Albian sedimentation of submarine autochthonous and allochthonous carbonates, east edge of the Valles-San Luis Potosi platform, Mexico. In: H.C. Cook. and P. Enos (Eds.), Deep-Water Carbonate Environments. SEPM Spec. Pub. No. 25, pp. 263-272.

Chilingar, G. V., 1956. Relationship between Ca/Mg ratio and geological age. Am. Assoc. Petrol. Geol. Bull., 40: 2256-2266.

Chimene, C. A., 1976. Upper Smackover reservoirs, Walker Creek field area, Lafayette and Columbia counties, Arkansas. In: J. Braunstein (Ed.), North American Oil and Gas Fields. AAPG Mem. 24, pp. 177-204.

Choquette, P. W. and Pray, L. C., 1970. Geologic nomenclature and classification of porosity in sedimentary carbonates. Am. Assoc. Petrol. Geol. Bull., 54: 207-250.

Choquette, P. W. and Steinen, R. P., 1980. Mississippian non-supratidal dolomite, Ste. Genevieve limestone, Illinois basin: evidence for mixed-water dolomitization. In: D. H. Zenger, J. B. Dunham and R. L. Ethington (Eds.), Concepts and Models of Dolomitization. SEPM Spec. Pub. No. 28, pp. 163-196.

Choquette, P. W. and Steinen, R. P., 1985. Mississippian oolite and non-supratidal dolomite reservoirs in the Ste. Genevieve Formation, North Bridgeport field, Illinois basin. In: P. O. Roehl and P. W. Choquette (Eds.), Carbonate Petroleum Reservoirs. Springer-Verlag/New York, pp. 209-225.

Choquette, P. W. and James, N. P., 1987. Diagenesis #12: Diagenesis in limestones - 3. The deep burial environment. Geosci. Can., 14: 3-35.

Chuber, S. and Pusey, W. C., 1985. Productive Permian carbonate cycles, San Andres Formation, Reeves field, West Texas. In: P. O. Roehl and P. W. Choquette (Eds.), Carbonate Petroleum Reservoirs. Springer-Verlag/New York, pp. 289-307.

Collins, A. G., 1975. Geochemistry of Oil Field Waters. Developments in Petroleum Science 1. Elsevier/ Amsterdam, 496 pp.

Coogan, A. H., 1970. Measurements of compaction in oolitic grainstones. J. Sediment. Petrol., 40: 921-929.

Coogan, A. H. and Manus, R. W., 1975. Compaction and diagenesis of carbonate sands. In: G. V. Chilingarian and K. H. Wolf (Eds.), Compaction of Coarse-Grained Sediments I. Developments in Sedimentology 18A. Elsevier/New York, pp. 79-166.

Coogan, A. H., Bebout, D. G. and Maggio, C., 1972. Depositional environments and geological history of Golden Lane and Poza Rica trends, Mexico, an alternative view. Am. Assoc. Petrol. Geol. Bull., 56: 1419-1447.

Craig, D. H., 1988. Caves and other features of the Permian karst in San Andres dolomite, Yates field reservoir, west Texas. In: N. P. James and P. W. Choquette (Eds), Paleokarst. Springer-Verlag/New York, pp. 342-363.

Crawford, M. L., Kraus, D. W. and Hollister, L. S., 1979. Petrologic and fluid inclusion study of calc-silicate rocks, Prince Rupert, British Columbia. Am. J. Sci., 279: 1135-1159.

Crevello, P. D., Harris, P. M., Stoudt, D. L. and Baria, L. R., 1985. Porosity evolution and burial diagenesis in a reef-debris reservoir, Smackover Formation (Jurassic), Hico Knowles field, Arkansas. In: P. O. Roehl and P. W. Choquette (Eds.), Carbonate Petroleum Reservoirs. Springer-Verlag/New York, pp. 385-406.

Curtis, C. D., Petrowski, C. and Ortel, G., 1972. Stable carbon isotope ratios within carbonate concretions: a clue to place and time of formation. Nature Phy. Sci., 235: 98-100.

Darwin, C., 1841. On a remarkable bar of sandstone off Pernambuco, on the coast of Brazil. London, Edinburgh and Dublin Philos. Mag. and J. of Sci., 19: 257-260.

Davies, G. R., 1975. Introduction. In: G. R. Davies (Ed.), Devonian Reef Complexes of Canada I Rainbow, Swan Hills. CSPG Reprint Series 1, pp. iii-ix.

Davies, G. R., 1977. Former magnesian calcite and aragonite submarine cements in Upper Paleozoic reefs of the Canadian Arctic: a summary. Geology, 5: 11-15.

Davies, G. R. and Krouse, H. R., 1975. Carbon and isotopic composition of Late Paleozoic calcite cements, Canadian Arctic Archipelago - preliminary results and interpretations. Can. Geol. Surv. Pap. 75-1, Pt. B, 215-220.

Deffeyes, K. S., Lucia, F. J. and Weyl, P. K., 1965. Dolomitization of Recent and Plio-Pleistocene sediments by marine evaporite waters on Bonaire, Netherlands Antilles. In: Dolomitization and Limestone Diagenesis. SEPM Spec. Pub. No. 13, pp. 71-88.

Dellwig, L. F., 1955. Origin of the Salina salt of Michigan. J. Sediment. Petrol., 25: 83-110.

Dravis, J. J., 1979. Rapid and widespread generation of Recent oolitic hardgrounds on a high energy Bahamian platform, Eleuthera Bank, Bahamas. J. Sediment. Petrol., 49: 195-208.

Druckman, Y. and Moore, C. H. 1985. Late subsurface porosity in a Jurassic grainstone reservoir, Smackover Formation, Mt. Vernon field, southern Arkansas. In: P. O. Roehl and P. W. Choquette (Eds.), Carbonate Petroleum Reservoirs. Springer-Verlag/New York, pp. 371-383.

Dullo, W. C., 1986. Variation in diagenetic sequences: an example from Pleistocene coral reefs, Red Sea, Saudi Arabia. In: J. H. Schroeder and B. H. Purser (Eds.), Reef Diagenesis. Springer-Verlag/Berlin, pp. 77-90.

Dunham, R. J., 1962. Classification of carbonate rocks according to depositional texture. In: W. E. Ham (Ed.), Classification of Carbonate Rocks. AAPG Mem. 1, pp. 108-121.

Dunham, R. J., 1971. Meniscus cement. In: O. P. Bricker (Ed.), Carbonate Cements. AAPG Stud. in Geol. No. 19. Johns Hopkins University Press/Baltimore, Maryland, pp. 297-300.

Dunham, R. J., 1972. Capitan Reef, New Mexico and Texas: Facts and Questions to Aid Interpretation and Group Discussion. SEPM Permian Basin Sec., Pub. No. 72-14.

Dunnington, H. V., 1967. Aspects of diagenesis and shape change in stylolitic limestone reservoirs. World Petrol. Cong. Proc., 7th, 2: 339-352.

Elliott, T. L., 1982. Carbonate facies, depositional cycles and the development of secondary porosity during burial diagenesis. In: J. E. Christopher and J. Kaldi (Eds.), Fourth International Williston Basin Symposium. Saskatchewan Geol. Soc., Spec. Pub. 6, pp. 131-151.

Enos, P., 1974. Reefs, platforms, and basins of Middle Cretaceous in northeast Mexico. Am. Assoc. Petrol. Geol. Bull., 58: 800-809.

Enos, P., 1977. Tamabra limestone of the Poza Rica trend, Cretaceous, Mexico. In: H. E. Cook and P. Enos (Eds.), Deep-water Carbonate Environments. SEPM Spec. Pub. No. 25, pp. 273-314.

Enos, P., 1986. Diagenesis of Mid-Cretaceous rudist reefs, Valles platform, Mexico. In: J. H. Schroeder and B. H. Purser (Eds.), Reef Diagenesis. Springer-Verlag/Berlin, pp. 160-185.

Enos, P., 1988. Evolution of pore space in the Poza Rica trend (Mid-Cretaceous), Mexico. Sedimentology, 35: 287-325.

Enos, P. and Sawatsky, L. H., 1981. Pore networks in Holocene carbonate sediments. J. Sediment. Petrol., 51: 961-985.

Enos, P. and Moore, C. H., 1983. Fore-reef slope environment. In: P. A. Scholle, D.G. Bebout and C. H. Moore (Eds.), Carbonate Depositional Environments. AAPG Mem. 33, pp. 507-537.

Esteban, M. and Klappa, C. F., 1983. Subaerial exposure environment. In: P. A. Scholle, D. G Bebout and C. H. Moore (Eds.), Carbonate Depositional Environments. AAPG Mem. 33, pp. 1-95.

Fairbridge, R. W., 1957. The dolomite question. In: R. J. Le Blanc and J. G. Breeding (Eds.), Regional Aspects of Carbonate Deposition. SEPM Spec. Pub. No. 5, pp. 124-178.

Feazel, C. T. and Schatzinger, R. A., 1985. Prevention of carbonate cementation in petroleum reservoirs. In: N. Schneidermann and P. M. Harris (Eds.), Carbonate Cements. SEPM Spec. Pub. No. 36, pp. 97-106.

Feazel, C.T. and Farrell, H.E., 1988. Chalk from the Ekofisk area North Sea: nannofossils + micropores = giant fields. In: A.J. Lamondo and P.M. Harris (Eds.), Giant Oil and Gas Fields. SEPM Core Workshop No. 12, 1: 155-178.

Feazel, C. T., Keany, J. and Peterson, R. M., 1985. Cretaceous and Tertiary chalk of the Ekofisk field area, central North Sea. In: P. O. Roehl and P. W. Choquette (Eds.), Carbonate Petroleum Reservoirs. Springer-Verlag/New York, pp. 497-507.

Fischer, A. G. and Garrison, R. E., 1967. Carbonate lithification on the sea floor. J. Geol., 75: 488-497.

Fishbuch, N. R., 1968. Stratigraphy, Devonian Swan Hills reef complexes of central Alberta. Bull. Can. Petrol. Geol., 16: 444-556.

Folk, R. L., 1954. The distinction between grain size and mineral composition in sedimentary rock nomenclature. J. Geol., 62: 344-359.

Folk, R. L., 1959. Practical petrographic classification of limestones. Am. Assoc. Petrol. Geol. Bull., 43: 1-38.

Folk, R. L., 1965. Spectral subdivision of limestone types. In: W. E. Ham (Ed.), Classification of Carbonate Rocks. AAPG Mem. 1, pp. 62-84.

Folk, R. L., 1968. Petrology of Sedimentary Rocks. Hemphills/Austin, Texas, 170 pp.

Folk, R. L., 1974. The natural history of crystalline calcium carbonate: effect of magnesium content and salinity. J. Sediment. Petrol., 44: 40-53.

Folk, R. L. and Robles, R., 1964. Carbonate sands of Isla Perez, Alacran reef complex, Yucatan. J. Geol., 72: 255-291.

Folk, R. L. and Land, L. S., 1975. Mg/Ca ratio and salinity; two controls over crystallization of dolomite. Am. Assoc. Petrol. Geol. Bull., 59: 60-68.

Ford, D., 1988. Characteristics of dissolutional cave systems in carbonate rocks. In: N. P. James and P. W. Choquette (Eds.), Paleokarst. Springer-Verlag/New York, pp. 25-57.

Friedman, G. M. and Sanders, J. E., 1978. Principles of Sedimentology. John Wiley and Sons/New York, 792 pp.

Friedman, G. M., Amiel, A. J. and Schneidermann, N., 1974. Submarine cementation in reefs: example from the Red Sea. J. Sediment. Petrol., 44: 816-825.

Friedman, I. and O'Neil, J.R., 1977. Compilation of stable isotope factors of geochemical interest. In: M. Fleischer (Ed.), Data of Geochemistry. U. S. Geol. Surv., Profess. Paper 440-KK.

Frost, S. H., Weiss, M. P. and Saunders, J. B., 1977. Reefs and related carbonates - ecology and sedimentology. AAPG Stud. in Geol. No. 4, 421 pp.

Fruth, Jr., L. S., Orme, G. R. and Donath, F.A., 1966. Experimental compaction effects in carbonate sediments. J. Sediment. Petrol., 36: 747-754.

Gaines, A., 1977. Protodolomite redefined. J. Sediment. Petrol., 47: 543-546.

Garrison, R. E., 1981. Diagenesis of oceanic carbonate sediments: a review of the DSDP perspective. SEPM Spec. Pub. No. 32, pp. 181-207.

Garven, G., 1985. The role of regional fluid flow in the genesis of the Pine Point deposit, western Canada sedimentary basin. Econ. Geol., 80: 307-324.

Gebelein, C. D., 1977. Mixing zone dolomitization of Holocene tidal flat sediments, southwest Andros Island, Bahamas (abst.). Am. Assoc. Petrol. Geol. Bull., 61: 787-788.

Gebelein, C. D., Steinen, R. P., Garrett, P., Hoffman, E. J., Queen, J. M. and Plummer, L. N., 1980. Subsurface dolomitization beneath the tidal flats of central west Andros Island, Bahamas. In: D. H. Zenger, J. B. Dunham and R. L. Ethington (Eds.), Concepts and Models of Dolomitization. SEPM Spec. Pub. No. 28, pp. 31-49.

Ginsburg, R. N., 1956. Environmental relationships of grain size and constituent particles in some south Florida carbonate sediments. Am. Assoc. Petrol. Geol. Bull., 40: 2384-2427.

Ginsburg, R. N. and Lowenstam, H. A., 1958. The influence of marine bottom communities on the depositional environment of sediments. J. Geol., 66: 310-318.

Ginsburg, R. N. and Schroeder, J., 1973. Growth and submarine fossilization of algal cup reefs, Bermuda. Sedimentology, 20: 575-614.

Ginsburg, R. N., Marszalek, D. S. and Schneidermann, N., 1971. Ultrastructure of carbonate cements in a Holocene algal reef of Bermuda. J. Sediment. Petrol., 41: 472-482.

Given, R. K. and Lohmann, K. C., 1985. Derivation of the original isotopic composition of Permian marine cements. J. Sediment. Petrol., 55: 430-439.

Given, R. K. and Lohmann, K. C., 1986. Isotopic evidence for the early meteoric diagenesis of the reef facies, Permian Reef complex of west Texas and New Mexico. J. Sediment. Petrol., 56: 183-193.

Given, R. K. and Wilkinson, B. H., 1985. Kinetic control of morphology, composition, and mineralogy of abiotic sedimentary carbonates. J. Sediment. Petrol., 55: 109-119.

Given, R. K. and Wilkinson, B. H., 1987. Perspectives: dolomite abundance and stratigraphic age: constraints on rates and mechanisms of Phanerozoic dolostone formation. J. Sediment. Petrol., 57: 1068-1078.

Gonzalez, L. A., and Lohmann, K. C., 1988. Controls on mineralogy and composition of spelean carbonates: Carlsbad Caverns, New Mexico. In: N. P. James and P. W. Choquette (Eds.), Paleokarst. Springer-Verlag/New York, pp. 81-101.

Goter, E. R., 1979. Depositional and diagenetic history of the windward reef of Enewetak Atoll during the mid-to-late Pleistocene and Holocene. Dissertation. Rensselaer Polytechnic Institute/Troy, New York, 240 pp.

Graton, L. C. and Fraser, H. J., 1935. Systematic packing of spheres - with particular relation to porosity and permeability. Jour. Geol., 43: 785-909.

Griffin, D. L., 1965. The Devonian Slave Point, Beaverhill Lake, and Muskwa formations of northeastern British Columbia and adjacent areas. Brit. Col. Dept. of Mines and Petrol. Res. Bull., 50: 1-90.

Grotzinger, J. P. and Read, J. F., 1983. Evidence for primary aragonite precipitation, lower Proterozoic (1.9 Ga) Rocknest dolomite, Wopmay orogen, Northwest Canada. Geology, 11: 710-713.

Grover, Jr., G. and Read, J. F. 1978. Fenestral and associated vadose diagenetic fabrics of tidal flat carbonates, middle Ordovician New Market limestone, southwestern Virginia. J. Sediment. Petrol., 48: 453-473.

Hallam, A., 1984. Pre-Quaternary sea level changes. Ann. Rev. Earth Planetary Sci., 12: 205-243.

Halley, R. B., 1987. Burial diagenesis of carbonate rocks. In: J. E. Warme and K. W. Shanley (Eds.), Carbonate Depositional Environments Modern and Ancient - Part 6: Diagenesis 2. Colorado School of Mines Press/Golden, Colorado. Colorado School of Mines Quarterly, pp. 1-15.

Halley, R. B. and Harris, P. M., 1979. Fresh-water cementation of a 1,000 year-old oolite. J. Sediment. Petrol., 49: 969-988.

Halley, R. B. and Evans, C. C., 1983. The Miami Limestone: a guide to selected outcrops and their interpretation. Miami Geol. Soc., 67 pp.

Halley, R. B. and Schmoker, J. W., 1983. High porosity Cenozoic carbonate rocks of south Florida: progressive loss of porosity with depth. Am. Assoc. Petrol. Geol. Bull., 67: 191-200.

Halley, R. B., Harris, P. M. and Hine, A. C., 1983. Bank margin environment. In: P. A. Scholle, D. G. Bebout and C. H. Moore (Eds.), Carbonate Depositional Environments. AAPG Mem. 33, pp. 463-506.

Hanor, J. S., 1978. Precipitation of beachrock cements: mixing of marine and meteoric waters vs CO_2-degassing. J. Sediment. Petrol., 48: 489-501.

Hanshaw, B. B. and Back, W., 1980. Chemical mass-wasting of the northern Yucatan Peninsula by groundwater dissolution. Geology, 8: 222-224.

Hanshaw, B. B., Back, W. and Deike, R. G., 1971. A geochemical hypothesis for dolomitization by groundwater. Econ. Geol., 66: 710-724.

Hardie, L. A., 1987. Dolomitization: a critical view of some current views. J. Sediment. Petrol., 57: 166-183.

Harper, M. L., 1971. Approximate geothermal gradients in the North Sea. Nature, 230: 235-236.

Harper, M. L. and Shaw, B. B., 1974. Cretaceous-Tertiary carbonate reservoirs in the North Sea. Stavanger, Norway Offshore Tech. Conf. Pap. G IV/4, 20 pp.

Harris, P. M. and Frost, S. H., 1984. Middle Cretaceous carbonate reservoirs, Fahud field and northwestern Oman. Am. Assoc. Petrol. Geol. Bull., 68: 649-658.

Harris, P. M. and Schneidermann, N. (Eds.), 1985. Carbonate Cements, SEPM Spec. Pub. No. 36, 379 pp.

Harris, W. H. and Matthews, R. K., 1968. Subaerial diagenesis of carbonate sediments: efficiency of the solution-reprecipitation process. Science, 160: 77-79.

Harrison, R. S., 1975. Porosity in Pleistocene grainstones from Barbados: some preliminary observations. Bull. Can. Petrol. Geol., 23: 383-392.

Havard, C. and Oldershaw, A., 1976. Early diagenesis in back-reef sedimentary cycles, Snipe Lake reef complex, Alberta. Bull. Can. Petrol. Geol., 24: 27-69.

Heydari, E. and Moore, C. H., 1988. Oxygen isotope evolution of the Smackover pore waters, southeast Mississippi salt basin. Geol. Soc. Amer., Accepted Abstr., Abstrs. with Program, 20: A261.

Heydari, E., Moore, C. H. and Sassen, R., 1988. Late burial diagenesis driven by thermal degradation of hydrocarbons and thermochemical sulfate reduction: Upper Smackover carbonates, southeast Mississippi salt basin. Amer. Assoc. Petrol. Geol. Bull., Abstr., 72: 197.

Hills, J. M., 1972. Late Paleozoic sedimentation in west Texas Permian basin. Am. Assoc. Petrol. Geol. Bull., 56: 2303-2322.

Horowitz, A. S. and Potter, P. E., 1971. Introductory Petrography of Fossils. Springer-Verlag/New York, 302 pp.

Hower, J., Eslinger, E. V., Hower, M. E. and Perry, E. A., 1976. Mechanism of burial metamorphism of argillaceous sediment: 1. Mineralogical and chemical evidence. Geol. Soc. Amer. Bull., 87: 725-737.

Hsü, K. J. and Siegenthaler, C., 1969. Preliminary experiments and hydrodynamic movement induced by evaporation and their bearing on the dolomite problem. Sedimentology, 12: 11-25.

Hudson, J. D., 1977. Stable isotopes and limestone lithification. Jour. Geol. Soc. London, 133: 637-660.

Humphrey, J. D., 1988. Late Pleistocene mixing zone dolomitization, southeastern Barbados, West Indies. Sedimentology, 35: 327-348.

Humphrey, J. D., Ransom, K. L. and Matthews, R. K., 1986. Early meteoric diagenetic control of Upper Smackover production, Oaks field, Louisiana. Am. Assoc. Petrol. Geol. Bull., 70: 70-85.

Illing, L. V., 1959. Deposition and diagenesis of some Upper Paleozoic carbonate sediments in western Canada. 5th World Petrol. Cong., New York Proc., pp. 23-52.

Illing, L. V., Wells, A. J. and Taylor, J. C. M., 1965. Penecontemporary dolomite in the Persian Gulf. In: L. C. Pray and R. C. Murray (Eds.), Dolomitization and Limestone Diagenesis - a Symposium. SEPM Spec. Pub. No. 13, pp. 89-111.

Inden, R. F. and Moore, C. H., 1983. Beach environment. In: P. A. Scholle, D. G. Bebout and C. H. Moore (Eds.), Carbonate Depositional Environments. AAPG Mem. 33, pp. 211-265.

Ingerson, E., 1962. Problems of the geochemistry of sedimentary carbonate rocks. Geochim. Cosmochim. Acta, 26: 815-847.

Jacka, A. D. and Brand, J. P., 1977. Biofacies and development and differential occlusion of porosity in a Lower Cretaceous (Edwards) reef. J. Sediment. Petrol., 47: 366-381.

Jacobson, R. L. and Usdowski, H. E., 1976. Partitioning of strontium between calcite, dolomite and liquids: an experimental study under higher temperature diagenetic conditions and the model for prediction of mineral pairs for geothermometry. Contrib. to Min. and Petrol., 59: 171-185.

James, N. P., 1979. Reefs. In: R. G. Walker (Ed.), Facies Models. Geosci. Can., Repr. Ser. 1, pp. 121-133.

James, N. P., 1983. Reef environment. In: P. A. Scholle, D. G. Bebout and C. H. Moore (Eds.), Carbonate Depositional Environments. AAPG Mem. 33, pp. 345-440.

James, N. P., 1984. Shallowing-upward sequences in carbonates. In: R. G. Walker (Ed.), Facies Models. Geosci. Can., Repr. Ser. 1, pp. 213-228.

James, N. P. and Ginsburg, R. N., 1979. The seaward margins of Belize barrier and atoll reefs. Inter. Assoc. Sed., Spec. Pub. 3, 197 pp.

James, N. P. and Choquette, P. W., 1983. Diagenesis 6. Limestones - The sea floor diagenetic environment. Geosci. Can., 10: 162-179.

James, N. P. and Klappa, C. F., 1983. Petrogenesis of Early Cambrian reef limestones, Labrador, Canada. J. Sediment. Petrol., 53: 1051-1096.

James, N. P. and Mountjoy, E. W., 1983. Shelf-slope break in fossil carbonate platforms: an overview. In: D. J. Stanley and G. T. Moore (Eds.), The Shelfbreak: Critical Interface on Continental Margins. SEPM Spec. Pub. No. 33, pp. 189-206.

James, N. P. and Choquette, P. W., 1984. Diagenesis 9 - Limestones - The meteoric diagenetic environment. Geosci. Can., 11: 161-194.

James, N. P. and Choquette, P. W., 1988. Introduction. In: N. P. James and P. W. Choquette (Eds.), Paleokarst. Springer-Verlag/New York, pp. 1-21.

James, N. P., Ginsburg, R. N., Marszalek, D. S. and Choquette, P. W., 1976. Facies and fabric specificity of early subsea cements in shallow Belize (British Honduras) reefs. J. Sediment. Petrol., 46: 523-544.

Jordan, Jr., C. F., Connally, Jr., T. C. and Vest, H. A., 1985. Middle Cretaceous carbonates of the Mishrif Formation, Fateh field, offshore Dubai, U. A. E. In: P. O. Roehl and P. W. Choquette (Eds.), Carbonate Petroleum Reservoirs. Springer-Verlag/New York, pp. 425-442.

Katz, A., 1973. The interaction of magnesium with calcite during crystal growth at 25°-90°C and one atmosphere. Geochim. Cosmochim. Acta, 37: 1563-1586.

Katz, A., Sass, E., Starinsky, A. and Holland, H. D., 1972. Strontium behavoir in the aragonite-calcite transformation: an experimental study at 40°-98°C. Geochem. Cosmochim. Acta, 36: 481-496.

Kendall, A. C., 1977a. Fascicular-optic calcite: a replacement of bundled acicular carbonate cements. J. Sediment. Petrol., 47: 1056-1062.

Kendall, A. C., 1977b. Origin of dolomite mottling in Ordovician limestones from Saskatchewan and Manitoba. Bull. Can. Petrol. Geol., 25: 480-504.

Kendall, A. C., 1984. Evaporites. In: R. G. Walker (Ed.), Facies Models. Geosci. Can., Repr. Ser. 1, pp. 259-298.

Kendall, A. C., 1985. Radiaxial fibrous calcite: a reappraisal. In: N. Schneidermann and P. M. Harris (Eds.), Carbonate Cements. SEPM Spec. Pub. No. 36, pp. 59-77.

Kendall, A. C. and Tucker, M. E., 1973. Radiaxial fibrous calcite: a replacement after acicular carbonate. Sedimentology, 20: 365-387.

Kendall, C. G. St. C., 1969. An environmental reinterpretation of the Permian evaporite/carbonate shelf sedimentation of the Guadalupe Mountains. Geol. Soc. Am. Bull., 80: 2503-2526.

Kerans, C., Hurley, N. F. and Playford, P. E., 1986. Marine diagenesis in Devonian reef complexes of the Canning basin, Western Australia. In: J. H. Schroeder and B. H. Purser (Eds.), Reef Diagenesis. Springer-Verlag/ Berlin, pp. 357-380.

Kinsman, D. J. J., 1966. Gypsum and anhydrite of recent age, Trucial Coast, Persian Gulf. In: 2nd Sym. on Salt. Northern Ohio Geol. Soc./Cleveland, Ohio, 1: pp. 302-306.

Kinsman, D. J. J., 1969. Interpretation of Sr +2 concentrations in carbonate minerals and rocks. J. Sediment. Petrol., 39: 486-508.

Kitano, Y. and Hood, D. W., 1965. The influence of organic material on the polymorphic crystallisation of calcium carbonate. Geochim. Cosmochim. Acta, 29: 29-41.

Klosterman, M. J., 1981. Applications of fluid inclusion techniques to burial diagenesis in carbonate rock sequences. Louisiana State University, Applied Carbonate Research Program, Contribution #7, 101 pp.

Klovan, J. E., 1964. Facies analysis of the Redwater reef complex, Alberta, Canada. Bull. Can. Petrol. Geol., 12: 1-100.

Klovan, J. E., 1974. Development of western Canadian Devonian reefs and comparison with Holocene analogues. Am. Assoc. Petrol. Geol. Bull., 58: 787-799.

Koepnick, R. B., 1984. Distribution and vertical permeability of stylolites within a Lower Cretaceous carbonate reservoir, Abu Dhabi, United Arab Emirates. In: Stylolites and Associated Phenomena Relevence to Hydrocarbon Reservoirs. Abu Dhabi National Reservoir Res. Found., Spec. Pub., pp. 261-278.

Kohout, F. A., 1965. A hypothesis concerning cyclic flow of salt water related to geothermal heating in the Floridian aquifer. Trans. New York Acad. Sci., Series 2, 28: 249-271.

Krynine, P. D., 1941. Petrographic studies of variations in cementing material in the Oriskany sand. Pa. Stat. Coll. Mineral Indus. Expt. Sta. Bull., 33: 108-116.

Lahann, R. W., 1978. A chemical model for calcite crystal growth and morphology control. J. Sediment. Petrol., 48: 337-344.

Lahann, R. W. and Siebert, R. M., 1982. A kinetic model for distribution coefficients and application to Mg-calcites. Geochim. Cosmochim. Acta, 46: 2229-2237.

Land, L. S., 1967. Diagenesis of skeletal carbonates. J. Sediment. Petrol., 37: 914-930.

Land, L. S., 1973a. Contemporaneous dolomitization of middle Pleistocene reefs by meteoric water, North Jamaica. Bull. Mar. Sci., 23: 64-92.

Land, L. S., 1973b. Holocene meteoric dolomitization of Pleistocene limestones, North Jamaica. Sedimentology, 20: 411-424.

Land. L. S., 1980. The isotopic and trace element geochemistry of dolomite: the state of the art. In: D. H. Ethington, J. B. Dunham and R.L. Zenger (Eds.), Concepts and Models of Dolomitization. SEPM Spec. Pub. No. 28, pp. 87-110.

Land, L. S., 1985. The origin of massive dolomite. J. Geol. Educ., 33: 112-125.

Land, L. S., 1986. Limestone diagenesis - some geochemical considerations. In: F. A. Mumpton (Ed.), Studies in Diagenesis. U. S. Geol. Surv. Bull., pp. 129-137.

Land, L. S. and Goreau, T. F., 1970. Submarine lithification of Jamaican reefs. J. Sediment. Petrol., 40: 457-462.

Land, L. S. and Hoops, G. K., 1973. Sodium in carbonate sediments and rocks: a possible index to the salinity of diagenetic solutions. J. Sediment. Petrol., 43: 614-617.

Land, L. S. and Moore, C. H., 1980. Lithification, micritization, and syndepositional diagenesis of biolithites on the Jamaican island slope. J. Sediment. Petrol., 50: 365-369.

Land, L. S. and Prezbindowski, P. R., 1981. The origin and evolution of saline formation water, Lower Cretaceous carbonates, south-central Texas, U. S. A. J. Hydrol., 54: 51-74.

Land, L. S., Salem, M. R. I. and Morrow, D. W., 1975. Paleohydrology of ancient dolomites: geochemical evidence. Am. Assoc. Petrol. Geol. Bull., 59: 1602-1625.

Laporte, L. F., 1968. Recent carbonate environments and their paleoecologic implications. In: E. T. Drake (Ed.), Evolution and Environment. Yale University Press/ New Haven, Conn., pp. 229-258.

Lees, A., 1975. Possible influence of salinity and temperature on modern shelf carbonate sedimentation. Marine Geol., 19: 159-198.

Lighty, R. G., 1985. Preservation of internal reef porosity and diagenetic sealing of submerged early Holocene barrier reef, southeast Florida shelf. In: N. Schneidermann and P. M. Harris (Eds.), Carbonate Cements. SEPM Spec. Pub. No. 36, pp. 123-151.

Lindsay, R. F. and Roth, M. S., 1982. Carbonate and evaporite facies, dolomitization and reservoir distribution of the Mission Canyon Formation, Little Knife field, North Dakota. 4th Intern. Williston Basin Symp., pp. 153-180.

Lindsay, R. F. and Kendall, C. G. St. C., 1985. Depositional facies, diagenesis, and reservoir character of Mississippian cyclic carbonates in the Mission Canyon Formation, Little Knife field, Williston basin, North Dakota. In: P. O. Roehl and P. W. Choquette (Eds.), Carbonate Petroleum Reservoirs. Springer-Verlag/New York, pp. 175-190.

Lippman, F., 1973. Sedimentary Carbonate Minerals. Springer-Verlag/Berlin, 228 pp.

Lloyd, E. R., 1929. Capitan limestone and associated formations of New Mexico and Texas. Am. Assoc. Petrol. Geol. Bull., 13: 645-658.

Lloyd, R. M., 1977. Porosity reduction by chemical compaction-stable isotope model. Amer. Assoc. Petroleum Geologists Bull., 61: 809.

Lockridge, J. P. and Scholle, P. A., 1978. Niobraru gas in eastern Colorado and northwestern Kansas. In: J. P. Pruit and P. E. Coffin (Eds.), Energy Resources of the Denver Basin. Rocky Mtn. Assoc. Geol., Symp. Guidebook., pp. 35-49.

Logan, B. W., 1987. The MacLeod Evaporite Basin, Western Australia. AAPG Mem. 44, 140 pp.

Lohmann, K. C., 1983. Unraveling the diagenetic history of carbonate reservoirs. In: J. L. Wilson, B. H. Wilkinsón, K. C. Lohmann and N. F. Hurley (Eds.), New Ideas and Methods for Exploration for Carbonate Reservoirs. Dallas Geol. Soc., Short Course.

Lohmann, K. C., 1988. Geochemical patterns of meteoric diagenetic systems and their application to studies of paleokarst. In: N. P. James and P. W. Choquette (Eds.), Paleokarst. Springer-Verlag/New York, pp. 58-80.

Lohmann, K. C. and Meyers, W. J., 1977. Microdolomite inclusions in cloudy prismatic calcites: a proposed criterion for former high magnesium calcites. J. Sediment. Petrol., 47: 1078-1088.

Longman, M. W., 1980. Carbonate diagenetic textures from nearsurface diagenetic environments. Am. Assoc. Petrol. Geol. Bull., 64: 461-487.

Longman, M. W., 1985. Fracture porosity in reef talus of a Miocene pinnacle-reef reservoir, Nido B. field, the Philippines. In: P. O. Roehl and P. W. Choquette (Eds.), Carbonate Petroleum Reservoirs. Springer-Verlag/New York, pp. 547-560.

Loreau, J. P. and Purser, B. H., 1973. Distribution and ultrastructure of Holocene ooids in the Persian Gulf. In: B. H. Purser (Ed.), The Persian Gulf - Holocene Carbonate Sedimentation and Diagenesis in a Shallow Epicontinental Sea. Springer-Verlag/Berlin, pp. 279-328.

Loucks, R. G. and Budd, D. A., 1981. Diagenesis and reservoir potential of the Upper Jurassic Smackover Formation of south Texas. Trans. Gulf Coast Assoc. Geol. Socs., 31: 339-346.

Loucks, R. G. and Longman, M. W., 1982. Lower Cretaceous Ferry Lake anhydrite, Fairway field, east Texas: product of shallow sub-tidal deposition. In: C. R. Handford, R. G. Loucks and G. R. Davies (Eds.), Depositional and Diagenetic Spectra of Evaporites. Calgary, Can., SEPM Core Workshop No. 3, pp. 130-173.

Loucks, R. G. and Anderson, J. H., 1985. Depositional facies, diagenetic terranes, and porosity development in lower Ordovician Ellenburger dolomite, Puckett field, west Texas. In: P. O. Roehl and P. W. Choquette (Eds.), Carbonate Petroleum Reservoirs. Springer-Verlag/New York, pp. 19-37.

Loucks, R. G. , Bebout, D. G. and Galloway, W. E., 1977. Relationship of porosity formation and preservation of sandstone consolidation history - Gulf Coast Lower Tertiary Frio Formation. Trans. Gulf Coast Assoc. Geol. Socs., 27: 109-120.

Loucks, R. G., Dodge, M. M. and Galloway, W. E., 1979. Sandstone consolidation analysis to delineate areas of high quality reservoirs suitable for production of geopressured geothermal energy along the Texas Gulf Coast. University of Texas, Bureau of Econ. Geol., Contract report for U. S. Dept. of Energy, EG-77-05-5554: 1-98.

Machel, H. G., 1985. Facies and diagenesis of the Upper Devonian Nisku Formation in the subsurface of central Alberta. Dissertation. McGill University/Montreal, Quebec, 392 pp.

Machel, H. G., 1986. Early lithification, dolomitization and anhydritization of Upper Devonian Nisku buildups, subsurface of Alberta, Canada. In: J. H. Schroeder and B. H. Purser (Eds.), Reef Diagenesis. Springer-Verlag/ Berlin, pp. 336-356.

Machel, H. G. and Mountjoy, E. W., 1986. Chemistry and environments of dolomitization - a reappraisal. Earth-Sci. Rev., 23: 175-222.

Macintyre, I. G., 1977. Distribution of submarine cements in a modern Caribbean fringing reef, Galeta Point, Panama. J. Sediment. Petrol., 47: 503-516.

Macintyre, I. G., 1985. Submarine cements - the peloidal question. In: N. Schneidermann and P. M. Harris (Eds.), Carbonate Cements. SEPM Spec. Pub. No. 36, pp. 109-116.

Mackenzie, F. T. and Pigott, J. D., 1981. Tectonic controls of Phanerozoic sedimentary rock cycling. Jour. Geol. Soc. London, 138: 183-196.

Maiklem, W. R., 1971. Evaporative drawdown - a mechanism for water-level lowering and diagenesis in the Elk Point basin. Bull. Can. Petrol. Geol., 19: 487-503.

Maiklem, W. R. and Bebout, D. G., 1973. Ancient anhydrite facies and environments, Middle Devonian Elk Point basin, Alberta. Bull. Can. Petrol. Geol., 21: 287-343.

Majewske, O. P. 1969. Recognition of Invertebrate Fossil Fragments in Rocks and Thin Sections. Brill/Leiden, 101 pp.

Major, R. P., 1984. The Midway Atoll Coral Cap: Meteoric Diagenesis, Amplitude of Sea-Level Fluctuation and Dolomitization. Dissertation. Brown University/ Providence, Rhode Island, 133 pp.

Malek-Aslani, M., 1977. Plate tectonics and sedimentary cycles in carbonates. Trans. Gulf Coast Assoc. Geol. Socs., 27: 125-133.

Margaritz, M., Groldenberg, L., Kafri, U. and Arad, A., 1980. Dolomite formation in the seawater-freshwater interface. Nature, 287: 622-624.

Marshall, J. F., 1986. Regional distribution of submarine cements with an epicontinental reef system: central Great Barrier Reef, Australia. In: J. H. Schroeder and B. H. Purser (Eds.), Reef Diagenesis. Springer-Verlag/ Berlin, pp. 8-26.

Marshall, J. F. and Davies, P. J., 1981. Submarine lithification on windward reef slopes: Capricorn-Bunker group, southern Great Barrier Reef. J. Sediment. Petrol., 51: 953-960.

Martin, G. D. , Wilkinson, B. H. and Lohmann, K. C., 1986. The role of skeletal porosity in aragonite neomorphism - *Strombus* and *Montastrea* from the Pleistocene, Key Largo limestone, Florida. J. Sediment. Petrol., 56: 194-203.

Matter, A., Douglas, R. G. and Perch-Nielsen, K., 1975. Fossil preservation, geochemistry, and diagenesis of pelagic carbonates from Shatsky Rise, northwest Pacific. In: J. V. Gardner (Ed.), Initial Reports of the Deep Sea Drilling Project. U. S. Govt. Print. Off., pp. 891-921.

Mattes, B. W. and Mountjoy, E. W., 1980. Burial dolomitization of the Upper Devonian Miette buildup, Jasper National Park, Alberta. In: D. H. Zenger, J. B. Dunham and R. L. Ethington (Eds.), Concepts and Models of Dolomitization. SEPM Spec. Pub. No. 28, pp. 259-297.

Matthews, R. K., 1968. Carbonate diagenesis: equilibration of sedimentary mineralogy to the subaerial environment; coral cap of Barbados, West Indies. J. Sediment. Petrol., 38: 1110-1119.

Matthews, R. K., 1974. A process approach to diagenesis of reefs and reef associated limestones. In: L. F. Laporte (Ed.), Reefs in Time and Space. SEPM Spec. Pub. No. 18, pp. 234-256.

Matthews, R. K., 1987. Eustatic controls on near-surface carbonate diagenesis. In: J. E. Warme and K. W. Shanley (Eds.), Carbonate Depositional Environments Modern and Ancient - Part 6: Diagenesis. Colorado School of Mines Press/ Golden, Colorado. Colorado School of Mines Quarterly, pp. 17-40.

Mazzullo, S. J., 1980. Calcite pseudospar replacive of marine acicular aragonite, and implications for araogonite cement diagenesis. J. Sediment. Petrol., 50: 409-422.

Mazzulo, S. J. and Reid, A. M., 1987. Sedimentary textures of recent Belizean peritidal dolomites. J. Sediment. Petrol., 58: 479-188.

McBride, E. F., 1964. Review of turbidite studies in the U. S. In: Bouma, A. H. and Brouwer, A. (Eds.), Turbidites. Developments in Sedimentology 3. Elsevier/ Amsterdam, pp. 93-105.

McCrosson, R. G. and Glaister, R. P. (Eds.), 1964. Geological History of Western Canada. Alberta Soc. Petrol. Geol., Spec. Vol., 232 pp.

McGillivray, J. G. and Mountjoy, E. W., 1975. Facies and related reservoir characteristics, Golden Spike reef complex, Alberta. Bull. Can. Petrol. Geol., 23: 753-809.

McGillis, K. A., 1984. Upper Jurassic stratigraphy and carbonate facies, northeastern east Texas basin. In: M. W. Presley (Ed.), The Jurassic of East Texas. E. Texas Geol. Soc., Spec. Pub./Tyler, Texas, pp. 63-66.

McKenzie, J. A., 1981. Holocene dolomitization of calcium carbonate sediments from the coastal sabkhas of Abu Dhabi, U.A.E.: a stable isotope study. J. Geol., 89: 185-198.

McKenzie, J. A., Hsü, K. J. and Schneider, J. F., 1980. Movement of subsurface waters under the sabkha, Abu Dhabi, UAE, and its relation to evaporative dolomite genesis. D. H. Zenger, J. B. Dunham, and R. L. Ethington (Eds.), Concepts and Models of Dolomitization. SEPM Spec. Pub. No. 28, pp. 11-30.

McQuillan, H., 1985. Fracture-controlled production from the Oligo-Miocene Asmari Formation in Gachsaran and Bibi Hakimeh fields, southwest Iran. In: P. O. Roehl and P. W. Choquette (Eds.), Carbonate Petroleum Reservoirs. Springer-Verlag/New York, pp. 511-523.

Meendsen, F. C., Moore, C. H., Heydari, E. and Sassen, R., 1987. Upper Jurassic depositional systems and hydrocarbon potential of southeast Mississippi. Trans. Gulf Coast Assoc. Geol. Socs., 37: 161-174.

Meissner, F. F., 1972. Cyclic sedimentation in Middle Permian strata of the Permian basin, west Texas and New Mexico. In: J. C. Elam and S. Chuber (Eds.), Cyclic Sedimentation in the Permian Basin. W. Texas Geol. Soc./Midland, Texas, pp. 203-232.

Meyerhoff, A. A., 1967. Future hydrocarbon provinces of Gulf of Mexico-Caribbean region. Trans. Gulf Coast Assoc. Geol. Socs., 17: 217-260.

Meyers, W. J., 1974. Carbonate cement stratigraphy of the Lake Valley Formation (Mississippian) Sacramento Mountains, New Mexico. J. Sediment. Petrol., 44: 837-861.

Meyers, W. J., 1978. Carbonate cements: their regional distribution and interpretation in Mississippian limestones of southwestern New Mexico. Sedimentology, 25: 371-400.

Meyers, W. J., 1980. Compaction in Mississippian skeletal limestones southwestern New Mexico. J. Sediment. Petrol., 50: 457-474.

Meyers, W. J., 1988. Paleokarst features in Mississippian limestones, New Mexico. In: N. P. James and P. W. Choquette (Eds.), Paleokarst. Springer-Verlag/New York, pp. 306-328.

Meyers, W. J. and Lohmann, K. C., 1978. Microdolomite-rich syntaxial cements: proposed meteoric-marine mixing zone phreatic cements from Mississippian limestones, New Mexico. J. Sediment. Petrol., 48: 475-488.

Meyers, W. J. and Hill, B. E., 1983. Quantitative studies of compaction in Mississippian skeletal limestones; New Mexico. J. Sediment. Petrol., 53: 231-242.

Meyers, W. J. and Lohmann, K. C., 1985. Isotope geochemistry of regionally extensive calcite cement zones and marine components in Mississippian limestones, New Mexico. In: N. Schneidermann and P. M. Harris (Eds.), Carbonate Cements. SEPM Spec. Pub. No. 36, pp. 223-239.

Meyers, W. J., Cowan, P. and Lohmann, K. C., 1982. Diagenesis of Mississippian skeletal limestones and bioherm muds, New Mexico. In: K. Bolton, H. R. Lane and D. Le Mone (Eds.), Symposium on the Paleoenvironmental Setting and Distribution of the Waulsortian Facies, El Paso, Texas. El Paso Geol. Soc./ El Paso, Texas, pp. 80-95.

Miller, J. A., 1985. Depositional and reservoir facies of the Mississippian Leadville Formation, northwest Lisbon field, Utah. In: P. O. Roehl and P. W. Choquette (Eds.), Carbonate Petroleum Reservoirs. Springer-Verlag/New York, pp. 161-173.

Milliman, J. D., 1971. The role of calcium carbonate in continental shelf sedimentation: in the new concepts of continental margin sedimentation. Amer. Geol. Inst., Washington, D. C., Application to the geological record (supplement), lecture no. 14, 35 pp.

Milliman, J. D., 1974a. Marine Carbonates. Springer-Verlag/ Berlin, 375 pp.

Milliman, J. D., 1974b. Precipitation and cementation of deep-sea carbonate sediments. In: A. L. Inderbitzen (Ed.), Deep Sea Sediments. Plenum/New York, pp. 463-476.

Miran, Y., 1977. Chalk deformation and large scale migration of calcium carbonate. Sedimentology, 24: 333-360.

Mitchell, J. T., Land, L. S. and Miser, D. E., 1987. Modern marine dolomite cement in a north Jamaican fringing reef. Geology, 15: 557-560.

Molenaar, C. M., 1977. The Pinedale oil seep - an exhumed stratigraphic trap in the southwestern San Juan basin. In: J. E. Fassett (Ed.), San Juan Basin III. New Mexico Geol. Soc. Guidebook, 28th Field Conf., pp. 243-246.

Moore, C. H., 1973. Intertidal carbonate cementation, Grand Cayman, West Indies. J. Sediment. Petrol., 43: 591-602.

Moore, C. H., 1977. Beach rock origin: some geochemical, mineralogical and petrographic considerations. Geosci. and Man, 18: 155-163.

Moore, C. H., 1984. Regional patterns of diagenesis, porosity evolution, and hydrocarbon production, upper Smackover of the Gulf rim. Trans. Gulf Coast Assoc. Geol. Socs., 34: 455.

Moore, C. H., 1985. Upper Jurassic subsurface cements: a case history. In: P. M. Harris and N. Schneidermann (Eds.), Carbonate Cements. SEPM Spec. Pub. No. 36, pp. 291-308.

Moore, C. H., 1987. Jurassic platform development, northwestern Gulf of Mexico. Am. Assoc. Petrol. Geol., 71: 595.

Moore, C. H. and Shedd, W. W., 1977. Effective rates of sponge bioerosion as a function of carbonate production. Proc. 3rd. Intern. Coral Reef Sym., Miami, Fla., 499-505.

Moore, C. H. and Druckman, Y., 1981. Burial diagenesis and porosity evolution, Upper Jurassic Smackover, Arkansas and Louisiana. Am. Assoc. Petrol. Geol. Bull., 65: 597-628.

Moore, C. H. and Brock, F. C., 1982. Discussion of paper by Wagner and Matthews. J. Sediment. Petrol., 52: 19-23.

Moore, C. H., Smitherman, J.M. and Allen, S. H., 1972. Pore system evolution in a Cretaceous carbonate beach sequence. 24th IGC, Sec. 6, 124-136.

Moore, C. H., Graham, E. A. and Land, L. S., 1976. Sediment transport and dispersal across the deep fore-reef and island slope (-55 m to -305 m), Discovery Bay, Jamaica. J. Sediment. Petrol., 46: 174-187.

Moore, C. H., Chowdhury, A. and Heydari, E., 1986. Variation of ooid mineralogy in Jurassic Smackover limestones as control of ultimate diagenetic potential. Am. Assoc. Petrol. Geol. Bull., 70: 622-623.

Moore, C. H., Chowdhury, A. and Chan, L., 1988 (in press). Upper Jurassic Smackover platform dolomitization northwestern Gulf of Mexico: a tale of two waters. In: V. Shukla and P. A. Baker (Eds.), Sedimentology and Geochemistry of Dolostones. SEPM Spec. Pub. No. 43.

Morrow, D. W., 1982a. Diagenesis 1. Dolomite - part 1: the chemistry of dolomitization and dolomite precipitation. Geosci. Can., 9: 5-13.

Morrow, D. W., 1982b. Diagenesis 2. Dolomite - part 2: dolomitization models and ancient dolostones. Geosci. Can., 9: 95-10.

Morse, J. W., Zullig, J. J., Bernstein, L. D., Millero, F. J., Milne, P., Mucci, A. and Choppin, G. R., 1985. Chemistry of calcium carbonate-rich shallow water sediments in the Bahamas. Am. Jour. Sci., 285: 147-185.

Moshier, S. O., 1987. On the Nature and Origin of Microporosity in Micritic Limestones. Dissertation. Louisiana State University/ Baton Rouge, Louisiana, 291pp.

Mountjoy, E. W., 1978. Upper Devonian reef trends and configuration of the western portion of the Alberta basin. In: I. A. McIlreath and P. C. Jackson (Eds.), The Fairholme Carbonate Complex. Can. Soc. Petrol. Geol. Guidebook, pp. 1-30.

Mucci, A. and Morse, J. W., 1983. The incorporation of Mg^{+2} and Sr^{+2} into calcite overgrowths: influences of growth rate and solution composition. Geochim. Cosmochim. Acta, 47: 217-233.

Mueller, H. W., 1975. Centrifugal progradation of carbonate banks: a model for deposition and early diagenesis, Ft. Terrett Formation, Edwards Group, Lower Cretaceous, central Texas. Dissertation. University of Texas/Austin, Texas, 300 pp.

Murray, R. C., 1960. Origin of porosity in carbonate rocks. J. Sediment. Petrol., 30: 59-84.

Murray, R. C., 1969. Hydrology of south Bonaire, Netherlands Antilles - a rock selective dolomitization model. J. Sediment. Petrol., 39: 1007-1013.

Mussman, W. J., Montanez, I. P. and Read, F. J., 1988. Ordovician Knox paleokarst unconformity, Appalachians. In: N. P. James and P. W. Choquette (Eds.), Paleokarst. Springer-Verlag/New York, pp. 211-228.

Neugebauer, J., 1973. The diagenetic problem of chalk - the role of pressure solution and pore fluid. N. Jb. Geol. Paläontol. Abhandl., 143: 223-245.

Neugebauer, J., 1974. Some aspects of cementation on chalk. In: K. J. Hsü and H. C. Jenkyns (Eds.), Pelagic Sediments on Land and Under the Sea: Inter. Assoc. Sed., Spec. Pub.1, pp. 149-176.

Neumann, A. C., Kofoed, J. W. and Keller, G. H., 1977. Lithoherms in the Straits of Florida. Geology, 5: 4-10.

Niemann, J. C. and Read, J. F., 1988. Regional cementation from unconformity - recharged aquifer and burial fluids, Mississippian Newman Limestone. J. Sediment. Petrol., 58: 688-705.

O'Hearn, T. C., 1984. A Fluid Inclusion Study of Diagenetic Mineral Phases, Upper Jurassic Smackover Formation, Southwest Arkansas and Northeast Texas. Thesis. Louisiana State University/Baton Rouge, Louisiana, 189 pp.

O'Hearn, T. C. and Moore, C. H., 1985. Fluid inclusion study of diagenetic mineral phases, Upper Jurassic Smackover Formation, southwest Arkansas and northeast Texas. Am. Assoc. Petrol. Geol. Bull. (Abstr.), 69: 294.

Oldershaw, A. E. and Scoffin, T. P., 1967. The source of ferroan and non-ferroan calcite cements in the Halkin and Wenlock limestones. Geol. J., 5: 309-320.

Patterson, R. J. and Kinsman, D. J. J., 1977. Marine and continental groundwater sources in a Persian Gulf coastal sabkha. AAPG Stud. in Geol. No. 4, 381-397.

Perkins, R. D. and Halsey, S. D., 1971. Geologic significance of microboring fungi and algae in Carolina shelf sediments. J. Sediment. Petrol., 41: 843-853.

Petta, T. J., 1977. Diagenesis and geochemistry of a Glen Rose patch reef complex, Bandera Co., Texas. In: D. G. Bebout and R. G. Loucks (Eds.), Cretaceous Carbonates of Texas and Mexico; Applications to Subsurface Exploration. Bureau of Econ. Geol., University of Texas, Austin, Texas. Inv. Report No. 89, pp. 138-167.

Pettijohn, F. J., 1957. Sedimentary Rocks. Harper & Row/New York, 718 pp.

Pingitore, N. E., 1976. Vadose and phreatic diagenesis: processes, products and their recognition in corals. J. Sediment. Petrol., 46: 985-1006.

Pittman, E. D., 1975. Porosity and permeability changes during diagenesis of Pleistocene corals, Barbados, West Indies. Geol. Soc. Am. Bull., 85: 1811-1820.

Pittman, III, W. C.,1978. Relationship between eustacy and stratigraphic sequences of passive margins. Geol. Soc. Am. Bull., 89: 1389-1403.

Playford, P. E., 1980. Devonian "Great Barrier Reef" of Canning basin, Western Australia. Am. Assoc. Petrol. Geol. Bull., 64: 814-840.

Plummer, L. N., 1975. Mixing of sea water with calcium carbonate ground water. In: E. H. T. Whitten (Ed.), Quantitative Studies in the Geological Sciences: A Memoir in Honor of William C. Krumbein. Geol. Soc. Am., Mem. No. 142, pp. 219-236.

Plummer, L. N., Vacher, H. L., Mackenzie, F. T., Bricker, O. P. and Land, L. S., 1976. Hydrogeochemistry of Bermuda: a case history of ground-water diagenesis of biocalcarenites. Geol. Soc. Am. Bull., 87: 1301-1316.

Prezbendowski, D. R., 1985. Burial cementation - is it important? A case study, Stuart City Trend, south-central Texas. In: N. Schneidermann and P. M. Harris (Eds.), Carbonate Cements. SEPM Spec. Pub. No. 36, pp. 241-264.

Prezbindowski, D. R. and Larese, R. E., 1987. Experimental stretching of fluid inclusions in calcite - implications for diagenetic studies. Geology, 15: 333-336.

Purser, B. H., 1969. Syn-sedimentary marine lithification of Middle Jurassic limestones in the Paris basin. Sedimentology, 12: 205-230.

Purser, B. H., 1978. Early diagenesis and the preservation of porosity in Jurassic limestones. J. Petrol. Geol., 1: 83-94.

Purser, B. H., 1985. Dedolomite porosity and reservoir properties of Middle Jurassic carbonates in the Paris Basin, France. In: P. O. Roehl and P. W. Choquette (Eds.), Carbonate Petroleum Reservoirs. Springer-Verlag/New York, pp. 341-355.

Purser, B. H. and Schroeder, J. H., 1986. Conclusions, the diagenesis of reefs: a brief review of our present understanding. In: J. H. Schroeder and B. H. Purser (Eds.), Reef Diagenesis. Springer-Verlag/Berlin, pp. 424-446.

Rabat, A. M., Negra, M. H., Purser, B. H., Sassi, S. and Ben Ayed, N., 1986. Micrite diagenesis in Senonian rudist build-ups in central Tunisia. In: J. H. Schroeder and B. H. Purser (Eds.), Reef Diagenesis. Springer-Verlag/Berlin, pp. 210-223.

Rassmann-McLaurin, B., 1983. Holocene and ancient hardgrounds: petrographic comparison. Am. Assoc. Petrol. Geol. Bull. (Abstr.), 65: 975-976.

Read, J. F., Grotzinger, J. P., Bova, J. A. and Koerschner, W. F., 1986. Models for generation of carbonate cycles. Geology, 14: 107-110.

Reeder, R. J., 1981. Electron optical investigation of sedimentary dolomites. Contrib. to Min. and Petrol., 76: 148-157.

Reeder, R. J., 1983. Crystal chemistry of the rhombohedral carbonates. In: R. J. Reeder (Ed.), Carbonates: Mineralogy and Chemistry. Mineralogical Society of America/Washington, D. C.: Reviews in Mineralogy, pp. 1-47.

Reinhardt, J., 1977. Cambrian off-shelf sedimentation, central Appalachians. In: H. E. Cook and P. Enos (Eds.), Deep-Water Carbonate Environments. SEPM Spec. Pub. No. 25, pp. 83-112.

Rittenhouse, G., 1971. Pore space reduction by solution and cementation. Amer. Assoc. Petrol. Geol. Bull., 55: 80-91.

Roberts, H. H. and Whelan, T., 1975. Methane-derived carbonate cements in barrier and beach sands of a subtropical delta complex. Geochim. Cosmochim. Acta, 39: 1085-1089.

Robinson, R. B., 1967. Diagenesis and porosity development in Recent and Pleistocene oolites from southern Florida and the Bahamas. J. Sed. Petrol., 37: 355-364.

Rodgers, K. A., Easton, A. J. and Downes, C. J., 1982. The chemistry of carbonate rocks of Niue Island, South Pacific. J. Geol., 90: 645-662.

Roedder, E., 1979. Fluid inclusion evidence on the environments of sedimentary diagenesis, a review. In: P. A. Scholle and P. R. Schluger (Eds.), Aspects of Diagenesis. SEPM Spec. Pub. No. 26, pp. 89-107.

Roedder, E. and Bodner, 1980. Geologic pressure determinations from fluid inclusion studies. Ann. Rev. Earth and Planetary. Sci., 8: 263-301.

Roehl, P. O., 1967. Stony Mountain (Ordovician) and Interlake (Silurian) facies analogs of recent low-energy marine and subaerial carbonates, Bahamas. Am. Assoc. Petrol. Geol. Bull., 51: 1979-2032.

Roehl, P. O., 1985. Depositional and diagenetic controls on reservoir rock development and petrophysics in Silurian tidalites, Interlake Formation, Cabin Creek field area, Montana. In: P. O. Roehl and P. W. Choquette (Eds.), Carbonate Petroleum Reservoirs. Springer-Verlag/New York, pp. 87-105.

Roehl, P. O. and Choquette, P. W. (Eds.), 1985. Carbonate Petroleum Reservoirs. Springer-Verlag/New York, 622 pp.

Roehl, P. O. and Weinbrandt, R. M., 1985. Geology and production characteristics of fractured reservoirs in the Miocene Monterey Formation, west Cat Canyon oil field, Santa Maria Valley, California. In: P. O. Roehl and P. W. Choquette (Eds.), Carbonate Petroleum Reservoirs. Springer-Verlag/New York, pp. 524-545.

Rose, P. R., 1963. Comparison of type El Abra of Mexico with "Edwards Reef Trend" of south-central Texas. In: Geology of Peregrina Canyon and Sierra de El Abra, Mexico. Corpus Christi Geol. Soc. Am., Ann. Fld. Trip Guidebook, pp. 57-64.

Runnells, D. D., 1969. Diagenesis, chemical sediments, and mixing of natural waters. J. Sediment. Petrol., 39: 1188-1201.

Rupke, N. A., 1978. Deep clastic seas. In: H. G. Reading (Ed.), Sedimentary Environments and Facies. Elsevier/New York, pp. 372-415.

Ruzyla, K. and Friedman, G. M., 1985. Factors controlling porosity in dolomite reservoirs of the Ordovician Red River Formation, Cabin Creek field, Montana. In: P. O. Roehl and P. W. Choquette (Eds.), Carbonate Petroleum Reservoirs. Springer-Verlag/New York, pp. 39-58.

Saller, A. H., 1984a. Diagenesis of Cenozoic Limestones on Enewetak Atoll. Dissertation. Louisiana State University/Baton Rouge, Louisiana, 363 pp.

Saller, A. H., 1984b. Petrologic and geochemical constraints on the origin of subsurface dolomite, Enewetak Atoll: an example of dolomitization by normal sea water. Geology, 12: 217-220.

Saller, A. H., 1986. Radiaxial calcite in Lower Miocene strata, subsurface Enewetak Atoll. J. Sediment. Petrol., 56: 743-762.

Sandberg, P. A., 1983. An oscillating trend in Phanerozoic non-skeletal carbonate mineralogy. Nature, 305: 19-22.

Sandberg, P. A., 1985. Aragonite cements and their occurrence in ancient limestones. In: N. Schneidermann and P. M. Harris (Eds.), Carbonate Cements. SEPM Spec. Pub. No. 36, pp. 33-57.

Sandberg, P. A., Schneiderman, N. and Wunder, S. J., 1973. Aragonitic ultra skeletal relics in calcite-replaced Pleistocene skeletons. Nature Phy. Sci., 245: 133-134.

Sarg, J. F., 1976. Sedimentology of the carbonate-evaporite facies transition of the Seven Rivers Formation (Guadalupian, Permian) in southeast New Mexico. Dissertation. The University of Wisconsin/Madison, Wisconson, 313 pp.

Sarg, J. F., 1977. Sedimentology of the carbonate-evaporite facies transition of the Seven Rivers Formation (Guadalupian, Permian) in southeast New Mexico. SEPM Field Conf. Guidebook, Pub. 77, 16 pp.

Sarg, J. F., 1981. Petrology of the carbonate-evaporite facies transition of the Seven Rivers Formation (Guadalupian, Permian) southeast New Mexico. J. Sediment. Petrol., 51: 73-96.

Sassen, R. and Moore, C. H., 1988. Framework of hydrocarbon generation and destruction in eastern Smackover trend. Am. Assoc. Petrol. Geol. Bull., 72: 649-663.

Sassen, R., Moore, C. H. and Meendsen, F. C., 1987. Distribution of hydrocarbon source potential in the Jurassic Smackover Formation. Org. Geochem., 11: 379-383.

Schlager, W., 1981. The paradox of drowned reefs and carbonate platforms. Geol. Soc. Am. Bull., Part 1, 92: 197-211.

Schlager, W. and James, N. P., 1978. Low-magnesian calcite limestones forming at the deep-sea floor, Tongue of the Ocean, Bahamas. Sedimentology, 25: 675-702.

Schlanger, S. O., 1965. Dolomite-evaporite relations on Pacific islands. Sci. Rep. Tohoku Univ., 2nd Ser. (Geol.), 378: 15-29.

Schlanger, S. O. and Douglas, R. G., 1974. Pelagic ooze-chalk limestone transition and its implications for marine stratigraphy. In: K. J. Hsü and H. C. Jenkyns (Eds.), Pelagic Sediments; on Land and Under the Sea. Inter. Assoc. Sed., Spec. Pub.1, pp. 117-148.

Schmidt, V. and McDonald, D. A., 1979. The role of secondary porosity in the course of sandstone diagenesis. In: P. A. Scholle and P. R. Schluger (Eds.), Aspects of Diagenesis. SEPM Spec. Pub. No. 26, pp. 175-207.

Schmidt, V., McIlreath, I. A. and Budwill, A. E., 1985. Origin and diagenesis of Middle Devonian pinnacle reefs encased in evaporites, "A" and "E" pools, Rainbow field, Alberta. In: P. O. Roehl and P. W. Choquette (Eds.), Carbonate Petroleum Reservoirs. Springer-Verlag/New York, pp. 140-160.

Schmoker, J. W., 1984. Empirical relation between carbonate porosity and thermal maturity: an approach to regional porosity prediction. Amer. Assoc. Petrol. Geol. Bull., 68: 1697-1703.

Schmoker, J. W. and Halley, R. B., 1982. Carbonate porosity versus depth: a predictable relation for south Florida. Am. Assoc. Petrol. Geol., 66: 2561-2570.

Schofield, J. C., 1967. Origin of radioactivity at Niue Island. New Zealand J. Geol. Geophys., 10: 1362-1371.

Schofield, J. C. and Nelson, C. S., 1978. Dolomitization and Quaternary climate of Niue Island, Pacific Ocean. Pacific Geol., 13:37-48.

Scholle, P. A., 1971. Diagenesis of deep-water carbonate turbidites, Upper Cretaceous Monte Antola flysch, northern Appennines, Italy. J. Sediment. Petrol., 41: 233-250.

Scholle, P. A., 1977. Chalk diagenesis and its relation to petroleum exploration: oil from chalks, a modern miracle? Am. Assoc. Petrol. Geol. Bull., 61: 982-1009.

Scholle, P. A. and Arthur, M. A., 1980. Carbon isotope fluctuations in Cretaceous pelagic limestones: potential stratigraphic and petroleum exploration tool. Am. Assoc. Petrol. Geol. Bull., 64: 67-87.

Scholle, P. A. and Halley, R. B., 1985. Burial diagenesis: out of sight, out of mind. In: N. Schneidermann and P. M. Harris (Eds.), Carbonate Cements. SEPM Spec. Pub. No. 36, pp. 309-334.

Scholle, P. A., Arthur, M. A. and Ekdale, A. A., 1983. Pelagic environment. In: P. A. Scholle, D. G. Bebout and C. H. Moore (Eds.), Carbonate Depositional Environments. AAPG Mem. 33, pp. 619-691.

Scholle, P. A., Bebout, D. G. and Moore, C. H. (Eds.), 1983. Carbonate Depositional Environments. AAPG Mem. 33, 708 pp.

Schreiber, B. C., Roth, M. S. and Helman, M. L., 1982. Recognition of primary facies characteristics of evaporites and the differentiation of these forms from diagenetic overprints. In: C. R. Handford, R. G. Loucks and G. R. Davies (Eds.), Depositional and Diagenetic Spectra of Evaporites. Calgary, Can., SEPM Core Workshop No. 3, pp. 1-32.

Schroeder, J. H., 1969. Experimental dissolution of calcium, magnesium and strontium from recent biogenic carbonates: a model of diagenesis. J. Sediment. Petrol., 39: 1057-1073.

Schroeder, J. H. and Zankl, H., 1974. Dynamic reef formation: a sedimentological concept based on studies of recent Bermuda and Bahama reefs. In: Proc. 2nd Intern. Coral Reef Sym., Brisbane, Australia, Great Barrier Reef Committee, pp. 413-428.

Schroeder, J. H. and Purser, B. H., 1986. Reef diagenesis: Introduction. In: J. H. Schroeder and B. H. Purser (Eds.), Reef Diagenesis. Springer-Verlag/Berlin, pp. 1-5.

Scruton, P. C., 1953. Deposition of evaporites. Am. Assoc. Petrol. Geol. Bull., 37: 2498-2512.

Sears, S. O. and Lucia, F. J., 1980. Dolomitization of northern Michigan Niagara Reefs by brine refluxion and freshwater/seawater mixing. In: D. H. Zenger, J. B. Dunham and R. L. Ethington (Eds.), Concepts and Models of Dolomitization. SEPM Spec. Pub. No. 28, pp. 215-236.

Selley, R. C., 1970. Ancient Sedimentary Environments. Chapman & Hall Ltd./London, 137 pp.

Sellwood, B. W., 1978. Shallow-water carbonate environments. In: H. G. Reading (Ed.), Sedimentary Environments and Facies. Elsevier/New York, pp. 259-313.

Shearman, D. J., 1966. Origin of marine evaporites by diagenesis. Trans. Insti. Mining and Metall., 75: 208-215.

Shearman, D. J. and Fuller, J. G., 1969. Anhydrite diagenesis, calcitization and organic laminites, Winnipegosis Formation, Middle Devonian, Saskatchewan. Bull. Can. Petrol. Geol., 17: 496-525.

Shinn, E. A., 1968. Practical significance of birdseye structures in carbonate rocks. J. Sediment. Petrol., 38: 215-223.

Shinn, E. A., 1969. Submarine lithification of Holocene carbonate sediments in the Persian Gulf. Sedimentology, 12: 109-144.

Shinn, E. A., 1983. Tidal flat. In: P. A. Scholle, D. G. Bebout and C. H. Moore (Eds.), Carbonate Depositional Environments. AAPG Mem. 33, pp. 171-210.

Shinn, E. A. and Robbin, D. M., 1983. Mechanical and chemical compaction in fine-grained shallow-water limestones. J. Sediment. Petrol., 53: 595-618.

Shinn, E. A., Ginsburg, R. N. and Lloyd, R. M., 1965. Recent supratidal dolomite from Andros Island, Bahamas. In: L. C. Pray and R. C. Murray (Eds.), Dolomitization and Limestone Diagenesis - a Symposium. SEPM Spec. Pub. 13, pp. 112-123.

Shinn, E. A., Lloyd, R. M. and Ginsburg, R. N., 1969. Anatomy of a modern carbonate tidal flat, Andros Island, Bahamas. J. Sediment. Petrol., 39: 1202-1228.

Shinn, E. A., Bloxsom, W. E. and Lloyd, R. M., 1974. Recognition of submarine cements in Cretaceous reef limestones from Texas and Mexico (Abstract). Am. Assoc. Petrol. Geol. Bull., 1: 82-83.

Shinn, E. A., Steinen, R. P., Lidz, B. H. and Halley, R. B., 1985. Bahamian Whitings - No Fish Story (Abstract). Am. Assoc. Petrol. Geol. Bull., 69: 307.

Shinn, E. A., Hudson, H. J., Halley, R. B., Lidz, B., Robbin, D. M. and Macintyre, I. G., 1982. Geology and sediment accumulation rates at Carrie Bow Cay, Belize. In: K. Rützler and I. G. Macintyre (Eds.), The Atlantic Barrier Reef Ecosystem at Carrie Bow Cay, Belize, I. Structure and Communities. Smithsonian Contrib. to the Mar. Sci., No. 12, pp. 63-75.

Shirley, K., 1987. Colorful history, odd geology - Yates Field celebrates 60 years. AAPG Explorer, 8: 4-5.

Sibley, D. F., 1980. Climatic control of dolomitization, Seroe Domi Formation (Pliocene), Bonaire, N. A. In: D. H. Zenger, J. B. Dunham and R. L. Ethington (Eds.), Concepts and Models of Dolomitization. SEPM Spec. Pub. No. 28, pp. 247-258.

Silver, B. S. and Todd, R. G., 1969. Permian cyclic strata, northern Midland and Delaware basins, west Texas and southeastern New Mexico. Am. Assoc. Petrol. Geol. Bull., 53: 2223-2251.

Simms, M., 1984. Dolomitization by groundwater-flow systems in carbonate platforms. Trans. Gulf Coast Assoc. Geol. Socs., 34: 411-420.

Skall, H., 1975. The paleoenvironment of the Pine Point lead-zinc district. Econ. Geol., 70: 22-47.

Steinen, R. P. and Matthews, R. K., 1973. Phreatic vs vadose diagenesis: stratigraphy and mineralogy of a cored bore hole on Barbados, W. I. J. Sediment. Petrol., 43: 1012-1020.

Stewart, S. K., 1984. Smackover and Haynesville facies relationships in north-central east Texas. In: M. W. Presley (Ed.), The Jurassic of East Texas. E. Texas Geol. Soc. Spec. Pub./Tyler, Texas, pp. 56-62.

Stoakes, F. A., 1979. Sea Level Control of Carbonate-Shale Deposition During Progradational Basin-Filling: The Upper Devonian Duvernay and Treton Formations of Alberta, Canada. Dissertation. The University of Calgary/Calgary, Alberta, 346 pp.

Stoessell, R. K. and Moore, C. H., 1983. Chemical constraints and origins of four groups of Gulf Coast reservoir fluids. Am. Assoc. Petrol. Geol. Bull., 67: 896-906.

Stueber, A. M., Pushkar, P. and Hetherington, E. A., 1984. A strontium isotopic study of Smackover brines and associated solids, southern Arkansas. Geochim. Cosmochim. Acta, 48: 1637-1649.

Surdam, R. C., Boese, S. W. and Crossey, L. J., 1984. The chemistry of secondary porosity. In: D. A. McDonald and R. C. Surdam (Eds.), Clastic Diagenesis. AAPG Mem. 37, pp. 127-149.

Swartz, J. H., 1958. Geothermal measurements on Eniwetok and Bikini atolls. U. S., Geol. Surv., Profess. Paper 260-U, 711-739.

Swartz, J. H., 1962. Some physical constraints for the Marshall Island area. U.S., Geol. Surv., Profess. Paper 260-AA, 953-989.

Swinchatt, J. P., 1965. Significance of constituent composition texture and skeletal breakdown in some Recent carbonate sediments. J. Sediment. Petrol., 35: 71-90.

Swirydczuk, K., 1988. Mineralogical control on porosity type in Upper Jurassic Smackover ooid grainstones, southern Arkansas and northern Louisiana. J. Sediment. Petrol., 58: 339-347.

Taft, W. H., Arrington, F., Haimovitz, A., Macdonald, C. and Woolheater, C., 1968. Lithification of modern carbonate sediments at Yellow Bank, Bahamas. Bull. Mar. Sci. Gulf Caribbean, 18: 762-828.

Taylor, S. R. and Lapré, J. F., 1987. North Sea chalk diagenesis: its effect on reservoir location and properties. In: J. Brooks and K. W. Glennie (Eds.), Petroleum Geology of North West Europe. Proceedings of the 3rd Conference on Petroleum Geology of North West Europe held at the Barbican Centre, London, 26-29 October, 1986. Graham and Trotman/London, pp. 483-495.

Thibodaux, B. L., 1972. Sedimentological Comparison Between Sombrero Key and Looe Key. Thesis. Louisiana State University/Baton Rouge, Louisiana, 83 pp.

Tissot, B. P. and Welte, D. H., 1978. Petroleum Formation and Occurrence: a New Approach to Oil and Gas Exploration. Springer-Verlag/Berlin, 538 pp.

Usiglio, J., 1849. Analyse de l'eau de la Mediterranée sur le Côtes de France. Ann. de Chem. Phys., 3rd series, 27: 92-107.

Vacher, H. L., 1974. Groundwater hydrology of Bermuda. Public Works Dept., Bermuda, 87 pp.

Vahrenkamp, V. C. and Swart, P. K., 1987a. Major and trace element sigatures of Late Tertiary platform dolomites from the Bahamas: indication for fluid evolution (Abstr.). SEPM Midyear Mtg., Abstr., 4: 85-86.

Vahrenkamp, V. C. and Swart, P. K., 1987b. Stable isotopes as tracers of fluid/rock interactions during massive platform dolomitization, Little Bahama Bank. Am. Assoc. Petrol. Geol. Bull. Abstr., 71: 624.

Vail, P. R. and Hardenbol, J., 1979. Sea level changes during the Tertiary. Oceanus, 22: 71-79.

Vail, P. R., Mitchum, Jr., R. M., Todd, R. G., Widmier, J. M., Thompson, III, S., Sangree, J. B., Bubb, J. N. and Hatlelid, W. G., 1977. Seismic stratigraphy and global changes of sea level. In: C. E. Payton (Ed.), Seismic Stratigraphy - Applications to Hydrocarbon Exploration. AAPG Mem. 26, pp. 49-212.

Van den Bark, E. and Thomas, O. D., 1981. EKOFISK; first of the giant oil fields in western Europe. In: M. T. Halbouty (Ed.), Giant Oil and Gas Fields of the Decade, 1968-1978. AAPG Mem. 30, pp. 195-224.

Veizer, J., 1983. Chemical diagenesis of carbonates: theory and application of trace element technique. In: M. A. Arthur (Ed.), Stable Isotopes in Sedimentary Geology. SEPM Short Course Notes No. 10, pp. 3.1-3.100.

Veizer, J. and Compston, W., 1974. $^{87}Sr/^{86}Sr$ composition of seawater during the Phanerozoic. Geochim. Cosmochim. Acta, 38: 1461-1484.

Veizer, J. and Hoefs, J., 1976. The nature of $^{18}O/^{16}O$ and C^{13}/C^{12} secular trends in sedimentary carbonate rocks. Geochim. Cosmochim. Acta, 40: 1387-1895.

von der Borch, C. C., Lock, D. E. and Schwebel, D., 1975. Ground-water formation of dolomite in the Coorong region of South Australia. Geology, 3: 283-285.

Wagner, P. D., 1983. Geochemical Characterization of Meteoric Diagenesis in Limestone: Development and Applications. Dissertation. Brown University/Providence, Rhode Island, 384 pp.

Wagner, P. D. and Matthews, R. K., 1982. Porosity preservation in the upper Smackover (Jurassic) carbonate grainstone, Walker Creek field, Arkansas: response of paleophreatic lenses to burial processes. J. Sediment. Petrol., 52: 3-18.

Walls, R. A., 1977. Cementation History and Porosity Development, Golden Spike Reef Complex (Devonian), Alberta. Dissertation. McGill University/Montreal, Quebec, 307 pp.

Walls, R. A. and Burrowes, G., 1985. The role of cementation in the diagenetic history of Devonian reefs, western Canada. In: N. Schneidermann and P. M. Harris (Eds.), Carbonate Cements. SEPM Spec. Pub. No. 36, pp. 185-220.

Walls, R. A., Mountjoy, E. W. and Fritz, P., 1979. Isotopic composition and diagenetic history of carbonate cements in Devonian Golden Spike reef, Alberta, Canada. Geol. Soc. Am. Bull., 90: 963-982.

Wanless, H. R., 1979. Limestone response to stress: pressure solution and dolomitization. J. Sediment. Petrol., 49: 437-462.

Ward, W. C., and Halley, R. B., 1985. Dolomitization in a mixing zone of near-seawater composition, Late Pleistocene, northeastern Yucatan Peninsula. J. Sediment. Petrol., 55: 407-420.

Wendte, J., 1974. Sedimentation and Diagenesis of the Cooking Lake Platform and Lower Leduc Reef Facies, Upper Devonian, Redwater, Alberta. Dissertation. University of California/Santa Cruz, California, 221 pp.

Weyl, P. K., 1960. Porosity through dolomitization: conservation-of-mass requirements. J. Sediment. Petrol., 30: 85-90.

Weyl, P. K., 1959. Pressure-solution and the force of crystallization: phenomenological theory. J. Geophys. Res., 64: 2001-2025.

Wilkinson, S., 1984. Upper Jurassic facies relationships and their interdependence on salt tectonism in Rains, Van Zandt, and adjacent counties, east Texas. In: M. W. Presley (Ed.), The Jurassic of East Texas. E. Texas Geol. Soc./Tyler Texas, pp. 153-156.

Wilkinson, B. H., Janecke, S. U. and Brett, C. E., 1982. Low-magnesian calcite marine cement in Middle Ordovician hardgrounds from Kirkland, Ontario. J. Sediment. Petrol., 52: 47-58.

Wilson, J. L., 1975. Carbonate Facies in Geologic History. Springer-Verlag/New York, 471 pp.

Wilson, J. L., 1977. Regional distribution of phylloid algal mounds in Late Pennsylvanian and Wolfcampian strata of southern New Mexico. Geology of the Sacramento Mountains, Otero County, New Mexico. W. Texas Geol. Soc. Field Trip Guidebook, pp. 1-7.

Wilson, J. L., 1980. Limestone and dolomite reservoirs. In: G. D. Hobsen (Ed.), Petroleum Geology - 2. Applied Sciences Publishers, Ltd./Essex, U. K., pp. 1-51.

Winker, C. D. and Buffler, R. T., 1988. Paleogeographic evolution of early deep-water Gulf of Mexico and margins, Jurassic to Middle Cretaceous (Comanchean). Am. Assoc. Petrol. Geol. Bull., 72: 318-346.

Winter, J. A., 1962. Fredericksburg and Washita strata (subsurface Lower Cretaceous) southwest Texas. In: Contributions to Geology of South Texas. S. Texas Geol. Soc./San Antonio, Texas, pp. 81-115.

Wollast, R., 1971. Kinetic aspects of the nucleation and growth of calcite from aqueous solutions. In: O. P. Bricker (Ed.), Carbonate Cements. AAPG Stud. in Geol. No. 19. Johns Hopkins University Press/Baltimore, Maryland, pp. 264-273.

Wong, P. K. and Oldershaw, A., 1981. Burial cementation in the Devonian Kaybob Reef complex, Alberta, Canada. J. Sediment. Petrol., 51: 507-520.

Wood, G. V. and Wolfe, M. J., 1960. Sabkha cycles in the Arab Darb Formation off the Trucial Coast of Arabia. Sedimentology, 12: 165-191.

Woronick, R. E. and Land, L. S., 1985. Late burial diagenesis, Lower Cretaceous Pearsall and Lower Glen Rose formations, south Texas. In: N. Schneidermann and P. M. Harris (Eds.), Carbonate Cements. SEPM Spec. Pub. No. 36, pp. 265-275.

Zenger, D. H., 1972a. Significance of supratidal dolomitization in the geologic record. Geol. Soc. Am. Bull., 83: 1-12.

Zenger, D. H., 1972b. Dolomitization and uniformitarianism. J. Geol. Educ., 20: 107-124.

Zenger, D. H. and Dunham, J. B., 1980. Concepts and models of dolomitization - an introduction. In: J. B. Dunham and R. L. Ethington (Eds), Concepts and Models of Dolomitization. SEPM Spec. Pub. No. 28, pp. 1-9.

INDEX

Abnormal pressure, 257
Abu Dhabi, 124, 142
__, dolomite isotopic composition, 231
Alabama, 151, 199
Alacran Reef, 3, 4
Alberta, 102, 103, 105, 106, 112, 155, 157, 254, 267, 268, 271, 272
__, shelf margin complex, 103
Albuskjell field, 254
Algae, coralline, 2, 6, 31, 32
__, green, 2, 3
__, microboring, 29
__, red, 3
Algal, laminations, 141, 142
__ mats, 9
__ sediments, 31
__ wackestones, 31
Allochems, 7
Alluvial fans, 136
Ancient cements, recognition of, 80-83
__, secular mineralogic trends, 82, 83
__, __, stable isotopes, 83
__, stabilization processes of, 80, 81
__, __ aragonite, 81
__, __ magnesian calcite, 80
__, types of, 80, 81
__, *see also* Cements
Andros Island, 125, 127, 222
Angle of repose, 16
Anhydrite, 24, 40, 123, 124, 125, 129, 132, 135, 152, 156, 157, 240, 263
Aquifer, gravity driven system, 193
Arab D Formation, 142

Aragonite, 2, 16, 34, 120, 121, 123, 133, 145, 162, 167, 168, 169, 194
__, ancient, criteria for recognition, 82
__ cements, 109
__ __, conversion to calcite, 183, 189
__, compensation depth, 44, 50, 96, 111
__ dissolution, 38, 109, 111
__ dissolution pathways, 170
__, *Halimeda* segments, 33
__ in caves, 162
__, lysocline, 76, 109
__, macroscale dissolution, 169, 170
__, metastable, 45, 56, 76
__, microscale dissolution, 169-197
__ needles, 4, 28
__, polar aragonite crystals, 29
__, porosity modification, 109
__, saturation state, 116, 194
__ sediments, 109
__ stability, 162
__ stabilization, 81
__, stable isotopes, 78, 90, 109
__, trace elements, 56, 90
__, __, distribution coefficient, 56, 57
__, __, original mineralogy, 59
__, *see also* Cement, aragonite
Arco Bodcaw #1, 189, 191, 192, 202, 249, 263, 267
__ log, 190
__, porosity, 190
Arkansas, 33, 35, 59, 87, 171, 187, 188, 191, 199, 200, 201, 202, 246, 249, 266
Arizona, 10
Artesia Group, 148

Artesian system, 173
Asmari Limestone, 40
Atascosa Formation, 97, 101
Australia, 87, 112, 145, 171, 230, 232, 234, 235
Autocementation, 44
Autochthonous, 258
Aux Vases Sandstone, 226

Baffin Bay, dolomite isotopic composition, 231
Bahamas, 13, 15, 53, 77, 87, 88, 109, 114, 125, 127, 180, 182, 183, 277
__, dolomite isotopic composition, 231
Barbados, 184, 191, 193, 222, 223
__, dolomite, 224
__, dolomite isotopic composition, 221
Barrier reef, see Reef
Basinal brines, 44, 45
Beach, 6, 16
__, lower shoreface, 10
__, shoreface sequences, 16
__, siliciclastic beach sediments, 16
__, upper shoreface, 9
Beachrock, 16, 76, 83-87
__, algal filaments 78, 84
__, cementation processes, 83, 87
__, cements, 53, 76, 77, 78, 84, 85, 87
__, depositional environments, 84, 85, 86, 87
__, porosity modification, 85, 86, 87
__, sediments, 83, 84, 86, 87
Beaverhill Lake, 155
Belize, 90, 91, 127
Bell Canyon Formation, 149
Bermuda, 32, 90, 182, 222
Besa River Formation, 273
Bigoray area, 271, 274
Bigoray Formation, 275

Bioherms, Late Paleozoic, 112
Bittern salts, 268
Bitumen, 38
__, solid bitumen, 40
Black Creek Basin, 155
Bliss Sandstone, 136, 139
__, depositional mode, 138
Bonaire, 145, 222, 223
__, dolomite isotopic composition, 231
Bone Spring Limestone, 149
Boring porosity, see Porosity types
Bossier Formation, 152, 200
Bouma sequence, 6
Boundstone, 8
Brachiopods, 28
Brazeau area, 271, 274
Brazil, 16
Breccia, 16
__ as hydrocarbon reservoir, 39
__, association with secondary porosity, 39
__, mineralization of, 39
__, porosity, see Porosity types
Brecciation, of carbonate rock sequences, 39
Breton Formation, 275
Bridgeport field, map, 227
__, see North Bridgeport field
Brine, 241, 242
British Columbia, 155
Brown University, 193
Brushy Canyon Formation, 149
Bryans Mill field, 37
Bryozoans, 2
Buckner Anhydrite, 13, 151, 152, 199, 200, 203
__ isopach map, 151
__ lithofacies map, 151
Burial cementation, 260
__, carbon isotopic composition, 265

—, geochemistry, 262
—, impact on reservoir porosity, 266
—, oxygen isotopic composition, 264, 265
—, petrography, 262
—, source of cement, 262
—, strontium isotopic composition, 265, 266
—, trace elements, 262, 264
Burial diagenesis,
—, burial diagenetic environment, 26, 237
—, susceptibility of carbonates, 18-19
—, burial diagenetic regimes, 238
Burrows, 9, 16
—, burrow porosity, see Porosity types

Cabin Creek field, 130, 131, 132, 133
Calcian dolomite, 121, 142
Calcite, 24, 34, 40, 45, 162, 172, 194
— as a diagenetic constituent, 22
— cement morphology, 49
— —, control by magnesium poisoning
— compensation depth, 44, 50, 75, 76, 96, 109, 111, 112
—, concurrent dissolution, 36, 37
— dissolution, 75, 113, 230
—, dissolution of relict calcite, 38
—, dominating pelagic sediment, 102
— growth habit, control, 47, 48, 49, 50
— —, control by magnesium calcium ratio, 48
— —, predicted by surface potential, 50
— stable isotopic composition, zoned, 208
—, luminescent zoned, 204, 206, 208
— lysocline, 50, 76, 114, 115
— nucleation rate step, 194
— precipitation, 109
— saturation, 51, 110, 194
— stability field 162, 164

—, stabilized terrain, 194
—, trace elements, 56
— —, distribution coefficent, 56, 57
—, see also Cement, calcite
Calcitic seas, 229
Calcium carbonate, high solubility, 18
—, rearrangement, 17
—, transport, 193
Calcrete, see Caliche
Calgary, 272
Caliche, 166, 177, 178
—, porosity occlusion by, 179, 186
—, profile sketch, 178
California, 40
Callovian, 152, 200
Cambrian, 3, 82, 83, 96, 136
Campanian, 10
Canada, 96, 102, 103, 104, 107, 155, 268, 271, 276, 283, 284
Canadian sedimentary basin, 102
Canning basin, 254
Capillary fringe, 179, 210
Capillary pressure, 22
Capitan Formation, 148
Carbon dioxide degassing, 40, 50, 83, 88, 90, 91, 182
—, generation, 241, 267
— in burial diagenesis pore fluids, 18
—, partial pressure in soil, 178, 179, 182, 191, 210
Carbon isotopic composition,
— of burial dolomite, 270
— of burial cements, 264, 265
Carbon reservoirs, mass and isotopic values, 65
Carbonate buildups, 14, 15
Carbonate, marine environment, 15, 16
—, potential nuclei, 15
— precipitation, 15, 16
— ramp, 11, 12

__ rock, classification, 7-9
__ sediments, 1-7, 18, 19
__ __, biological influence, 1-6, 19
__ __, controlling environmental parameters, 1, 2, 18, 19
__ __, distribution, 1,2, 18, 19
__ __, grain composition, 5, 6, 19
__ __, grain size, 3, 19
__ __, grain supported, 30
__ __, mud dominated, 31
__ __, mud supported, 30
__ __, origin, 1, 2, 19
__ __, primary porosity in, 27-32
__ __, roundness, 4
__ __, sorting, 3
__ __, texture and fabric, 3
__ shelf, 9, 12, 13, 14
__ __ margin, 11, 15
__ __ relief, 13
__ __, rimmed, 12, 13
Carbonate mud, 8
Carbonate platform, 10, 13-15, 17
__, emergent, 13
__, epicontinental, 14,15
__, temporal distribution of types, 14,15
Carbonate saturation versus water depth, 76, 110
Carbonates and siliciclastics, differences, 19
Carlsbad New Mexico, 148, 149
Carnalite, 268
Carnarvon, 145
Caribbean, 215
Castille Formation, 157
Catagenesis, 240, 241
Cathodoluminescence, 61
__, see Luminescent zonation
Cave, 34, 210, 211, 212, 214
__, phreatic development, 24
__ porosity, see Porosity, cave

__ waters, 162,
__ __, calcium carbonate, 163
__ __, magnesium calcium ratio, 163
Cavernous porosity, see Porosity types
Cedar Creek anticline, 130, 131, 133
Celestite, 40
Cement, 2, 15
__, anhydrite, replacement, 54, 55
__, aragonite,
__, __, ancient, recognition of, 82
__, __, beachrock, 76, 77
__, __, botryoidal, 78, 81, 90, 94, 96
__, __ dissolution, 81
__, __, epitaxial overgrowths, 90
__, __, fibrous, 47, 50, 88, 90
__, __, fibrous circumgranular, 77
__, __, ghosts, 81, 84
__, __, marine, 45
__, __ mesh, 78
__, __ needle, 77
__, __ reefal, 90
__, __ stabilization, 81
__, burial, carbon isotopic composition, 264, 265
__, calcite,
__, __ associated with algae, 87
__, __, beachrock, 77, 85
__, __, bladed, 49, 50, 77, 80, 87, 111
__, __, circumgranular, 49, 52, 53, 55, 77
__, __, concentrations of, 179
__, __, equant, 47, 48, 49, 50, 180, 183, 220
__, __, fasicular optic, 94
__, __, fibrous, 47, 49, 50, 77, 80, 87
__, __, __, ancient, 80, 81, 85, 94
__, __, __, pendant, 85
__, __, fresh water, 47
__, __, hardground, 87, 88
__, __, isopachous, 54, 89
__, __, magnesium calcium ratio, 47, 115

__, __, marine, 47, 77, 82
__, __, meniscus, 51, 52, 53
__, __, micrite, 77, 78, 109
__, __, microcrystalline cements, 85
__, __, microstalactitic, 51, 52, 53, 84
__, __, pelleted micrite, 78, 91
__, __, peloidal, reefal 90, 91
__, __, poikilotopic, 50, 54, 55, 71, 191, 262, 263
__, __, polygonal structure, 49, 77, 81
__, __, polyhedral, 109
__, __, post-compaction subsurface, 55
__, __, radial fibrous, 94, 95, 96
__, __, radiaxial, 78, 79, 94, 95, 102, 111, 112
__, __, reef, 76, 77, 90
__, __, rhombic, 49
__, __, spalled, 54, 55
__, __, whisker crystal, 50
__, circumgranular crust, 183, 220
__, compaction features, 53, 54, 55
__, distribution by diagenetic environment, 52, 53
__, dolomite, 113
__, geopetal crusts, 92
__, growth habit controls,
__, __, magnesium poisoning, 47, 48
__, __, rate of precipitation, 50
__, __, surface potential, 48, 39
__, intergranular, 92, 200
__, isopachous, 183, 246
__, marine, 16, 18, 19, 32
__, meniscus cementation pattern, 180
__, microstalactitic, 180
__, porosity modification by, 25
__, pressure solution, relation to, 198
__, subsurface Jurassic, trace elements in, 69
__, __, beachrock, 87
__, vadose, isotopic composition, 180

__, __, petrography, 179-180
__, __, strontium isotopic composition, 181
__, __, trace element composition, 180
__, zoned, isotopic composition, 208
__, __, luminescent, 204, 206, 208
__, see also Ancient cements
Cementation, marine, 44
__, passive, 45
__, shallow marine cementation processes, 76-80
__, __, cements, 76, 77
__, __, environments, 76, 77, 80
__, __, kinetics, 79
__, __, relation to sedimentary environments, 80
__, __, stable isotopes, 78, 79
__, __, trace elements, 78
Cenozoic, 96
Central basin platform, 212, 213
Central Graben, 254, 257, 258, 259
Central Montana uplift, 156
Chalk, 78, 240, 244, 251, 256, 260, 279
__, isotopes, 78, 709
__, porosity, 256
Chalky texture, 178
Channels, 34
Channel porosity see Porosity types
Charles Formation, 36
Chemical compaction, 44, 247
__, factors affecting, 251, 252
Chemical potential, 247
Chemical reactivity, 15-18
__, of siliciclasitcs, 16-18
Cherry Canyon Formation, 149
Chesterian, 204, 266
Christ Church Ridge, hydrologic setting, 193
Chicago, 155
Chlorite, 268

Choquette and Pray porosity classification, 21-27
Clionid borings, galleries, 93, 94
Coelenterates, 4
Compaction, 238, 243, 238
__, chemical, 243
__, dewatering, 243
__ in sandstones, 22
__, mechanical, 243
Composite grains, 28
Conchs, 3
Conglomerates, 16
Coniacian, 10
Continental margins, 14
Convection, 25
Cooking Lake Platform, 102
Coorong, 230, 232, 233, 234, 235
__, cross section, 233
__, hydrologic setting, 233,
__, lagoon waters, 162
Coralline algae, 2, 6
Corals, 2-4, 6
__, coral polyps, 28
__, scleractinian, 3, 30
__, septum, 5
__, ultrastructure, 5
Cordilleran tectonics, 102
Core analysis, 22
Coulommes field, 38
Cretaceous, 13, 39, 70, 84, 86, 87, 94, 96, 97, 98, 99, 100, 101, 102, 112, 115, 150, 155, 212, 252, 253, 254, 255, 260, 274
Crystal silt, 179
Cyclicity, 9-11, 18
Cynthia Formation, 275

Dakota shelf, 155
Dan field, 254
Danian, 255
Debris flow, 258, 259
Decarboxilation, 267
Dedolomitization, 37
Deep burial pore fluids, 241
Deep marine diagenetic environments, 108-116
__, dolomitization below calcite compensation depth, 112-114
__, sediments, 108
__, thermal convection model of marine water dolomitization, 114-116
__, zones of aragonite dissolution, 109-112
Deep marine environment,
__, calcite/aragonite saturation state, 100
__, depositional environments, 110, 114
__ sediments, 108
Delaware basin, 137, 148, 149, 213
Density stratification, 13
Depositional environment of carbonates, 5
__, restricted, 3
__, see also Deep marine
__, see also Lagoon
__, see also Reef
__, see also Sabkha
__, see also Shallow marine
Depositional porosity, see Porosity types
Depositional stage of primary porosity formation, 29
Devonian, 96, 102, 104, 107, 112, 131, 226, 251, 254, 270, 271, 272, 283, 284
Devonian reefs, 96, 103, 104, 107
Diagenesis, compaction during, 26
__, environments of, 24-26
__, influence in carbonates, 22
__, influence in sandstones, 22
Diagenetic environments of porosity modification, 43-46
__, marine, 44, 45

__, meteoric, 45
__, subsurface, 45
__, tools for recognition, 43-74
Diagenetic events, timing by petrographic relationships, 54, 55
Diagenetic overprint, 16
__, early, 16-19
__, meteoric, 186
Diagenetic terrain, 130, 131, 137, 140
Diffusive flow, 172
Dissolution, bioclastic molds, 14
__, carbonate grains, 17
__, concurrent, 36-37
__, incongruence, 34
__, meteoric, 37
__, porosity generation, 33-35
__, porosity modifications, 25
__, relict calcite or aragonite, 38
__, sedimentation, 37
Directed pressure, 239
Dolinen, 210
Dolomite, 34, 35-38, 40, 45, 119-121, 162, 164
__, adjacent to palm hammocks, 125, 127
__, burial, carbon isotope composition, 270
__ crust, 125, 127
__, disordered, 232
__ distribution coefficient, 56, 57
__ distribution through time, 128-129
__, __, cyclic, 128
__, __, exponential increase, 128
__ cement as diagenetic constituent, 22
__, in cave environment, 164
__, in natural levees, 125
__, in telogenetic zone, 26
__, isotopic composition, 231
__, kinetics of precipitation, 120
__, __, Mg/Ca, 120
__, __, salinity, 120

__, microdolomite, 81, 96
__, mixed water, 165
__, non-stoichiometric, 232
__, ordered, 232
__, Persian Gulf, 124
__, __, Abu Dhubi, 124
__, protodolomite, 64, 121
__, recrystallization, 122, 130, 142, 154
__, relationship to porosity, 36
__, replacement, 75
__, reservoir seal, 234
__, rhombs, 36, 38
__, solubility constant, 221
__, stability field, 162, 164
__, stable isotopes, 64, 112-114
__, __, Abu Dhabi, 122
__, __, Baffin Bay, 122
__, __, Buckner, 122
__, __, Sugarloaf Key, 122
__, stable isotopic composition, 220-222, 224, 225, 228, 229, 235
__, sucrosic dolomite reservoir, 38
__, sucrosic, 129, 130
__, synergetic, 124
__, trace elements, 114
__, water composition, 232
__, zoned, 223, 229
__, *see also* Mixed water dolomitization
Dolomitization, 21, 31, 33, 35-38, 44, 45, 87, 240, 268
__, continental waters, 230-235
__, criteria for recognition
__, __, fresh, 145
__, __, marine water, 145
__, __, mixed, 145
__, deep marine, 112
__, Dorag *see* mixed water dolomitization
__, evaporative marine, 125
__, evaporative setting, 230
__, hydrothermal, 224

__, mixed water or mixing zone, *see* Mixed water dolomitization
__, patchy, 220
__, platform, 220
__, porosity modification by, 25
__, reflux, 128, 143-145, 148
__, relationship to carbonate levels, 234
__, Ste. Genevieve Limestone model, 230
__, subsurface, 268
__, tidal pumping, 125
__, marine dominated, 224
Dolomitization below calcite compensation depth, 112-114
__, related to porosity, 112
__, related to stable isotope compositions, 113, 114
Dolomitized grainstone reservoir, 221
Dorag dolomitization, 165
__, *see also* Mixed water dolomitization
Dorag zone, 164
Dorcheat field, 201
Downey's Bluff Limestone, 226
Dubai, 212
Duricrust, 178
Dune, 6
Dunham classification, 8
Duperow Evaporite, 102-104
Duvernay Formation, 273

Eagle Formation, 10
Edmonton, 271, 272
Edwards Formation, 53, 70, 85, 86
__, porosity modification in, 87
__, strontium isotopes in, 70
Eifelian, 103, 155
Ekofisk Field, 254, 256
__ cross-section, 255
__ depositional model, 258
__ lithology, 259

__ location, 254
__, permeability within, 258
__, porosity within, 255, 258
__, pressure-depth relationship, 257
Ekofisk Formation, 255
El Abra Formation, 97, 99, 100, 102, 115
El Abra shelf margin, 102
Elat, 145
Elk Point basin, 155-157
__ conceptual model, 156
__, map of, 155
__ stratigraphic cross-section, 155
Elk Point Group, 155
Ellenburger Formation, 39, 136
__ depositional environment, 136, 140
__ depositional model, 136, 138
__ lithology, 140
__ porosity and permeability, 139, 140
__, porosity generation within, 141
__ solution collapse brecciation, 140, 141
__ structure map, 137
Endolithic algae, 78
Enewetak Atoll, 33, 67, 79, 96, 110-113, 115, 192, 221, 222, 246, 251
__ cements, 111
__ diagenesis, 112, 115
__ facies, 111
__ stable isotopes, 111
__ porosity, 112
__ stratigraphy, 110-112
__ structure, 110
England, 38, 253
Environmental reconstruction, 5
Eocene, 69, 111, 112, 113
Eogenetic, 23, 24, 25, 26
Erasian, 155
Euphotic zone, 13
Europe, 157, 244, 256, 279
Evaporative lagoon, 143
__, density stratification within, 144

__, pycnocline, 143, 144
__, reflux dolomitization associated with, 144
Evaporite deposits, 13
__, solution collapse, 39
Evaporites, 12
Exposure surface, 9

Fabric, 24
__, definition, 22
Fabric selectivity, 21-24, 34
__, fabric selective porosity, *see* Porosity types
Facies tracts, 11-13
__, lagoonal, 13
__, shorezone, 13
Fairholme Formation, 273
Falmouth Formation, 221, 222
Faults, 35
__, association with brecciation, 39
Fenestral porosity, *see* Porosity types
Ferry Lake Anhydrite, 13, 150
__ depositional model, 150
Floating meteoric lens, 173, 174
Florida, 6, 90, 151, 199, 220, 277, 278, 281
__, dolomite isotopic composition, 231
Florida Bay, 245
Florida Peninsula, 276
Fluid inclusions, 40, 71-73
__, description, 71
__, homogenization temperature, 71
__, salinity of precipitating fluid, 72
__, two-phase, 71, 141, 152
Fluorite, 40
Folk classification, 7-8
Foraminifera, 2, 4, 6, 24, 28
__, pelagic, 29
Forereef, 103
Fort Simpson Formation, 273

Fracture, 35
__, fracture fills, 40
__, fracture porosity, *see* Porosity types
__, fracturing, 33, 39, 40
__, influence in carbonates, 22
__, influence in sandstones, 22
Framework porosity, *see* Porosity types
Frasnian, 102, 103, 271
Fredricksburg Group, 97
Free flow, 172
Fresh-marine water mixing, 83, 114
Freshwater, *see* Meteoric water
Freshwater flushing, 129, 133, 136, 142
Fungi, microboring, 28

Galena, 40
Gaschsaran field, Iran, 40
Gastropod, 3, 28, 31
Geopressure, 239, 240, 280
Geothermal gradient, 239, 240
Gilmer Formation, 152, 200
Givetian, 102, 103, 155
Glen Rose Formation, 98, 99
Golden Lane, 97, 99, 100, 102, 212
Golden Lane Platform, 102
Grainsize, ultrastructural control, 4
Grainstone, 8, 9, 49, 86, 111, 267
__, primary depositional porosity, 28
Grain types, in Florida environments, 6
Grand Cayman, 32, 84, 222
Gravity driven flow, 172
__, confined, 173
__, unconfined, 173
Grayburg Formation, 212, 213
Great Bahama Bank, 13, 225
Great Barrier Reef, 90
Great Britain, 244
Great Oolite, Bath, 38
Green algae, 2, 3

Grosmont Formation, 273
Guadalupe Mountains, 148
Guadalupian, 148, 149, 215
Gulf of Aqaba, 145
Gulf of Mexico, 94, 96-102, 107, 187, 195, 199, 200, 253
___ Cretaceous reef bound margin, 97-102
___, Jurassic source rocks, 13
___, Lower Cretaceous carbonate platform, 15
___, porosity-burial depth relationships, 17
Gulf of Mexico Basin, 97, 150, 151
Gypsum, 24, 120, 121, 123-125, 129, 144, 145, 148, 157, 240

Halimeda, 3, 4, 6, 28, 31, 33, 111, 246
Halite, 123-125, 134, 145, 152, 156, 157
Hardground, 9, 76, 87, 88, 258, 260
___, ancient, criteria for recognition, 88-89
___, cements, 77, 87
___, distribution, 87, 88
___, dolomite, 114
___, lithology, 88
___, pelagic, 109
Hardpan, 178, 179
Hare Indian Shale Basin, 102
Haynesville Field, 187, 201
Haynesville Formation, 152, 200
Hay River Barrier, 155
High-Mg calcite, *see* Magnesian calcite
Highstand, *see* Sea level
Holocene, 32, 51, 79, 83, 111, 145, 219
___, dolomite isotopic composition, 231
___, porosity permeability plot, 30
Hope Gate Formation, 222, 223, 229, 231
Horsetails, 247, 249
Hydrocarbon
___, association with fracture fills, 40
___, maturation, 33, 35, 241, 267

___, source facies, 13
___, thermal degradation, 18, 35, 45, 267
Hydrocorallines, 2
Hydrofracturing, 240
Hydrogen sulfide, 18
___, generation, 241, 267
Hydrology of meteoric diagenetic environment, diffuse flow, 172, 173
___, free flow, 172, 173
___, mixed zone, 172, 173
___, phreatic, 172, 173
___, vadose, 172, 173
Hydrology of subsurface waters, 243
Hydroseal, 147
Hydrostatic pressure, 239

Illinois, 226, 227
Illinois basin, 225
Illite, 240
Imperial Formation, 273
Incongruent dissolution, *see* Dissolution
Indian Ocean, 145
Indiana, 226
Infiltration zone, 210
Intercrystalline porosity, *see* Porosity types
Interlake, 131
___ sequence, 131
Internal drainage, 210
Internal sediments, 23, 211
___ in reef, *see* Reef
Interparticle porosity, *see* Porosity types
Interstitial water, 114
Intertidal zone, 9, 45, 76, 83-87
___ cementation, *see* Beachrock
___ depositional environments, 86, 87
___ mud flats, 15
___ pore types, 85, 87
___ porosity modifications, 86, 87
___ sediments, 86, 87

Intraclasts, 7
Intracratonic basins, 14-15
Intraparticle porosity, see Porosity
Iran, 40
Ireton Formation, 102, 273
Isotopes, see Stable isotopes
Israel, dolomitization, 222

Jamaica, 32, 78, 79, 90, 91, 93, 114,
__, Discovery Bay, 222
__, dolomite, 219, 221, 223, 231
Joint systems, 24
Joulters Cay, 180, 182, 183, 222
Jurassic, 37, 38, 49, 55, 59, 60, 69, 71, 72, 81, 87, 94, 96, 151, 181, 187, 188, 191, 195, 199, 202, 207, 246, 256, 257, 262, 263, 264, 265, 266, 267, 273
__ formation water chemical composition, 58

Karst, 12, 211, 214
__ processes and products, 210-211
__ __, humid environment, 210
__ profile, 210
Karstification, 13, 44, 102, 209, 214
Keg River Formation, 155, 156, 273
__ reef, 106
Kentucky, 226
Keystone vugs, 16
Kimmeridgian, 151, 152, 199, 200, 256
Kinderhookian, 204
Kohout thermal convection, 114, 115, 238, 276, 284

Laborcita Formation, 81
Lag deposit, 9
Lagoon, 9, 12

__, mud-dominated, 3
Lake Valley Formation, 204, 206, 208
__ porosity evolution, 209
__ stratigraphic setting, 205
__ timing of cement zones, 207
Laramide tectonics, 102
LaSalle, arch, 225
__ paleoshoal, 230
Leduc reefs, 96, 102, 103, 104, 106, 107, 271, 273, 275
__, diagenesis, 106
Lenticular zone,
__ karst profile, 210
Leonardian, 149
Limestone, euxinic, 13
__ in telogenetic zone, 26
__, solution collapse, 39
Lithoherms, 109
Lithostatic pressure, 239, 256
Little Knife field, 134, 135, 136
Lobstick Formation, 275
Local island model diagenesis, 185-186
__, sea level changes, 186
Louann Salt, 151, 200
Louisiana, 199, 200, 201, 253, 284
Luminescent zonation in calcite cements, 203, 204, 205, 206
Lysocline, 259

Maastrichtian, 255
Macedonia field, 201
MacLeod salt basin, 145, 157
__ evolution, 147
__ location map, 146
__ topography, 146
Macroscale dissolution, 169, 170
Madison Group, 135
Magnesian calcite, 2, 16, 34, 80, 162, 167, 168, 170, 192

__, algal related, 77, 90
__, distribution of, 76, 77, 90, 109
__, growth habit of, 48
__, internal sediment, 77, 91
__, macroscale dissolution of, 171
__, metastable, 45, 56, 76
__, original mineralogy of, 59
__, stabilization of, 80, 81, 109, 111, 170, 171
__ stable isotopes, 78, 80, 90
__ trace elements, 80
Manitoba, 131, 135, 156
Manitoba shelf, 155
Marginal marine evaporative environments, 119
__, evaporative lagoon, 119, 143-145
__, sabkha, *see* Sabkha
Marine cementation, 102, 107, 192
Marine corrosion surface, 9
Marine diagenetic environment, 44, 45, 50
__ cements, 48, 84, 99
__, deep water, 44, 50, 75, 96, 108, 110, 111, 113, 114
__, evaporative, 45, 114
__, metastable carbonate minerals in, 56, 80
__, stable isotopes in, 62, 63, 82, 90
__, strontium isotopes in, 68, 69
__ surface water, 51, 75, 79, 88, 89, 90, 91, 92, 94
Marine hardgrounds, 179
Marine phreatic zone, 53, 277
__ cements, 53, 84
__ circulation, 111, 112
Marine vadose cements, 53, 84
Marine waters, 25, 119
__, density of, 120
__ evaporation sequence, 119, 120
Mean high tide, 25
Mechanical compaction, 44, 240, 244

__ fabrics, 53-56
Mediterranean sea water, 120
Megapore, 23, 26
Meniscus cement, 228
Meramec, 226
Mesogenetic, 23-26, 33, 35, 238
Mesopore, 23
Mesozoic, 49, 59, 60, 72, 82, 98, 252, 254
Metamorphic regimen, 238
Metastable carbonate, 45, 110
__, stabilization of, 57, 80
__ cements, 76
Meteoric aquifer, 193, 194, 200, 238
__, cement luminescence in, 197, 198
__ diagenetic model, 194, 195
__ geochemical trends, 196-199
__, porosity development in, 196
__, porosity loss in, 204
__, stable isotope values in, 199
Meteoric calcite line, 166, 167
Meteoric carbonate, 166
__, stable isotope compositions, 166
__, strontium and magnesium signatures, 166, 167
Meteoric diagenetic environment, 12, 17, 45, 50, 161
__, cements, 48
__, diagenetic fabrics of, 224
__, fluids, 50, 57, 161
__, stable isotopes, 62, 63, 65, 66
__, strontium isotopes, 69
Meteoric lens, 173, 174, 182, 183, 184-186, 194, 214, 215, 220
__, cyclicity in, 182
__ diagenetic patterns, 185
__, dimensions of, 182
__, effect of sea level changes on, 186
__ geochemical trends, 185
__, mixing zone, 183
__, porosity development in, 185

__, Walker Creek field, 192
Meteoric phreatic, 13, 26, 44, 50, 99
__ cements, 51, 52, 53
__ __, petrography of, 183-184
__ __, stable isotope composition of, 66, 184
__ __, trace elements of, 184
__ __, zonation in, 184
__ dissolution, 37
__ environment, 26, 34
__ __, stable isotope trends in, 66
__ fluids, 45
__, metastable carbonate sequence, 181-209
Meteoric vadose, 13, 26, 30, 44, 50
__, carbon dioxide degassing, 50
__ cements, 51, 52, 53
__ environment, 177-217
__ fluids, 45, 50
__ soil zone, 50
__, stable isotope signature, 65, 66
__ zone of capillarity, 51
Meteoric water, 17, 25, 34, 37, 38, 39, 161
__, carbon dioxide content of, 162
__, carbon isotopic composition of, 165
__, dolomitization by, 220
__, geochemistry of, 161
__, hydrologic setting, 172
__, oxygen isotopic composition, 165
__, porosity evolution associated with, 168
Methane generation, 241, 267
Mexico, 3, 4, 97, 99, 100, 101, 102, 112, 115, 151, 212, 215
Michigan Basin, 157, 251
Micrite, 7
Microcodium, 179
Microporosity, 23, 26, 252
Microscale dissolution, 169
Microstalactitic cement, 228, 180
Microstylolite, 248

Midale beds, *see* Charles Formation
Middle Devonian reef trends of Western Canada, 102-108
__, distribution, 103
__, paleogeography of, 103
__, permeability within, 104
__, porosity within, 104
__, reservoir quality in, 104
__, stratigraphy of, 103
__, subsurface cross-section of, 103
Middle East, 252
Midland Basin, 212, 213
Midway field, 201
Miette reef, 284
Mineral stabilization, 24, 25, 34
Miocene, 33, 111, 112, 246
Mission Canyon Formation, 134
__ depositional environments, 135
__, diagenetic modification in, 136
__, dolomitization of, 135
__, fresh water flushing of, 136
__, gamma-ray/sonic log of, 136
__, log profile of, 135
Mississippi, 151, 199, 246, 249, 263
Mississippian, 3, 69, 82, 96, 134, 135, 196, 204, 205, 225, 226, 229, 251, 263, 267, 272
Mixed meteoric-marine water, 17, 37, 38
Mixed water dolomitization, 183, 219, 222, 235
__, criteria for recognition, 220, 225
__ in lagoonal mudstones, 228
__, modern examples, 222
__, __, Andros Island, 222
__, __, Israel, 222
__, __, Jamaica, 222
__, __, Yucatan, 222
__ in relation to sea level, 223, 224
__ as reservoir, 225, 226
__, validity of, 220-225

___, water composition in, 219, 224
Mixed zone, ocean, 75
Mixing zone, 277
___, *see also* Mixed water dolomitization
Moldic porosity, *see* Porosity types
Molluscs, 2, 6
Monroe uplift, 187
Montana, 131, 135, 155
Monterey Formation, 40, 240
Moon milk, 210
Mt. Vernon field, 201
Mudcracks, 9, 142
Mudstone, 8, 103
Murphy Giffco #1, 202, 246
Mururoa Atoll, 90
Muskeg Evaporites, 102, 103, 107, 155, 156, 157, 273

Nebraska, 131
Neomorphism, 196
New Albany Shale, 226
___, regional structure on, 226
New Mexico, 10, 13, 137, 148, 149, 204, 205
Niagaran, 157
Niobrara Formation, 10, 281
Nisku Formation, 103, 271, 272, 284
___ burial history, 272, 276
___ geochemistry, 274
___ location, 274
___ reef development, 275
Nisku shelf, 274
Niue Atoll, 115
Nodules, 178
Non-skeletal grains *see* Allochems
Normal marine diagenetic environments, 75-117
___, deep water, 108-117
___, shallow water, 76-108

North America, 253, 256, 279
North Bridgeport field, 225, 226, 227, 229, 230
North Dakota, 131, 134, 135, 155
North Louisiana salt basin, 187
North Sea, 240, 244, 253, 254, 257
North Sea chalk porosity curve, 256

Oceanic atolls, 110, 111
Offlap, 11, 12, 13
Oil-field water salinity variation, 242
Oil generation, 241
Oligocene, 40
Oncoids, 4
On lap, 11, 13
Ooid grainstone reservoir, 187, 199, 200, 203, 225, 226, 227
___ sands, 16
___ shoals, 16
Ooids, 4, 7, 22, 24, 28, 49, 55, 56, 59, 78, 81, 87, 88, 96, 246, 249, 266, 267
___, aragonitic, 24, 203
___, calcitic, 192, 203
___, dissolution of, 182
___, formation of, 15
___, Shark Bay, 171
___, Smackover, 171
___, stable isotopic composition of, 78, 79
Oolites, *see* Ooids
Ooze, *see* Pelagic ooze
Ordovician, 130, 131, 136, 219
Ore mineralization, 211
Organic acids, 18, 241, 267
Organic framework, 24
Organic matter, thermal degradation, 18
Osagean, 204
Overpressuring, 239, 256
Oxfordian, 152, 199, 200
Oxidizing environment, 180

Oxygen isotopic composition of burial cements, 264, 265
__, of burial dolomites, 270

Packstones 8, 9, 86, 111
__, porosity in, 28, 29
Paleozoic, 81, 82, 112, 128, 267, 273, 276, 284
Paragenesis, 55
Paris Basin, 37, 38
Partial pressure of carbon dioxide, 33, 34
Passive margin, 15
Peace River arch, 102, 103, 156
Pearsall Shale, 97
Pekelmeer, 145
Pelagic foraminiferal limestones, 97
Pelagic ooze, 29, 109, 244, 245, 259
Pellets, 24
Peloids, 25
Pembina area, 271, 274
Pennsylvanian, 3, 81
Perched water table, 210
Percolation zone, 210
Peri-platform ooze, *see* Pelagic ooze
Permeability, 30, 31
__ in carbonates, 22
__ in dolomite reservoirs, 229
__ in meteoric lens, 185
__ in ooid reservoirs, 229
__ in sandstones, 22
__ in vadose zone, 181
__ loss in caliche formation, 186
Permian, 81, 144, 149, 157, 212, 215, 216, 255, 259
Permian Reef complex, 13
Persian Gulf, 88, 124, 128
Petrography, of cement distribution patterns, 51-53
__ of cement morphology, 47-51
__ of grain-cement relationships relative to compaction, 53-56
pH-Eh gradient, 198
Phreatic, *see* Meteoric phreatic
__, see Marine phreatic
Pierre Shale, 10
Pinnacle reef, 156
Pisoids, 178, 179
Plate tectonics, 15
Pleistocene, 51, 111, 112, 115, 145, 171, 186, 191, 192, 193, 222, 223, 224
__ dolomite isotopic composition, 221
__ dolomitized reef, 219
Pliocene, 112, 222
Plio-Pleistocene, 223
__, dolomite isotopic composition, 231
Polyhalite, 268
Pore fluids, chemistry of, 33
__ in burial diagenesis, 18
__, marine, 34
__, partial pressure of carbon dioxide in, 33
__, salinity of, 33
__, temperature of, 33
Pore pressure, 256
Pore systems, 40
__ in carbonates, 21
__, secondary, 24
__, size classes, 23, 26
Pore types, 85, 91
Porosity and cement variation with pressure solution, 250
__, classification, 21-27
__, nature of carbonate porosity, 21-41
Porosity development associated with regional unconformity, 211, 214
__ in meteoric aquifer, 196
__ in meteoric lens, 185
__ in vadose environment, 181
Porosity dimensions, shapes, 22, 180

__, sizes, 22
Porosity enhancement, by dissolution, 33, 35, 277
__ by dolomitization, 271
Porosity evolution, 25
__, effects of burial processes, 277
__, effects of surface processes, 277
__ in Devonian reefs, 102
__ in Smackover Formation, 199
__ with depth, 245, 279, 280
Porosity in carbonates, 21-41, 229
__ in deep marine sediments, 109
Porosity modification, 25-26, 43-74
__, diagenetic environments of, 43-46
__, tools for recognition of, 46-74
Porosity preservation, 192, 193
__ by oil migration, 257
__ by overpressuring, 256
__ in carbonates, 17-18, 22
__ in siliciclastics, 17-18, 22
Porosity reduction, 203, 204 205
__ by caliche formation, 186
__ by cementation 44,, 87, 45, 277
__ by chemical compaction, 244, 246
__ by compaction, 53, 191, 244
__ by pressure solution, 189, 192, 248
Porosity types, 21-41
__, biomoldic, 33
__, boring, 23
__, breccia, 23, 39, 41
__, burrow, 23
__, cavernous, 23, 34, 35, 102
__, channel, 23
__, depositional, 24, 27-32
__, fenestral, 23, 30-32
__, fracture, 23, 39-41
__, framework, 21, 23, 30-32
__, growth, 91, 98
__, intercrystalline, 23, 37, 226
__, intergranular, 87, 91, 98, 202, 203

__, interparticle, 22, 23, 27, 28
__, intragranular, 87, 91, 98, 202, 203
__, intraparticle, 21, 22, 23, 28, 29
__, moldic, 12, 23, 24, 27, 33, 34, 37, 38, 128, 181, 183, 196, 220, 226, 228, 274
__, oomoldic, 33, 200, 202, 203
__, primary, 17, 22-24, 27-32, 35, 41
__, secondary, 17, 21, 23-26, 33-41
__, shelter, 23
__, shrinkage, 23
__, solution-enlarged, 35, 212
__, vuggy, 34, 23, 35, 200, 203, 274
Poza Rica, Mexico, 115
Precambrian, 82, 96, 128, 136
Precambrian Shield, 102, 103, 156
Presquile reef, 102, 103, 104, 107, 155, 156
__, diagenesis of, 105
__ reservoir history, 105
Presquile shelf margin, 102
Pressure solution, 18, 19, 45, 206, 247, 248
__, cement overgrowth during, 198
Primary grains, *see* Allochems
Primary porosity, *see* Porosity types
Puckett field, Texas, 39, 137, 138
__ cross-section, 137
__, porosity-permeability relationships in, 138-139
Pycnocline, 143, 144, 150
Pyrobitumen, 263, 267, 268

Queen Formation, 148, 149, 212, 213

Rainbow reefs, 96, 102, 103, 104, 106, 107, 155, 157
__, diagenesis of, 106
__, resevoir history of, 106
Recrystallization, 22

Red algae, 3
__, coralline, 2, 6, 31, 32
Red River Formation, 130
__ depositional environments, 132-133
__, diagenetic features of, 131-133
__, fresh water flushing in, 133
__, gamma ray-neutron logs from, 132
__, oil saturation in, 132
__, pore size in, 132
__, pore system development in, 134
__ porosity, 131, 132
__, total saturation in, 132
__, water saturation in, 132
Red Sea, 90
Reef 2, 3, 16, 18, 76, 80, 89-93
__, algal cup, 32
__, barrier, 12, 13
__ bioeroders, 2
__ building organisms, 30
__ cementation, 2, 76, 77, 78, 90
__, closed framework, 31
__, definition of, 2
__ detrital fill, 2
__ development, 31
__ diagenesis, 89-93
__ __, ancient, 93-97
__ __, bioerosion, 91, 92, 93
__ __, cements, 90, 91, 92, 94, 95, 96
__ __, distribution, 89, 90, 96
__ __, internal sedimentation, 91, 92, 93
__ __, pelletization, 92
__ __, porosity evolution, 91, 92, 93, 94
__ __, stable isotopes, 90
__ __, trace elements, 90
__ framework, 78, 92, 93
__ __ organisms, 2
__ internal structure, 2, 31, 32
__, modern, 2-3, 32
__, open framework, 30, 31
__, patch, 13

__, pinnacle, 13
__ structure, 2
__ zonation, 78, 89
__ __, back reef, 111
__ __, fore reef, 78, 90, 111, 96
__ __, island slope, 78
__ __, lagoon margin, 111
__ __, reef crest, 111
Reef bound shelf margin, Lower-Middle Cretaceous Gulf of Mexico, 97-102
__, burial curve, 99
__, depositional environments, sedimentary facies, 97, 98, 99, 100, 101
__, diagenesis, 99, 101, 102
__, hydrocarbon production, 97, 98, 100
__, porosity evolution, 98, 99, 102
__, structure, 97, 98, 100, 101
Reef trend, western Canada, 102-108
Renault Sandstone, 226
Reservoir seal, dolomite, 234
Reservoir properties, oolite versus dolomite, 228, 229
__, semi-log plots of, 229
Reservoir compartmentalization, 87
Reservoir hydrocarbon production, 97, 98, 99, 100, 102
Restricted carbonate environment, 3
Rhizoids, 178
Rhodoliths, 4
Rip-up clasts, 9, 142
Rock-water interaction, 35
Rudist, 3, 28
__ reefs, 97, 99, 100, 150

Sabine Uplift, 187
Sabkha, 38
__, ancient, 127
__, __, criteria for recognition, 141
__,__, dolomitization, 141

___, isotopes, 62
___, modern, 123
___, ___ Bahamas, 123
___, ___ humid, 125
___, ___ hydrology, 123-124
___, ___ Persian Gulf, 124
___, ___ pore fluid chemistry, 123, 124, 125
___, ___ pore fluid origin, 123
Sacramento Mountains, 204, 205
Saddle dolomite, 262, 263
___, Devonian, 270
___, Jurassic, 270
___, stable isotopes, 270
Salinity, of pore fluid, 33
San Andres Formation, 149, 212, 213, 214, 215
San Andres Mountains, 204, 205
San Andres unconformity, paleotopography on, 125
San Juan Basin, 10
San Salvador, dolomite isotopic composition, 231
Sand flat, 103
Sandstones, compositional classes, 7
___, aspects of porosity, 22
Santonian, 10
Saskatchewan, 36, 103, 131, 155
Schooner Cay, 182, 183, 184
Sea level fall, 12, 13
___ fluctuation rate, 11-12
___ fluctuations, 10-13
___ ___, carbonate sedimentation response, 12
___ highstand, 12, 13
___, Permian, 216
___, rapid rise, 10-13
___ relationship to dolomitization, 223
___, upper Jurassic, Gulf Coast, 200
Secondary porosity, 161, 165, 169, 170, 171, 241, 277

___, bioerosion, 91, 92
___, cavernous, 102
___, dissolution, 45, 87
___, dolomitization, 45
___, *see* Porosity types
Sedimentation rate, 12
Seroe Domi Formation, 222
Seven Rivers Formation, 148, 149, 212, 213, 214, 215
Shale dewatering, 33, 34
Shallowing-upward sequence, 134, 136
Shallow marine depositional environments, 80
___, lagoon, 80
___, platform, 80, 88
___, shallow basin, 80
___ ___, lithoherms associated with marine cements, 80
___, shelf margin, 80, 88, 89, 92, 96, 97, 102, 110
___ ___, sand shoals, 80
___, shoreface, 84
___, strandline, 80
Shallow siliciclastic shelf sediments, 16
Shallow water, normal marine diagenetic environments, 76
___, ancient cements, recognition, 80-83
___, ancient hardgrounds, 88-89
___, ancient reef diagenesis, 93-97
___, Canadian Middle Devonian reefs, 102-108
___, cementation process, 76-80
___, diagenetic setting in the intertidal zone, 83-87
___, Gulf Coast Lower-Middle Cretaceous shelf margins, 97-102
___, modern hardgrounds, 87-88
___, modern reefs, 89-93
Shark Bay, 87
___, Pleistocene ooids, 171

Shekilie Barrier, 155
Shelf margin, Lower Cretaceous, 97, 101, 112
__, see also Carbonate shelf margin
Shelter porostiy, see Porosity
Shelter voids, 91, 98
Shoaling upward, cycles, 188
__, sequence, 9-11, 19
Shoreline complexes, 6
__, high energy, 16
Shore zone, 12, 13
Shrinkage porosity, see porosity types
Sierra Madre Oriental, 102
Siliciclastic, 2-4
__, early burial diagenesis, 18
__ grain composition, 5, 19
__ progradational shoreline sequences, 10
__ provenance, 5, 19
__ rock classification, 7
__, roundness, 4
__, sediment size, 3
__ shelfal evolution, 11
__ sorting, 3
__ terrain, 196, 198, 199
Silurian, 131, 157
Sinkholes, 210, 214
Sioux Arch, 155
Slave Point Formation, 103, 104, 155
Sligo Formation, 97
Slump, 258, 259
Smackover Formation, 13, 49, 55, 59, 60, 69, 71, 72, 81, 87, 187, 188, 191, 192, 195, 199, 200, 201, 202, 204, 207, 246, 249, 267
__, Bryans Mill field, 37
__, Clarke County, Mississippi, 39
__ diagenetic fabrics and porosity, 202
__ diagenetic zones and gradients, 201
__ dolomitization in East Texas, 151, 159
__ __, dolomite geochemistry 152-154

__ __, dolomitization model, 151-153
__ __, recrystallization, 154
__ __, stratigraphic setting, 152
__ __, time of dolomitization, 152
__ geologic setting, 200
__ in Arkansas, southern, 33, 35
__, luminescence, 203
__ ooids, 171
__ porosity evolution, 199
__ porosity-permeability relationships, 201
__, relationship to sea level changes, 200
__ reservoir porosity, 208
__, stable isotopic composition of cements, 202
Smectite, 240, 268
Soil formation and brecciation, 39
Solubility contrast, calcite and aragonite, 194
Solution collapse breccia, 129, 140, 141, 142, 211
__, ore mineralization, 211
__, reservoirs, 211
Solution cavities, 210
Solution, cementation and porosity evolution in diagenetically mature systems, 211, 212
Solution seam, 248, 249
Solution towers, 210, 214
Source of carbonate cement, 260
South Dakota, 131, 135
Speleothems, 50, 162, 210
Sphalerite, 40
Sponges, 31
__, clionid, 31, 32
Spontaneous nucleation, 168
St. Croix, 222
St. Louis Limestone, 226
Ste. Genevieve Formation, 225, 226, 228, 229
__, associated formations, 226

__ dolomite isotopic composition, 231
__ dolomite, 231
Stabilization of unstable minerals , 168, 169
Stable isotopes, 61-68
__, carbon, 61, 64, 65, 66
__, __, organic fractionation, 64, 65
__, definition, 61, 62
__, environmental distribution, 63
__, oxygen, 61, 62, 63, 64, 65, 66
__, uses, 65, 66
__, variation, burial, 60
__, __, shifts across meteoric interface, 66
__, __, temporal, 66, 67
Stable isotopic composition, 198, 199
__, carbon, 191
__, __ in burial cements, 264, 265
__, __ in burial dolomite, 270
__, dolomite, 221, 231
__, meteoric phreatic cement, 184
__, mixed water dolomite, 220, 221, 222, 224, 225, 228, 229, 235
__, Smackover, 202
__, vadose cements, 180
__, zoned cement, 205, 208
Starved basin 12, 13
Stony Mountain Formation, 132
Stromatoporoids, 3, 31
__, coral framework, 103
Strontianite, 40
Strontium isotopes, 68-71
__, burial cements, 265, 266
__, dating diagenetic events, 69
__, Edwards Formation, 70
__, __, limestones and brines, 70
__, formation waters, 266
__, saddle dolomite, 270
__, sea water, 68
__, __, Jurassic, 181
__, __, temporal variations, 68

Strontium levels, 203
Stuart City reef, 97, 98, 99, 101, 102
Stylolites, 35, 247, 248, 249, 269
Stylolitization, 18, 207
Subaerial exposure, 9, 39, 136, 148, 228
Submarine canyon, fill and slope deposits, 6
Submarine cements, *see* Marine cements and Cements
Submarine hardground, 104
Subsidence, 13
Subsurface cement, 107
Subsurface diagenetic environment, 45, 46, 50
__, cements, 51, 52
__, pore fluids, 45, 57, 59
__, porosity destruction, 45
__, __, pressure solution, 45
__, secondary porosity, 45
__, stable isotopes, 62, 63, 64, 65
__, strontium isotopes, 69
Subsurface dissolution, 267
Subsurface dolomitization, 107, 268
__ geochemistry, 269
__ impact on reservoir porosity, 270
__ petrography, 269
__ stable isotopic composition, 270
__ strontium isotopic composition, 270
Subsurface waters, 242
__, hydrology, 243
__, magnesium to calcium ratio, 242
__, oxygen isotope composition, 242
Subtidal, 9, 11, 25
Sucrosic dolomite, 129, 130
Sugarloaf Key, Florida, 32, 114
Sulfate cement, 22
__, replacement , 22
Sulfur Point Formation, 155, 273
Supratidal, 9, 11, 25

__, algal related sediments and wackestones, 31
__ cap, 131
__ mud flats, 15
__ sequence, 103
Surface active anions, 50
Surface active cations, 50
Swan Hill Formation, 254, 271, 273
Swan Hill reefs, 96, 102, 103, 104, 105, 107
__, diagenesis, 105
__, reservoir history, 105
Swift Formation, 281

Tamabra Limestone, 115
Tamaulipas Formation, 100
Tansill Formation, 148, 149
Tectonic sill, 150
Tectonic stress, 253
Teepee structures, 178, 179
Telegraph Creek Formation, 10
Terra rosa, 211
Tertiary, 82, 114, 254, 255, 274
Texas, 13, 37, 39, 94, 97, 98, 99, 101, 102, 112, 136, 137, 144, 148, 149, 150, 151, 157, 171, 212, 213, 246, 253
Thermal convection, 114-116
__ fluids, 114, 115
__ model of marine water dolomitization, 114-116
__ __, stable isotopes, 114, 115
__ __, trace elements, 114, 115
__, see also Kohout convection
Thermal degradation of hydrocarbons, see Hydrocarbons
Thermal maturity, 281
Thermochemical sulfate reduction, 241
Thermocline, 75, 108
Tidal bar complexes, 6

__ channel, 6, 16
__ __ migration, 16
__ delta, 6
__ flat, 3, 6, 9, 38
__ __ dolomite, 37
__ pumping, 125
Tithonian, 152, 200
Tongue-of-the-Ocean, 109
Tools for recognition of diagenetic environments in the geologic record, 46-74
Tor field, 254
Trace element, 56-61
__ composition, 198
__ distribution coefficients, 56, 57, 58, 59
__ __, commonly accepted values, 57
__ __, controls over, 58
__ __, case study of, Jurassic, 59, 60
__ __, original mineralogy, 59
__ fluid concentration, 57, 58
__ geochemistry of cements and dolomites, 56-61, 180, 231
__, paleo-fluid flow, determination of, 61
__, redox conditions involving, 61
Transgressive-regressive cycles, 10
Transgressive processes, 6
Trucial Coast, 123, 142
Tunisia, 94
Turbidites, 258, 259
__, basinal, 7
__, carbonate, 6
Tuxpan Platform, 97

Unconfined aquifer, 173
Unconformity, 26, 34
United States, 267
Upper Jurassic source rocks, 13

Vadose cement, 141

—, diagenesis in metastable sequences, 177-181
—, environment, 177-182
— fabrics, 220
—, flow, 172
—, porosity development, 181
—, seepage, 172
— zone, 172, 177, 277
— —, lower, 177, 179, 186
— —, upper (soil), 177-179, 186
—, *see also* Meteoric vadose
Val Verde basin, 137
Valles-San Luis Potosi Platform, 97, 100
Vitrinite reflectance, 241
Vug, 34
—, keystone, 16
—, vuggy porosity, *see* Porosity types

Wabamun Formation, 273
Wackestone, 8, 9, 92, 103
—, primary depositional porosity in, 28, 29
—, supratidal, algal-related, 31
Walker Creek field, 187, 188, 191, 192, 201, 202, 249, 267
— as island model, 187-193
— cross section, 189
— location map, 187
— porosity isopach, 188
— reserves, 187
— reservoir character, 187
— structure map, 188
Walnut Canyon, 148, 149
Water table, 172, 173
Watt Mountain, 155
Weathering, 25
West Cat Canyon field, 40
Whitings, 15
Williston basin, 130, 131, 134, 135, 267

Winterburn basin, 271
Winterburn Shale, 273, 275
Wisconsin, 219
Wispy seam, 248, 249
Wolf Lake Formation, 275
Woodbend Shale, 273
Woodbend shelf, 103

Xel Ha, 215

Yankeetown Sandstone, 226
Yates field, 39, 212, 213
— geologic setting, 213
— wells with caves, 214
— cave abundance, 216
Yates Formation, 148, 149
Yucatan, 215
— Pleistocene dolomites, 221, 222, 223, 224, 229
Yukon, 103

Zama reefs, 157
Zechstein Basin, 157
Zetalk reef, 275
Zoning in calcite cement, 203, 204, 205, 206, 207, 208
— in dolomite, 229
— in dolomite cement, 223
Zuni basin, 10